An Introduction to Microprocessors: Experiments in Digital Technology

NOEL T. SMITH

HAYDEN BOOK COMPANY, INC.
Rochelle Park, New Jersey

Library of Congress Cataloging in Publication Data

Smith, Noel T.
An introduction to microprocessors.

Includes index.
1. Integrated circuits. 2. Microprocessors.
I. Title

TK7874.S62	621.381'73	81-6568
ISBN 0-8104-0867-8		AACR2

Printed in the United States of America

1	2	3	4	5	6	7	8	9	PRINTING
81	82	83	84	85	86	87	88	89	YEAR

Preface

When one stands before some of the mammoth machines of our day, he is impressed with the technology that has created them. Equally amazing, perhaps even more so, are the tiny machines that are operating now in almost every sphere of life—the microprocessors. Man's advance in his technological capability has been coupled with his increase in numbers. Perhaps it was the press of a burgeoning population that caused man to think small. Whatever the reason, man has compressed his technical functioning into a space smaller than his fingernail, and in the process has revolutionized the world he lives in.

This book is designed, in a learn by doing style, to be a guide for learning how to use the tiny electronic devices that are collectively called integrated circuits (IC). Starting with the simplest gates, this text goes beneath the plastic cases of the integrated circuits and investigates the elemental parts that give ICs their capabilities. The characteristics of the devices are explored and this knowledge is used to explain how to combine them into working and useful circuits. The book progresses in logical order from the simple gates, through increasingly complex devices, to the ultimate in integrated complexity—the microprocessor. Together, the chapters form a course of study, complete with experiments, which will provide the student with an understanding of integrated circuitry and a capability in electronic design using integrated circuits. The final two chapters discuss the use of microprocessors in microcomputers. Construction, functioning, and programming of microcomputers are covered in these chapters.

This book is the product of long hours and hard work. Much of that toil was undertaken by Judy. For her help, a simple thanks is inadequate. Information and help came from several sources. My thanks go to each of the manufacturers who have willingly permitted the use of their IC data sheets. To the fellows at Sterling Electronics in Austin, Texas, and to Irwin Carrel and Lyle Sapp at Motorola, a special word of appreciation is due. Also, thanks to James Melton, who loaned the Melton Special for inclusion in the book.

Finally, this volume is dedicated to my children—Amy, Christi, and Carrie. It is their world that will reap the full fruit of integrated technology.

Contents

ONE

An Introduction to Digital Technology

The Rise of Digital Technology

Advancements in technological capabilities seldom occur in isolation. Advancements usually result from a beneficial confluence of concepts and techniques that have been developed at earlier times. Quantum leaps within technology occur when some perceptive mind visualizes the value that comes from combining existing capabilities. The advancement that has been coined "digital technology" is no exception.

Digital technology in its current form is the result of several propitious developments. Indeed, almost all of current digital technology is reapplication of preexisting concepts. Digital is as elemental as man's fingers and toes. The abacus at 5000 B.C. was applying digital techniques to mathematical problems. The mathematical principles used by modern computers bares the name of their developer George Boole. Boole developed Boolean algebra in the 1850s. Why, then, if the concepts were available, did digital techniques wait until recent years to gain their preeminent status within technology? Digital technology waited until that unique set of circumstances would cause it to be more versatile, more economical, and more advantageous than its rivals. A brief inquiry into the ancestry of modern technology will show this to be true.

Digital manipulation was at first accomplished by mechanical devices that used gears to count and store the numbers. Mechanical devices were capable of handling decimal calculations, which was an efficient numbering system for these devices. The impetus to convert from decimal to binary numbering systems[1] had to wait on a complementary technology. That technology took several forms. The binary math system was a natural choice for electromechanical relays that were either on or off. However, relays in the quantities required by computers and other devices represented excessive cost and massive bulk. In addition, they were far from reliable. The introduction of vacuum tubes represented a significant improvement over relays. But even vacuum tubes possessed their own problems of bulk, complicated circuitry, heat generation, and high power requirements. Nor was the reliability of tube type systems at a desired level.

It was the advent of the transistor that heralded the true beginnings of digital techniques. The transistor performed the function of a data switch. By interconnecting numbers of transistors and using their on and off states to produce meaningful codes, logical functions were accomplished and with great speed. Addition, subtraction, logical combination of data, and data storage are but a few of the functions that were permitted by transistor digital circuitry. As might be expected, there were drawbacks to transistor logic circuits. One was cost. Even using inexpensive transistors, the vast numbers required to perform logical functions resulted in great cost. Complexity also was a problem. Thousands of transistors made maintenance and construction difficult. Using discrete devices such as transistors and diodes for logical operations pointed toward a new day in electronics, but the fullness of that day waited upon yet another technological advance—integration.

The transistor was a useful device. It could perform all of the operations required for digital functions. But transistors were single devices. They had to be constructed one at a time. Each one had to have its own housing and its own leads. Within circuits, each one required its own support components such as resistors and capacitors. The time and materials required for construction caused the base cost of digital equipment using transistors to stay beyond the level required for mass popularity. A multitran-

[1]The binary numbering system uses only two digits: "0" and "1." This method of representing numerical values will be discussed more fully in later sections.

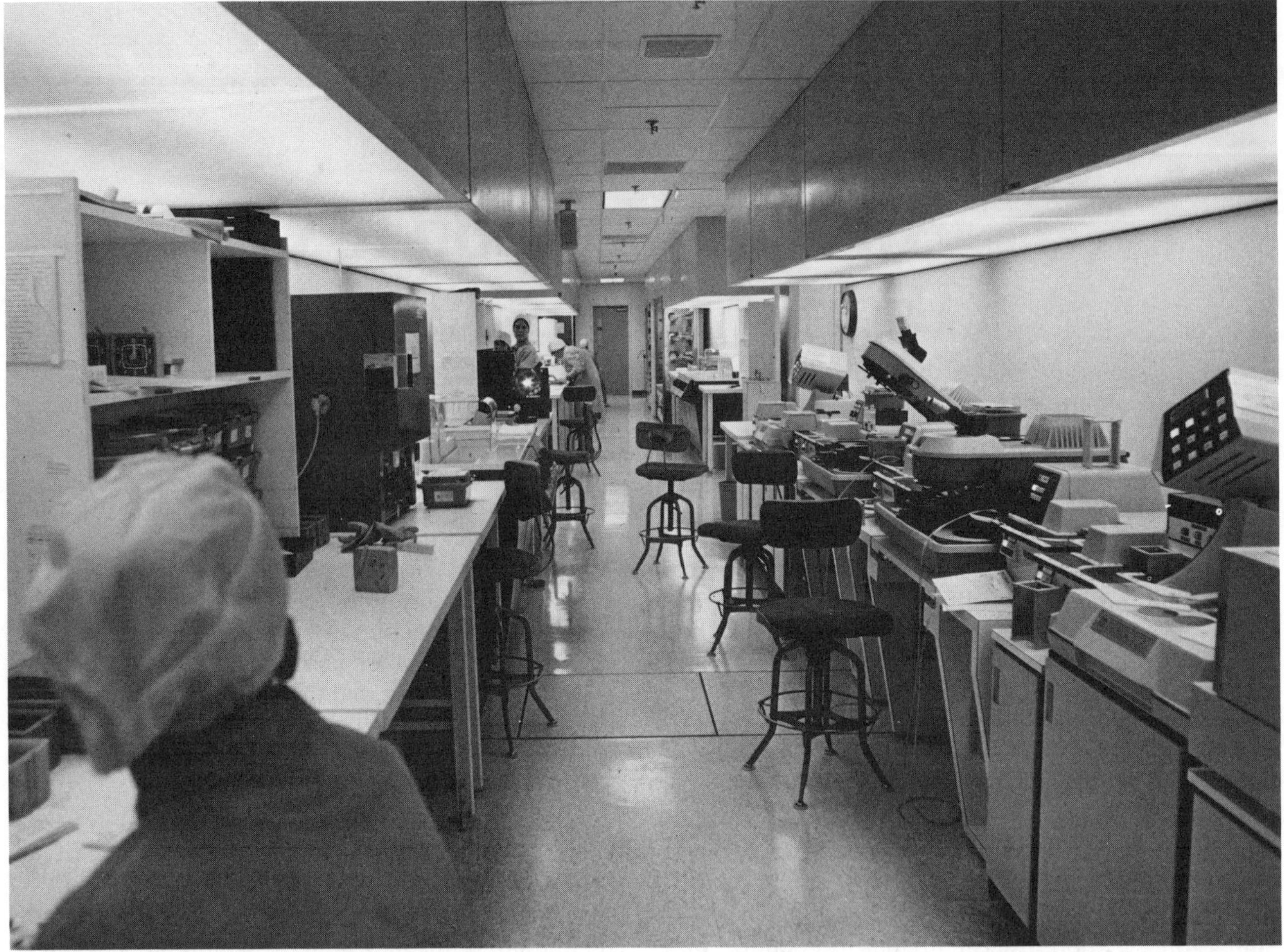

Fig. 1-1 Integrated circuit fabrication in a clean room environment. (Courtesy of Motorola, Inc.)

sistor circuit module that was mass producible, small in size, low in cost, and flexible was needed. While this list of requirements seemed impossible to satisfy, a tiny electronic miracle referred to as the integrated circuit met that need.

Fabrication of Integrated Circuits

The integrated circuit is the offspring of the transistor. Semiconductor materials such as silicon and germanium were used in the manufacture of transistors. It was refinement of transistor fabrication techniques that resulted in the integrated circuit. The name *integrated circuit* is significant. An integrated circuit is an electronic circuit consisting of the necessary transistors, diodes, resistors, and conductors, all of which have been fabricated (integrated) on a single piece of semiconductor material. The process of fabricating components on a single "chip" is exacting and laborious during the design phase. Once mass production starts, however, the circuits can be produced at a rapid rate, and for many ICs, the user cost approaches a negligible level. The use of computers greatly assists the design process.

The actual fabrication of microelectronic circuitry (integrated circuits are classified as microelectronic circuits) takes place in clean room factories by personnel wearing smocks and working in air streams filtered of all particulate matter by laminar flow filtration systems such as the one shown in Fig. 1-1. While these "clean factories" produce the finished products, the process begins long before. Figure 1-2 shows the production processing steps required for integrated circuit fabrication. An integrated circuit begins with a need. Someone has a job that can best be accomplished using microelectronics. The job may be as specific as the control of a photocopy machine or as general as providing a number of AND gates[2] for use in any circuit requiring them. Whatever the job requirements may be, a circuit specification is developed that will satisfy

[2]AND gates are discussed in Chap. 2.

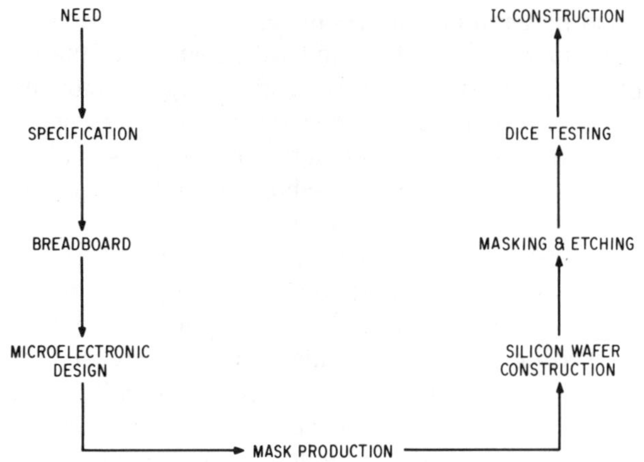

Fig. 1-2 Integrated circuit fabrication flowchart.

those requirements. Depending on the nature of the requirements, the circuit may be designed using discrete components in a breadboard arrangement or, if the circuit has previously been designed using microelectronic techniques, it may immediately be fabricated using microelectronic design principles. To facilitate design, the circuit is considered in many small sections. The final design is drawn and is then checked by trained personnel. Once the design is accepted, it is photographed and reduced for construction of the master masks. The masks are round pieces of glass on which the circuit pattern is deposited by photolithographic methods. The mask may use either photographic emulsion or a chromium film on the glass plate to create the required pattern. The chromium film master is more durable, but it also is more expensive. Photographic emulsion masks are capable of only a few uses before minute scratches and dust particles render them worthless. The chromium masks are capable of many more uses before they require cleaning or replacement.

The basic element of IC construction is silicon. The pure silicon is grown into a round cylinder. This cylinder is then cut into thin wafers using a diamond saw. The wafers are ground and polished leaving a final thickness of approximately ½ mm. This silicon wafer, highly polished on one side, is the raw material from which the IC will be built.

In processing the wafer, silicon dioxide is created on its surface and used as an electrically insulating medium. This silicon dioxide is selectively removed using a photolithography process. The masks are used to create the desired patterns in the silicon dioxide. Dopant or impurity atoms are then added to create the electron-rich "n" regions or the positive ion-rich "p" regions. Using a succession of masks, as many as seven or more exposures will be made in completing a wafer. Each separate mask will provide the pattern required for creating n and p regions or for connections within the microelectronic circuit. Once the wafer processing is completed, each microcircuit on a wafer (up to 250 on a 100-mm wafer) is tested. Those circuits having defects that render them useless are marked and discarded. In most cases, if 40 out of 250 circuits pass the test, the wafer is considered profitable.

The next step is to cut the IC circuits (called *dice*) apart and place them in packages. Packages are usually constructed of plastic or ceramic. The leads of the package are attached to metal "pods" that were formed on the dice during fabrication. The package is sealed and tested, and the finished IC is ready for the user.

Digital/Analog Comparison

In this volume, the emphasis is on digital circuitry. Digital circuitry is best understood by comparing it with analog circuitry. A typical analog signal is the sine wave, which is shown in Fig. 1-3. It can be seen that the voltage level of this wave moves smoothly from one value to the next and that all voltage levels of the signal from minimum to maximum are represented briefly at some point in the cycle. This continuously varying voltage level is the primary characteristic of analog signals. Figure 1-4 shows a simple analog circuit, an amplifier. If a sine wave voltage generator is connected to the transistor base (B), a similar sine wave signal is found at the collector output (C). However, this output sine wave is larger than the input due to the amplifier action of the circuit.

In contrast to analog circuits, digital circuits do not provide smoothly varying signals. The action of digital circuits is more analogous to a switch. Digital circuits are either on or off. Digital signals assume a

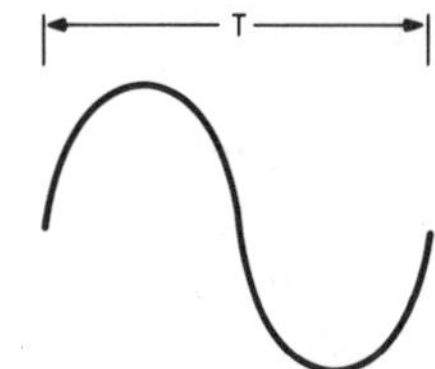

Fig. 1-3 Sine wave typical analog signal.

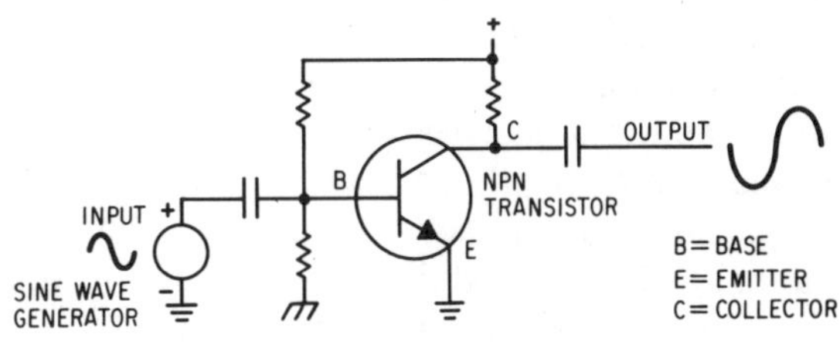

Fig. 1-4 Analog amplifier circuit.

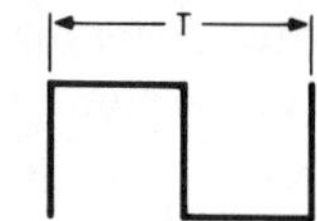

Fig. 1-5 Square wave typical digital signal.

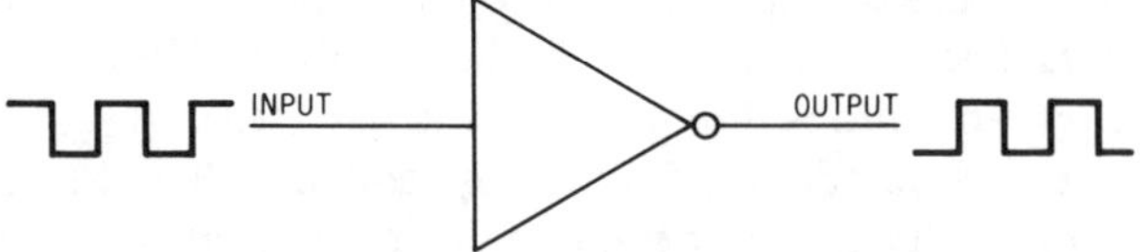

Fig. 1-6 NOT gate symbol.

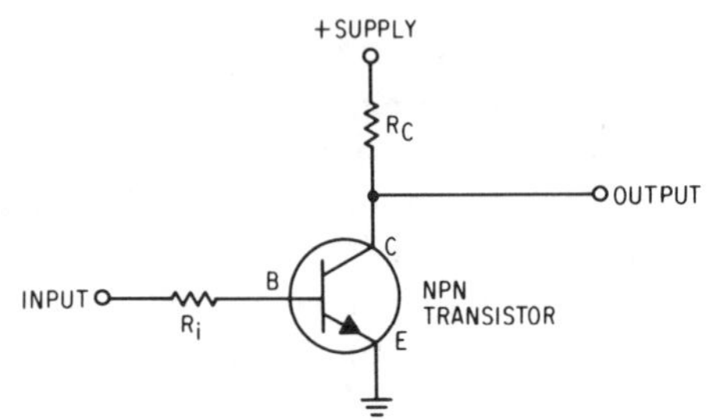

Fig. 1-7 Transistorized inverter.

finite number of values. The intervening voltage levels are passed through so quickly that they can be ignored. If this digital circuit action is plotted over a time course similar to the time course of Fig. 1-3, the resulting graph would be the one shown in Fig. 1-5. The resultant signal is called a square wave. The *square wave* is a typical waveform found in digital circuitry. Notice that the square wave has essentially two voltage levels. The transition from one to the other is very rapid. The voltage levels that characterize digital circuits are the low and the high levels. They reflect the "on/off" action of the circuits. The two voltage levels or states are referred to in several different ways: low or high, 0 or 1, absence or presence of signal.

Figure 1-6 shows the symbol for one type of digital circuit called an *inverter* or a *NOT gate*. As shown, the symbol is composed of a triangle, a small circle, and input and output leads. The triangle represents the active elements of the device. The circle means that the output is an inverted input signal. This circuit will change its output from minimum to maximum depending on the voltage applied to its input. If the input assumes a low voltage, the output will assume a high voltage level. If a positive voltage of proper level is applied to its input, the output will be minimum. Notice that the input and output square wave signals are of the same magnitude.

Only two voltage levels are permitted in digital circuitry and these voltage levels must be maintained by each circuit. As indicated before, the output square wave is inverted with respect to the input. This inversion process gives the circuit its name—inverter. The alternate name, NOT gate, which also is applied to this circuit, is a description of the circuit action; i.e., when the input signal is high, the output signal is not high (or low).

A symbol for the inverter was used in Fig. 1-6. The symbol is adequate to describe the action of the inverter but does not explain the component action required to accomplish this function. Figure 1-7 shows the schematic of a typical inverter. The device uses only one transistor to accomplish the inversion process. Actually, IC inverters are not commonly available as single devices. Typical ICs usually have six inverters in one package. All six of the inverters will be constructed on the same semiconductor chip. Even so, each inverter operates independently just as though it were a discrete device. In this circuit, as with all integrated circuits, standard transistor action is observed.

Transistor Theory Review

A brief review of transistor action may be helpful at this point. There are of course two types of semiconductor materials, p types and n types. P-type semiconductor material is usually composed of silicon or germanium that has been "doped" with minute quantities of an impurity element such as indium or gallium. The result is a chemical structure that provides an excess of positive ions. N-type material is doped with an element such as arsenic or antimony, which results in an excess of electrons. The importance of this semiconductor structure is that when voltage is applied to it, electromagnetic actions are initiated that cause the positive ions in the p-type material to drift away from a positive voltage source and toward a negative voltage source. Electrons react to voltage polarity in the opposite way, drifting toward the positive and away from the negative.

When p-type and n-type materials are placed next to one another, a p-n junction is formed. The characteristics of this junction constitute the basis upon which all semiconductor devices operate. When the junction is first formed, the electrons and ions begin to cross it. The negative electrons cross the junction and enter the positive p-type material. The positive ions cross the junction in the opposite direction, entering the n region. Soon, the excess charges are depleted near the junction, and the drifting of the carriers (ions and electrons) ceases. (See Fig. 1-8a.) The junction will remain in this stable condition unless external voltages are applied to the two regions creating the junction. If the positive

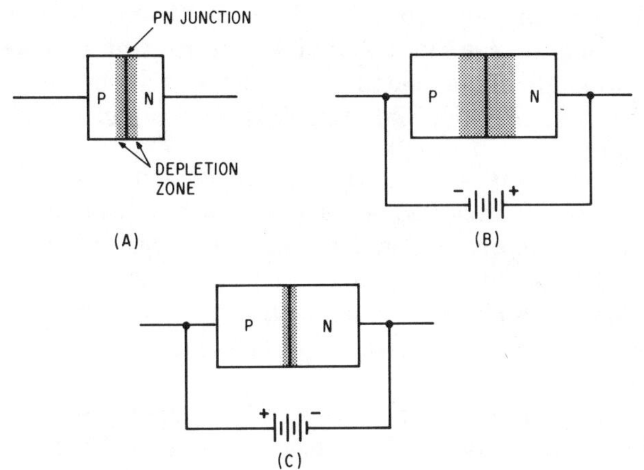

Fig. 1-8 P-n junction: (a) depletion zone, (b) reverse bias, and (c) forward bias.

voltage source is connected to the n region and the negative voltage source to the p region, the electrons and ions will flow into the voltage source, and the depletion zone will be increased in width, as shown in Fig. 1-8b. Increasing the applied voltage causes an increase in the depletion zone with a resultant increase in the effective resistance of the junction. When a junction is connected in this fashion, it is said to be *reverse biased.*

If the voltage source is connected with the positive to the p region and the negative to the n region, electrons and ions are forced toward the junction. This causes the depletion zone to become more narrow. Indeed, the carriers cross the junction and flow in a complete circuit; i.e., from the source, through the semiconductor and back to the source. (See Fig. 1-8c.) Since the depletion zone is now quite narrow, very little resistance is encountered and the current flows easily. In this situation, the junction is said to be *forward biased.*

A semiconductor diode is simply a p-n junction. The characteristics of the junction apply to the diode. When the external voltages forward bias the diode, current flows easily through the diode. When the voltages reverse bias the diode, it represents a high resistance to the flow of current.[3]

A bipolar transistor, as shown in Fig. 1-9, is composed of two p-n junctions that share a common element, the base. If the base is constructed of n-type material, the transistor is a pnp device. If p-type material is used for the base, the device is an npn transistor. The sharing of the base element and the design of the base as a very thin element result in important interaction between external circuits of the two junctions. The junctions are biased so that one junction is forward biased and the other is reverse biased. This is shown in Fig. 1-10a. Figure 1-10b shows the symbols used to represent transistors. Notice that the arrowhead of the symbol for the transistor points toward the n-type material. Signal voltages may be combined with the bias voltage to produce amplification. Various configurations are shown in Fig. 1-11. The flow of current through the emitter junction varies as the input signal adds to or reduces the bias voltage. Because the base material is thin, the emitter current flow results in charge carrier "diffusion" into the other junction. The charges cross the second junction and flow out to the

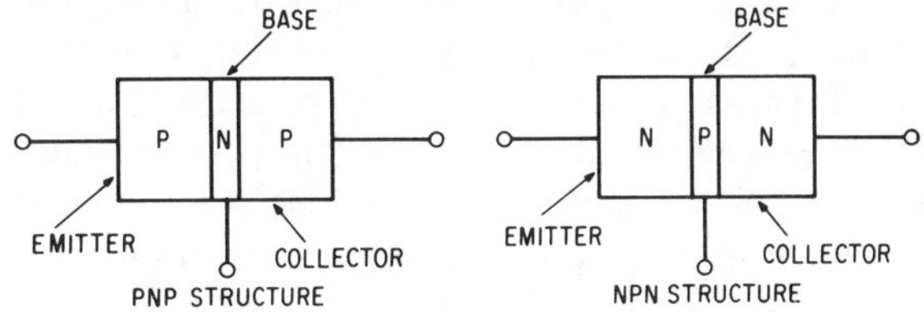

Fig. 1-9 Junction transistor types.

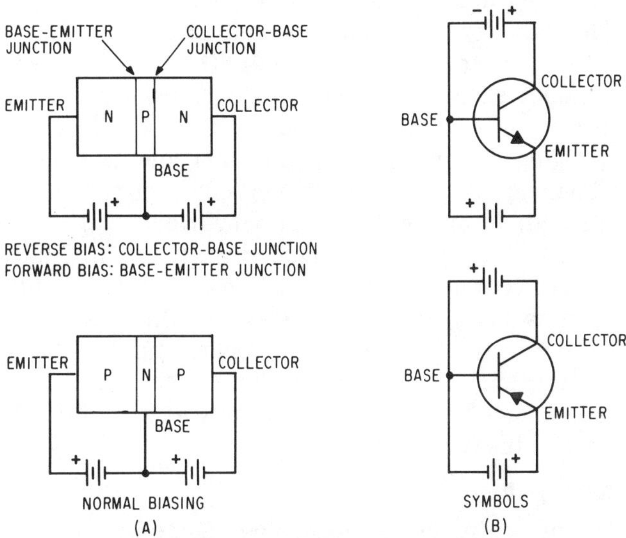

Fig. 1-10 Bipolar transistors showing typical biasing: (a) symbolic drawing and (b) schematic drawing.

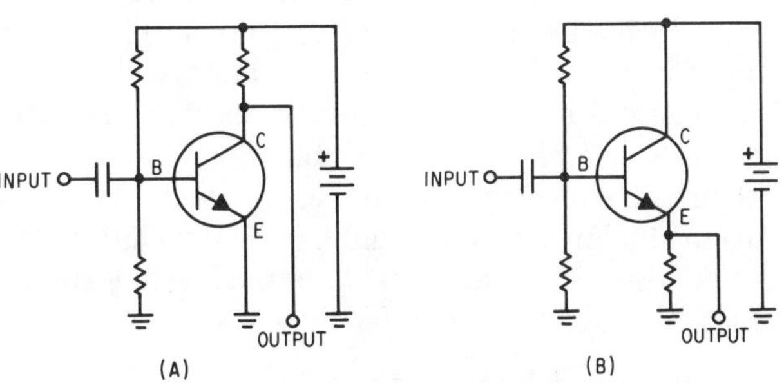

Fig. 1-11 Current amplification configuration: (a) common emitter and (b) common collector.

[3]This discussion of transistor theory is abbreviated. Details such as major and minor carriers have not been included in the interest of space. The purpose of this section is to provide a foundation for understanding the following circuit discussions. The reader is encourged to consult a transistor design text for a more detailed treatment.

collector. This second junction is reverse biased. When these charges cross the collector junction, they are in a high resistance area. Because of this high resistance, the relatively small current flow results in relatively high output power. The explanation for this is the power formula, which says that a small current flowing through a large resistance will result in an output power proportionate to the current squared times resistance ($P = I^2R$). In analog circuits this is used to obtain amplification from the transistor. Digital circuits simply turn the transistor on and off to obtain either a "full on" output signal or "no signal."

The npn transistor with *no* input is turned *off*. In this off condition, no current is drawn through the collector resistor (see Fig. 1-7) resulting in no voltage drop across the resistor. With no voltage drop, the output displays a positive value approximately equal to the power supply voltage. This positive level is called a "high" logic level. The same circuit action results if the input is grounded; i.e., brought to a "low" logic level.

The transistor will act differently, however, if a positive voltage is connected to the input resistor, Ri. This will forward bias the emitter circuit and reverse bias the collector circuit. The transistor will conduct current through collector resistor, Rc, resulting in a voltage drop across Rc and effectively bringing the output to ground or a low logic level. The inversion process is evident in this circuit. As indicated, the circuit is called an inverter.

The inverter is a simple digital circuit. There are, of course, many other types of digital circuits, such as AND gates, NOR gates, and flip-flops. Most of these circuits are simply combinations of basic circuits such as the inverter. This aspect of IC technology will be discussed more fully as it pertains to each of the IC families.

IC Families

Integrated circuit technology has undergone an evolution of its own. The milestones of the field mark progress in both fabrication techniques and semiconductor design. Several types of semiconductor technologies have been developed. Devices that are constructed under the same technology display a fundamental resemblance to one another; i.e., they are built from the same basic building blocks. It has become customary, therefore, to refer to these various technologies as families of ICs. While different families are often capable of performing similar functions, they do so in different ways.

Resistor-Transistor Logic

The inverter described in the preceding section is characteristic of a family of ICs called *Resistor-Transistor Logic* or *RTL*. It is apparent from the schematic (see Fig. 1-7) that the circuit contains only two kinds of components—resistors and transistors. The name is derived from this component complement.

RTL was developed early in the era of IC technology. It heralded a new age in electronics and began the transition from discrete components to integrated circuitry. As with any new field, this early IC technology was destined for improvement, refinement, and virtual replacement by later developments.

A family of ICs share common characteristics, just as other types of families have family characteristics. These common traits are helpful in understanding the capabilities and limitations of the IC family. It has already been stated that the inverter (see Fig. 1-7) is the basic building block of RTL circuits. The circuit operation of the inverter has been discussed in a general way. However, a more detailed consideration of the circuit will assist in understanding the RTL family.

RTL is referred to as a saturated logic family. This means that for the circuit to provide the on/off, high/low, output signal, it is necessary that the transistor alternate between fully conducting and fully cutoff states. The saturated condition results in a low logic level output, the cutoff state in a logical high output. To saturate a bipolar transistor, its emitter, base, and collector voltage must be adjusted so that the transistor turns on and conducts heavily. As explained earlier, when a sufficiently positive voltage is applied to the base of the transistor, the transistor turns on or saturates. In this state the transistor provides a relatively low resistance path from collector to ground. Collector current results in a small voltage across the transistor (a low resistance). The collector voltage is close to 0 V, the low logic level. In the cutoff state, no current flows in the collector circuit and the output rises to a positive level approximately equivalent to the power supply level. This high output voltage corresponds to a high logic level.

Standard supply voltage for RTL is 3.6 V. RTL devices will operate on 5 V, providing compatibility with TTL[4] devices that operate on 5 V. However, increasing the voltage does not improve the operation of RTL devices but simply increases the power dissipation. Power dissipation is an important consideration in choosing a logic family. For a single RTL device, the power dissipation will be in the milliwatt range. However, if hundreds of RTL gates and circuits are connected together to form a sophisticated electronic machine such as a computer, the ac-

[4]TTL stands for Transistor-Transistor Logic, which will be covered later in this chapter.

cumulated power dissipation can reach many amperes and create excessive heat. This relatively large power dissipation is a major limitation of RTL.

An important characteristic of logic devices is the number of other devices (of the same family) that can be driven by a given logic gate. This capacity is called the *fan-out*. Two elements of the design of a given logic family determine the driving capacity or fan-out. The first parameter is the amount of current needed to drive the input of a logic circuit. The second parameter is the amount of current available at the output of a logic gate (without exceeding the limits of the defined logic levels). The ratio of the available output (called *sinking* capability) to the required input current determines the fan-out.

For instance, a typical TTL logic output will sink approximately 16 mA of current. A typical input will require approximately 1.6 mA to place it in the saturated mode. By simple calculation, it is apparent that the TTL output is capable of driving ten TTL inputs. If too many inputs are placed on an output, the output will be unable to sink sufficient current to activate all of the circuits, and output voltages will no longer correspond to the logic levels of a given family. To reduce the difficulty of logic design, a convention has been established that refers to logic device drive and load requirements as relative units rather than precise current levels. The output drive is referred to as *fan-out* and the input load as *fan-in*. Fan-in and fan-out are rated in compatible units. If a device is listed as having a fan-out of three, it will be capable of driving a device with a fan-in of three, or three devices that each have a fan-in of one. A typical RTL fan-out is five. Typical fan-in for RTL is two or three. Compared with other logic families, RTL fan-out is low. When multiple gates must be driven, a supplementary device called a *buffer* will often be necessary. The buffer provides a larger fan-out capacity. There are, however, problems in using buffers as will be explained in a later section. Low fan-out is a shortcoming of RTL.

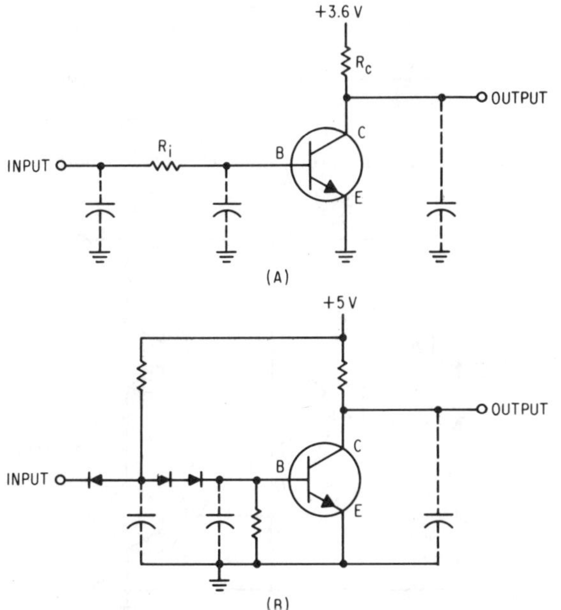

Fig. 1-12 Stray capacitances of (a) RTL inverter and (b) DTL inverter.

The construction of RTL devices contribute to two additional problems. One is speed of operation, and the second is noise immunity. Operating speeds of logic devices are normally measured in nanoseconds (billionths of seconds). The speed of operation of RTL devices is slower than the speed of some other IC families. The reason for the slower speed is the nature of the circuit and the circuit action of RTL. While RTL circuits only use resistors and transistors, the proximity of the components and printed wiring adds stray capacitance to each circuit. Figure 1-12 shows this stray capacitance connected as actual circuit components. Dashed lines have been used to indicate that it is stray circuit capacitance. Notice that the capacitance is across the input and output of the circuit. Visualized in this way, resistance/capacitance combinations can easily be seen in the circuits.

When the transistor is in the unsaturated state, the output capacitance will charge to a voltage level equal to the power supply. If the input is now made high, the input capacitance will have to be charged up to a level that will turn the transistor on. Since the input resistance and the input capacitance form an RC time constant, the turn-on speed of the transistor will be slowed. In a similar fashion, the speed of the output level change will be reduced by the output capacitance discharging through the transistor. When the transistor is turned off once again (by driving to a low voltage level), the input capacitance is discharged through the input resistance, and the output capacitance is charged through the load resistance. The two RC time constants will again limit the speed at which the circuit can change levels. The speed of the RTL circuit is not subject to change by external power supplies or circuitry. RTL speed represents a design decision made at the time of manufacture of the circuit. The decision must be made between circuit speed and circuit power consumption. If greater speed is required, a different family of ICs must be used.

The noise immunity of a logic family relates to logic levels. As indicated previously, there are only two logic levels, high and low. For RTL, a logic high is between 2 and 3.6 V.[5] An RTL low is between 0 and 0.4 V. Between 0.4 and 2 V, the circuit action may be erratic. For this reason, the voltage corridor

[5]RTL can use 5 V if increased power dissipation can be tolerated.

between low and high levels is called a forbidden region and must be avoided when designing and operating circuits. If the input to a logic gate rises above 0.4 V for a logical zero (low) the gate may not function properly. Thus, the noise immunity or signal that this family will reject without error is 0.4 V.

Noise immunity is less for RTL circuits than for some other IC families. Again, the reason is circuit construction. As shown in Fig. 1-12a, the base of the transistor connects directly to the input resistor. If a voltage is applied to the input, it must overcome the diode drop of the transistor base emitter circuit before it will turn the transistor on. Since the normal voltage drop of a silicon junction diode is approximately 0.4 V,[6] the circuit will consistently reject noise spikes less than 0.4 V. If the noise results in a level greater than 0.4 V, the transistor may be placed into a state of conduction. This low threshold results in the decreased noise immunity characteristic of RTL.

RTL is available in two basic types. One type is standard RTL, which conforms to what has been discussed. A second type retains all of the family characteristics of RTL but significantly reduces power dissipation (1/8 the dissipation of standard RTL). This power dissipation reduction is obtained by increasing the value of the collector resistor. Dissipation is effectively reduced, but speed also is reduced. The reason for the speed reduction is the circuit capacitance. The capacitance must still be charged and discharged, and with the larger resistors, the time constant is increased and the state change time is increased.

Diode-Transistor Logic

Closely allied with RTL is an IC family called *Diode-Transistor Logic* or *DTL*. DTL succeeded RTL. Indeed, in many instances DTL supplanted RTL. The reason for DTL's popularity over RTL is increased noise immunity. The circuit shown in Fig. 1-12b is a DTL inverter. Comparison of this circuit with that of the RTL inverter (Fig. 1-12a) indicates that the primary difference between the two is the nature of the base-driving circuit of the transistor. The circuit still operates by bringing the output low during transistor saturation and high when the transistor is off. Since DTL is saturated logic, power drain will be greater than is true of families such as MECL (to be discussed later). Due to the construction of DTL, there is an advantage in using 5 V as the supply voltage. This supply voltage, greater than the 3.6 V of RTL circuitry, permits a greater distinction between a logic high and a logic low. At the same time, the two series diodes raise the signal threshold from the single diode drop of RTL to a double diode drop or 0.8 V. This means that for noise to turn the transistor on, it would have to be a minimum of 0.8 V. This feature effectively doubles the noise immunity of DTL over that of RTL. A side benefit of DTL construction is increased fan-out. DTL typically has a fan-out of eight. DTL also enjoys increased speed over RTL. The reason for this is that the distributed input capacitance is dissipated through a forward biased diode that represents a very low resistance. The capacitance can therefore be discharged much more quickly, and the change from one state to the other is accomplished more rapidly.

Transistor-Transistor Logic (TTL)

DTL provided some advantage over RTL. However, DTL was somewhat eclipsed by a later development called *Transistor-Transistor Logic (TTL)*. TTL has become an extremely popular logic family. Preeminent among the factors contributing to TTL's popularity is its speed of operation. The secret of TTL's increased speed is an output configuration that is referred to as a "totem pole" (see Fig. 1-13). The two output transistors complement

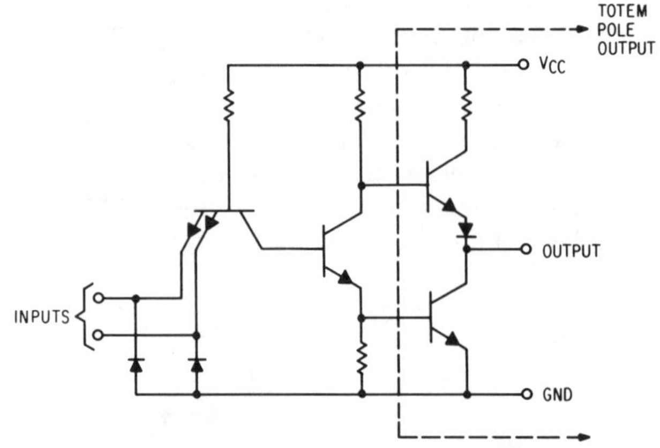

Fig. 1-13 TTL gate circuit showing totem pole output.

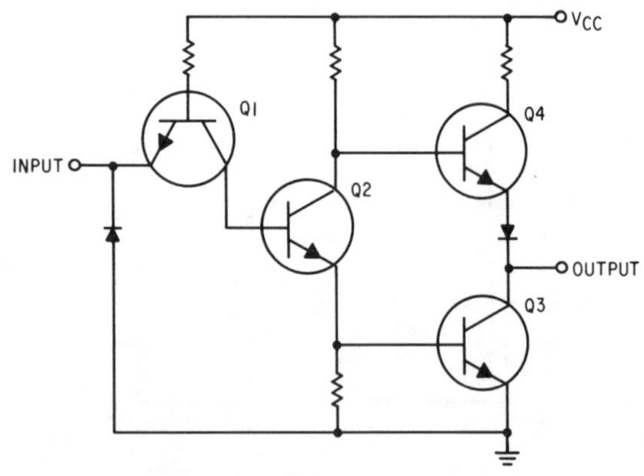

Fig. 1-14 TTL inverter.

[6]Silicon-diode voltage drop is often given as 0.6 V. Using this figure will alter the noise immunity figures given in the text.

each other's action. The TTL inverter shown in Fig. 1-14 demonstrates the fundamental operating rules of TTL. The input is connected to the emitter of the input transistor rather than to the base, as is true of RTL. If the emitter of the input transistor is attached to ground or a logic level low, the transistor will be correctly biased for operation in a saturated "on" condition. This will effectively ground the base of the transistor Q2 and turn it off. With Q2 off, the collector will be positive and Q4 will turn on. Q3, the complement of Q4 in the totem pole, will turn off simultaneous with Q4 turning on. The result of the process is that the output will be positive, a logic level high. The inversion process is easily seen: low input, high output. If the input is now connected to a logic high, the input transistor emitter base junction will be reverse biased. The base collector junction of Q1 will now function as a forward-biased diode and will supply a positive level to the base of Q2, which in turn will provide a positive level to the base of Q3. Q2 will turn on bringing its collector and the base of Q4 to ground. This will turn Q4 off while Q3 will turn on. The output will go low and be capable of sinking considerable current from outside circuits. The circuit action of the inverter will be the same if the input is left unconnected since the emitter base junction is still not forward biased.

The speed of TTL results primarily from the complementary action of the two totem pole transistors. As indicated previously, when one of the totem pole transistors turns on, the other turns off. These transistors do not change immediately from one state to the other, however. As they change from one condition to the other, each conducts heavily. This results in greatly increased switching speed, but it also produces a large current spike that can produce noise in other TTL circuits. This problem is normally reduced by liberal use of bypass capacitors.

The transistor-to-transistor construction of TTL circuits also promotes rapid switching. The input and output stages employ no resistors that would create slow time constants. The intermediate stage is designed to ensure fast operation at this point in the circuit. Speed is enhanced still further through the interaction of Q1 and Q2. When Q1 is turned on, it forces Q2 off. Q1 operates as a dynamic switch, not only turning Q2 base bias off but actively forcing the base bias to ground at a more rapid rate.

TTL is faster than RTL and DTL IC families, but it is still a saturated logic technology. Saturated logic design has an inherent speed limitation. The change from a saturated to an unsaturated condition requires the dissipation of junction charges that accumulate during saturation. In addition to this speed limitation, saturated circuitry results in current spikes that can cause erratic operation of TTL circuits. As will be discussed shortly, other technologies offer improvement in these limitations. Even with its limitations, however, TTL is still the most popular IC family for small-scale integration projects.

Normal fan-in for standard TTL circuits is one. TTL fan-out is usually ten. TTL is referred to as *current-sinking logic*. TTL outputs must accept and short to ground sufficient current from external circuits to turn on the input transistor. When the TTL output is high, current is not exchanged with other circuits since the positive voltage level reverse biases the input transistor of the driven circuit. (Only a small leakage current has to be supplied.)

Noise immunity of TTL is similar to DTL circuits. TTL logic low is from 0 to 0.8 V. Logic high is from 2 to 5 V positive. As explained previously, TTL does have the problem of current spikes, which can create erratic operation if not included in the design decisions.

TTL is available in several forms. Standard TTL has just been described. As indicated, this technology is very popular even with its limitations of power consumption and speed. Because of TTL's popularity, other forms were created to overcome these specific problems. *Low-power TTL* was devised to reduce power consumption and fan-in. This is achieved by increasing the base resistance of Q1 (see Fig. 1-14). With a larger resistance, the current that must be sunk through Q2 and the input transistor (Q1) is reduced by a factor of ten. Unfortunately, low-power operation is achieved at the expense of speed. The larger resistance reduces power consumption and at the same time slows the time constant of the RC circuit formed by the resistor and the stray capacitance of the circuit. As an example of the low-power option, typical power consumption per gate for standard TTL is 10 mW and for low-power TTL is 1 mW. The speed of standard TTL is approximately three times that of low-power TTL.

Figure 1-15 shows the schematic of a high-speed TTL device. As can be seen, the primary change is the addition of an extra amplifier Q5, between Q2 and the totem pole. The function of this transistor is to force the output transistors to change states more quickly. In addition to the amplifier, resistor values have been modified to ensure rapid circuit action. The result is a TTL device with switching speeds almost twice that of standard TTL and concomitantly, more than twice the power dissipation. These special devices are designed to achieve specific goals. Unless high-speed or low-power dissipation is a necessity, the advisability of using specialized devices is questionable and probably not cost effective.

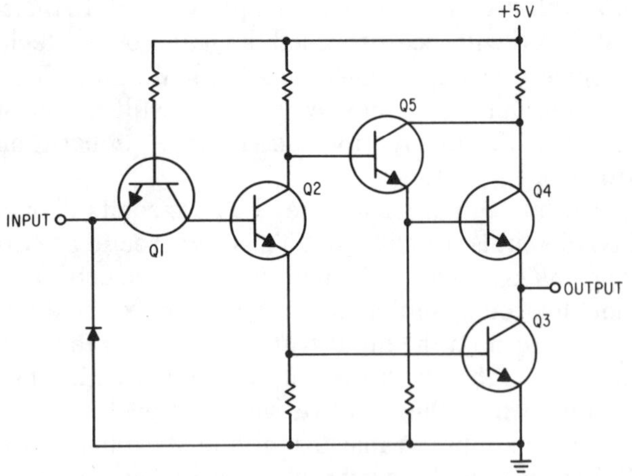

Fig. 1-15 High-speed TTL device.

There are other types of TTL devices available. These are the Schottky and low-power Schottky TTLs. The Schottky design offers significant speed advantages with less power dissipation than that of high-speed TTL. In the low-power version, the speed is approximately equal to standard TTL, but the power consumption is only one-fifth as much.

Emitter-Coupled Logic (ECL)

From the previous discussion of RTL, DTL, and TTL, it should be apparent that speed is an important characteristic of an IC device. It was indicated that RTL is relatively slow. TTL is faster, and high-speed TTL faster yet. However, even the speed of high-speed TTL and Schottky TTL is inadequate for some applications. If a digital frequency counter is to count a high frequency signal, for example, the digital circuit must be capable of switching at an RF frequency rate. If the switching rate is exceeded, errors will occur. To provide for those applications that require very fast switching, other technologies were created. *Emitter-Coupled Logic (ECL)* is a technology developed for use in applications requiring increased speed (see Fig. 1-16).

ECL achieves its speed by using nonsaturated techniques. Rather than defining highs and lows as full-on or full-off, ECL defines them as off or conducting. Drive signals are maintained at a level that will not place the transistors into saturation. The problem of the charge accumulation between the base and collector of saturated transistors was discussed previously. This charge accumulation is an inherent speed limitation for saturated logic technologies. ECL avoids this by not saturating its transistors. As with most advantages, some disadvantages result from ECL's method of operation. To preclude saturation, ECL uses logic levels that are significantly different from the other families we have discussed. An ECL high is approximately − 0.9 V, and an ECL low is in the − 1.7 V range. Notice that both levels are negative and that separation between high and low is 0.8 V. Considering that

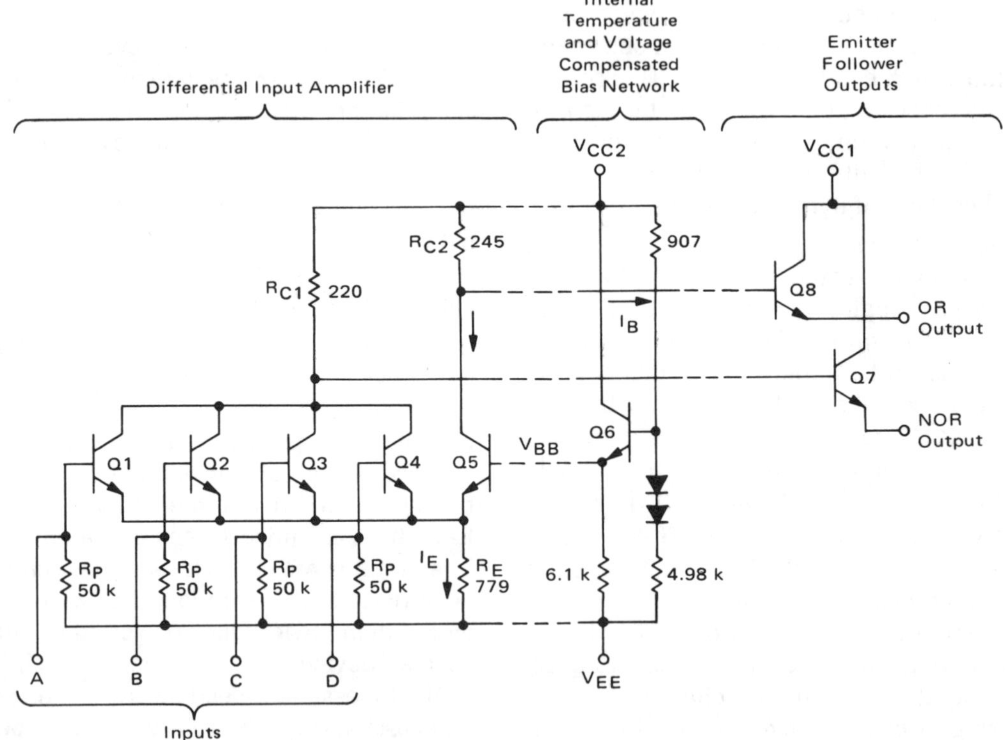

Fig. 1-16 Basic MECL gate. (Courtesy of Motorola, Inc.)

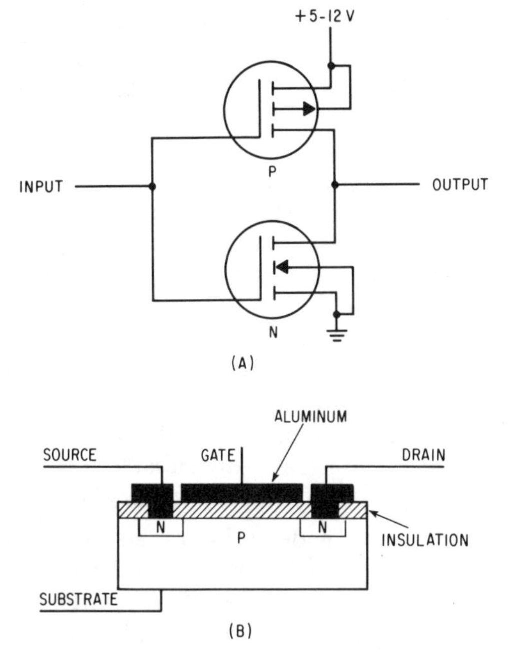

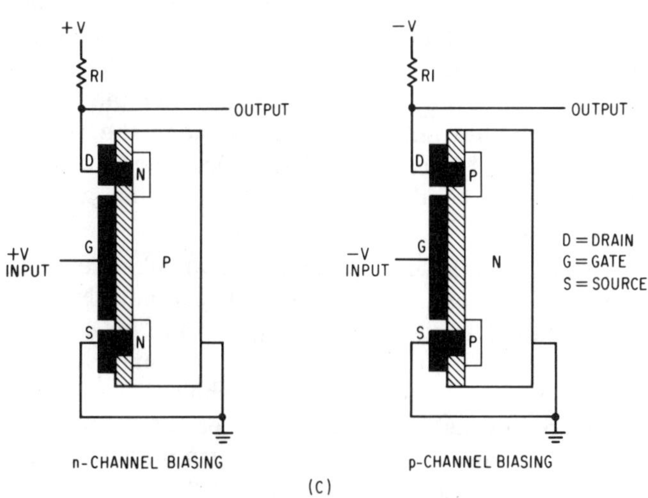

Fig. 1-17 MOS construction details: (a) CMOS circuit, (b) n-channel MOS showing lead designations, and (c) typical MOS biasing arrangements.

saturated logic we have discussed will not tolerate negative inputs and that the logic levels spread is much larger with saturated logic than is true with ECL, it is evident that ECL is not readily compatible with the saturated logic families. The smaller voltage swing does promote more rapid operation for ECL and special level shifters are available for interface of ECL and TTL and other families.

Unlike the saturated logic families that are driven into saturation by their external inputs, ECL compares the input with an internally derived voltage using a differential amplifier (Q5 and Q6 in Fig. 1-16). If the input is greater than the reference, the output will be high (OR output of Fig. 1-16). If the input is less than the reference, the output will be low. ECL often employs complementary outputs such that when one is high the other is low and vice versa.

One particular advantage of ECL is that it maintains a constant load on the power supply regardless of its state. This results in no current spikes, which eliminates noise problems.

Complementary Metal-Oxide-Silicon (CMOS)

All of the foregoing IC families have used traditional bipolar transistor construction. CMOS employs a different technology. For many applications, there are indications that CMOS may replace saturated logic in popularity. To understand the appeal of this new design, it is necessary to consider some preliminary information concerning the construction and operation of metal-oxide-silicon devices.

CMOS stands for Complementary Metal-Oxide-Silicon.[7] The MOS sandwich is the secret to advantages of drive requirements. The complementary construction results in decreased power consumption. A CMOS device is constructed with at least two transistors (see Fig. 1-17a). One transistor of this complementary pair will be of n-channel construction (lower transistor symbol in Fig. 1-17a) and the second will be p-channel (upper transistor in Fig. 1-17a).

The process of fabricating an n-channel transistor begins with the doping of a silicon substrate with a chemical impurity that produces a deficiency of electrons. This lack of electrons can be visualized as holes that are devoid of electrons. The process continues by doping two small n regions within the p substrate. Connections are made to the substrate and to the two n regions. The next fabrication step is to lay down a layer of silicon dioxide over the p substrate between the n regions. Finally, this silicon dioxide layer is plated with metal and a contact is attached to the metal. The n-channel transistor is now complete (see Fig. 1-17b). Names have been assigned to the connections made to the transistor. The connection to the silicon bar is referred to as the *substrate* lead. The connection to one n region is called the *source*, and the second n region connection is called the *drain*. The connection to the metal placed over the oxide layer is called the *gate*.

To place the n-channel transistor into operation, it must be biased properly. The p-type substrate and the two n regions form two p-n junctions. If the substrate lead is connected to ground and the drain is connected to a positive voltage source (via resistor R1 in Fig. 1-17c), the p-n junction they form will be

[7] CMOS has varying names among the literature—e.g., Complementary Symmetry Metal-Oxide-Silicon, Complementary Metal-Oxide-Semiconductor, etc.

reverse biased. With the source grounded, the p-n junction formed by it and the substrate is essentially shorted. If the gate was not included on the transistor, it would be a useless device. No conduction would take place. Using the gate, however, the n-channel transistor becomes a very useful device. The gate, constructed as a sandwich of metal, oxide, and silicon alters the operation of the p-n junction. Consider that the oxide layer is very thin and is a good insulator. With metal on one side and a silicon conductor on the other, the gate region behaves as a capacitor. If the gate is grounded, nothing happens. The negative force of electrons on the metal connector fosters the electron deficiency of the substrate between the source and drain. If, however, the gate is connected to the positive voltage source, the positive metal plate attracts electrons to the substrate between the source and drain. This accumulation of electrons causes the p region to take on the characteristics of an n region. Biased as indicated previously, the drain/substrate p-n junction will change from reverse to forward bias, and the source/substrate junction also will be conducting. As a result, the transistor will conduct through the channel between the drain and the source bringing the drain to a low condition due to a voltage drop through load resistor R1 (see Fig. 1-17c). When the gate is grounded, once again, the current flow through the channel ceases and the output goes to a high or positive level.

This same MOS technology can be used with an n-type substrate. In this case, the transistor will be called a p-channel transistor. The source and drains will, of course, be doped as p-type regions. The biases will vary to bring about conduction when a negative voltage is applied to the gate. Typical connections for MOS transistors are shown in Fig. 1-17c.

The circuit action described for both the n- and p-channel transistors is characteristic of enhancement mode MOS field-effect transistors. The designation *enhancement mode* simply means that the transistor must be acted upon to bring about conduction. In the case of these transistors, the action is the application of an appropriate voltage to the gate of the transistor. If no voltage is applied to the gate, the transistor will not conduct. A transistor that is designed to conduct with no voltage applied is called a *depletion mode device*. To cut the transistor off, the conducting channel must be depleted of the current carriers through the application of an appropriate voltage to the transistor gate. The devices (both enhancement and depletion) are called *field-effect transistors* because they react to the electric field imposed on their gates.

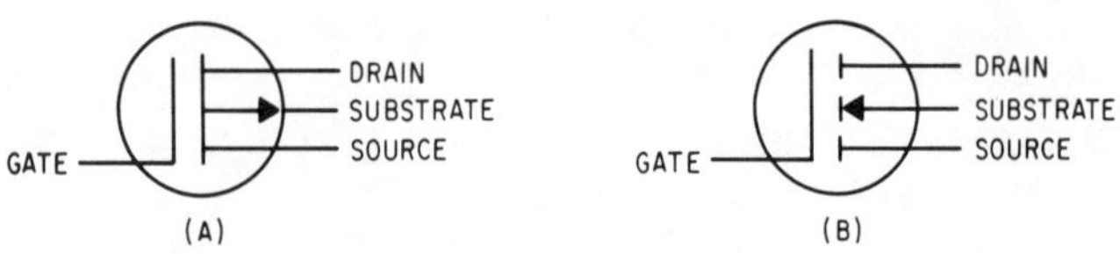

Fig. 1-18 MOS transistor symbols: (a) p-channel depletion type and (b) n-channel enhancement type.

The schematic representation of the MOS transistors is shown in Figs. 1-18a and 1-18b. The arrow on the substrate lead indicates whether the substrate is p- or n-type material. If the arrow points toward the substrate, the substrate is p type and the transistor is an n-channel MOS. If the arrow points away from the substrate, the substrate is of n-type material and the transistor is p-channel MOS.

The operating mode of the transistor also is indicated by the schematic diagram. In Figs. 1-18a and 1-18b notice the bars between the gates and the drain, substrate, and source leads. In Fig. 1-18a the bar is continuous between the three leads. This represents a depletion mode device that will conduct unless prevented from doing so by an applied gate bias. Figure 1-18b shows a device in which the bar is segmented and not continuous between the leads. This is indicative of an enhancement mode device that will not conduct unless biased by the proper gate voltage.

The circuit in Fig. 1-17a shows two enhancement mode MOS transistors connected together. Notice that the top transistor is PMOS (p-channel MOS) and the bottom transistor is NMOS (n-channel MOS). With voltage connections made as shown, the circuit action is interesting. If the two gates are connected to ground, the top transistor will turn on due to the voltage difference between the substrate and gate while the bottom transistor will turn off due to the zero voltage difference between its gate and substrate. The lead marked output will effectively be connected to the positive supply voltage through the top transistor. If the gates are now connected to the positive voltage source, the top transistor will have a zero voltage difference between the gate and substrate resulting in an off condition. At the same time the bottom transistor will be biased on, and the output, effectively connected to ground through the bottom transistor will be low. Examination of this circuit action; i.e., low input = high output and high input = low output, will reflect an inverting action. This circuit is a CMOS inverter.

The "C" of CMOS stands for complementary. Complementary-metal-oxide-silicon devices use PMOS and NMOS transistors to complement one another's circuit action, which was done in the inverter. This complementary action results in several advantages. Rather than the tug-of-war that the TTL totem pole provided, the CMOS complementary pair work in concert. When one is on, the other is off. Reverse the gate bias and the transistors reverse their status. There is no competition between them and therefore no wasted current. Each transistor performs independently, and this independent action takes place in a complementary fashion.

Another benefit of CMOS circuitry is the resulting noise immunity. CMOS circuits switch without producing voltage spikes. This helps keep the circuit free of noise. In addition, CMOS does not depend on diode drops for its switch threshold but rather switches halfway between ground and the positive voltage level. Add to this the very low current flow in CMOS circuits, the very low resulting voltage drops, and a voltage almost equal to ground and the positive supply characterize CMOS low and high levels respectively. Using a typical supply of + 12 V, the noise immunity of the device is significantly improved over that of other families.

Much of the low-power dissipation of CMOS is due to the capacitor input. This represents an extremely high impedance that draws only leakage current when not switching. During switch transitions, the capacitor charge must be dissipated and then reformed at the opposite polarity. Current will flow during these transition periods. However, unless the switching is occurring at a rapid rate, the current drain will be miniscule in comparison with bipolar technologies.

CMOS is not without shortcomings. The most significant CMOS limitation is speed. Typically, standard CMOS circuits are limited to maximum speeds of 5 MHz. Some premium circuits operate at multiples of this speed, but the cost also is increased. The primary culprits that decrease the speed of CMOS circuits are the input and stray capacitances. If the circuit does not require excessive speed, however, CMOS is a very good choice.

A second CMOS shortcoming is the fragility of the input capacitor. The silicon dioxide dielectric layer is easily punctured by static electricity. For this reason, CMOS must be stored, handled, and installed with care. Some of these safeguards are described in the section on construction techniques. Protective input diodes are usually provided for CMOS circuits to decrease static charge problems. Even with these few limitations, CMOS promises to become very popular because of its benefits of low power requirements and high noise immunity. Table 1-1 summarizes some of the important characteristics of the IC families. Using the table, comparison is easily made.

The Data Sheet

Description of every integrated circuit that is currently available would require a massive volume. It becomes important, therefore, that a prospective IC designer be able to find information on any specific IC. This type of information is usually available in the form of manufacturer's data sheets. Data sheets contain a lot of useful information about the ICs. Among other things, the data sheet provides technical specifications for the IC. Voltages, current drain, speed of operation, and basic construction details are in the data sheet. Practical information such as "pin-out" details (what connects to which pin), application suggestions, and fan-in/fan-out considerations help in the design of working circuits. So becoming familiar with the IC data sheet is well worth the effort.

The data sheet for the Motorola MC14011 NAND Gate is shown in Fig. 1-19. This data sheet is typical and will be used to familiarize the reader with the data sheet.

Notice that at the top of Fig. 1-19a, the function of the device is listed, which is "NAND" Gate. As will be discussed in the next chapter, the NAND gate is an elemental logic device that is used in the construction of more sophisticated logic devices. The NAND gate operates by outputting a low only when both inputs are high. Data sheet designations of ICs normally follow positive logic action. Thus the

Table 1-1 IC family characteristics.

	Typical Supply Voltage	*Logic High Levels**	*Logic Low Levels**	*Typical Fan-out*	*Typical Power (per gate)*	*Frequency Limit*
RTL	+ 3.5 V	2.4 to 3.5 V	0 to 0.4 V	5	0.025 W	8 MHz
DTL	+ 5 V	2.4 to 5 V	0 to 0.8 V	8	0.011 W	30 MHz
TTL	+ 5 V	2.4 to 5 V	0 to 0.8 V	10	0.01 W	50 MHz
ECL	− 5 V	− 0.75 V	− 1.6 V	25	0.025 W	100 MHz
CMOS	+ 5 to + 18 V	8 V to 12 V (for 12 = V supply)	0 to + 4 V (for 12 = V supply)	50	0.001 W	10 MHz

*These levels are necessary for consistent logic switching.

"NAND" GATE

MC14011AL
MC14011CL
MC14011CP

QUAD 2-INPUT "NAND" GATE

The MC14011 quad 2-input NAND gate finds primary use where low power dissipation and/or high noise immunity is desired.

- Quiescent Power Dissipation = 2.5 nW/package typical @ 5 Vdc
- Noise Immunity = 45% of V_{DD} typical
- Diode Protection on All Inputs
- Supply Voltage Range = 3.0 Vdc to 18 Vdc (MC14011AL)
  3.0 Vdc to 16 Vdc (MC14011CL/CP)
- Single Supply Operation – Positive or Negative
- High Fanout – > 50
- Input Impedance = 10^{12} ohms typical
- Logic Swing Independent of Fanout
- Pin-for-Pin Replacement for CD4011A

McMOS

(LOW-POWER COMPLEMENTARY MOS)

QUAD 2-INPUT "NAND" GATE

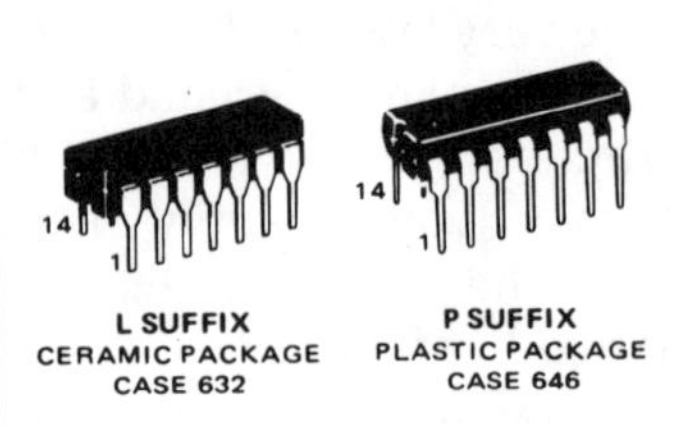

L SUFFIX
CERAMIC PACKAGE
CASE 632

P SUFFIX
PLASTIC PACKAGE
CASE 646

MAXIMUM RATINGS (Voltages referenced to V_{SS}, Pin 8.)

Rating		Symbol	Value	Unit
DC Supply Voltage	MC14011AL MC14011CL/CP	V_{DD}	+18 to -0.5 +16 to -0.5	Vdc
Input Voltage, All Inputs		V_{in}	V_{DD} to -0.5	Vdc
DC Current Drain per Pin		I	10	mAdc
Operating Temperature Range	MC14011AL MC14011CL/CP	T_A	-55 to +125 -40 to +85	°C
Storage Temperature Range		T_{stg}	-65 to +150	°C

This device contains circuitry to protect the inputs against damage due to high static voltages or electric fields; however, it is advised that normal precautions be taken to avoid application of any voltage higher than maximum rated voltages to this high impedance circuit. For proper operation it is recommended that V_{in} and V_{out} be constrained to the range $V_{SS} \leq (V_{in}$ or $V_{out}) \leq V_{DD}$.

Unused inputs must always be tied to an appropriate logic voltage level (e.g., either V_{SS} or V_{DD}).

CIRCUIT SCHEMATIC

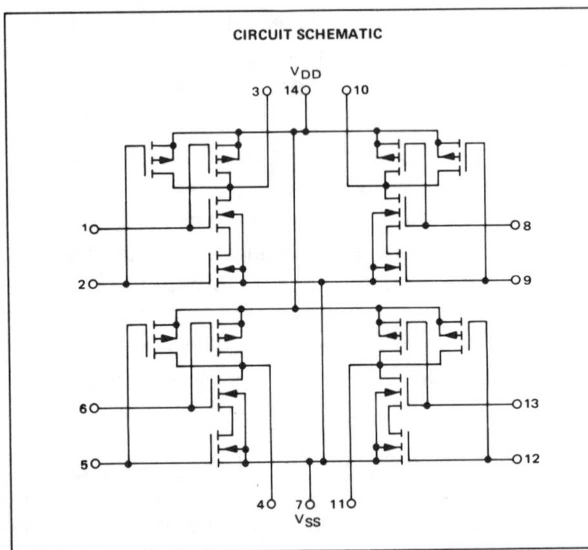

LOGIC DIAGRAM
POSITIVE LOGIC

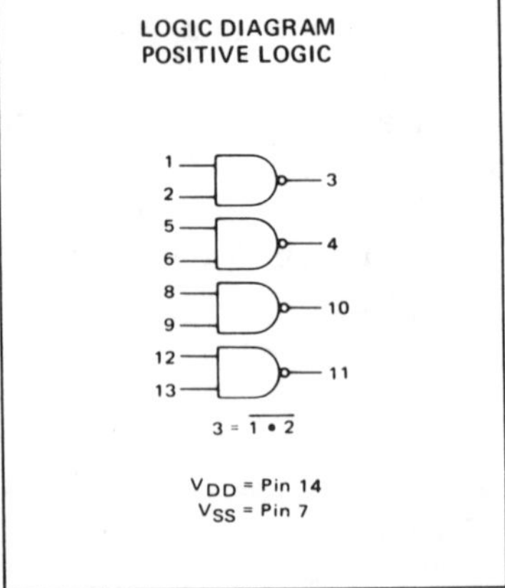

(A)

Fig. 1-19 Data sheet for the Motorola MC14011 NAND Gate. (Courtesy of Motorola, Inc.)

NAND gate will AND the two logic high inputs to obtain a logic high output but will then invert that high output to provide a logic low output. The "N" of the NAND means "not" and acts as an inverter. Gate functions will be discussed in Chap. 2. In the left-hand box of the data sheet (see Fig. 1-19a), the device is described as a "Quad 2-Input NAND Gate." This means that the 14-pin dual-in-line package (DIP) contains four independent NAND gates. The pin connections of the DIP are shown in the schematic and

ELECTRICAL CHARACTERISTICS

Characteristic	Symbol	V_{DD} Vdc	T_{low}* Min	T_{low}* Max	25°C Min	25°C Typ	25°C Max	T_{high}* Min	T_{high}* Max	Unit
Output Voltage "0" Level	V_{out}	5.0	–	0.01	–	0	0.01	–	0.05	Vdc
		10	–	0.01	–	0	0.01	–	0.05	
		15	–	0.05	–	0	0.05	–	0.25	
"1" Level		5.0	4.99	–	4.99	5.0	–	4.95	–	Vdc
		10	9.99	–	9.99	10	–	9.95	–	
		15	14.95	–	14.95	15	–	14.75	–	
Noise Immunity#	V_{NL}									Vdc
($\Delta V_{out} \leq$ 1.4 Vdc)		5.0	1.5	–	1.5	2.25	–	1.4	–	
($\Delta V_{out} \leq$ 2.8 Vdc)		10	3.0	–	3.0	4.50	–	2.9	–	
($\Delta V_{out} \leq$ 3.5 Vdc)		15	3.75	–	3.75	6.75	–	3.6	–	
($\Delta V_{out} \leq$ 1.4 Vdc)	V_{NH}	5.0	1.4	–	1.5	2.25	–	1.5	–	Vdc
($\Delta V_{out} \leq$ 2.8 Vdc)		10	2.9	–	3.0	4.50	–	3.0	–	
($\Delta V_{out} \leq$ 3.5 Vdc)		15	3.6	–	3.75	6.75	–	3.75	–	
Output Drive Current (AL Device)	I_{OH}									mAdc
(V_{OH} = 2.5 Vdc) Source		5.0	-0.62	–	-0.50	-1.7	–	-0.35	–	
(V_{OH} = 9.5 Vdc)		10	-0.62	–	-0.50	-0.9	–	-0.35	–	
(V_{OH} = 13.5 Vdc)		15	-1.8	–	-1.5	-3.5	–	-1.1	–	
(V_{OL} = 0.4 Vdc) Sink	I_{OL}	5.0	0.50	–	0.40	0.78	–	0.28	–	mAdc
(V_{OL} = 0.5 Vdc)		10	1.1	–	0.90	2.0	–	0.65	–	
(V_{OL} = 1.5 Vdc)		15	4.2	–	3.4	7.8	–	2.4	–	
Output Drive Current (CL/CP Device)	I_{OH}									mAdc
(V_{OH} = 2.5 Vdc) Source		5.0	-0.23	–	-0.20	-1.7	–	-0.16	–	
(V_{OH} = 9.5 Vdc)		10	-0.23	–	-0.20	-0.9	–	-0.16	–	
(V_{OH} = 13.5 Vdc)		15	-0.69	–	-0.60	-3.5	–	-0.48	–	
(V_{OL} = 0.4 Vdc) Sink	I_{OL}	5.0	0.23	–	0.20	0.78	–	0.16	–	mAdc
(V_{OL} = 0.5 Vdc)		10	0.60	–	0.50	2.0	–	0.40	–	
(V_{OL} = 1.5 Vdc)		15	1.8	–	1.5	7.8	–	1.2	–	
Input Current	I_{in}	–	–	–	–	10	–	–	–	pAdc
Input Capacitance (V_{in} = 0)	C_{in}	–	–	–	–	5.0	–	–	–	pF
Quiescent Dissipation (AL Device) (Per Package)	P_Q	5.0	–	0.00025	–	0.0000025	0.00025	–	0.015	mW
		10	–	0.001	–	0.000010	0.001	–	0.06	
		15	–	0.003	–	0.000023	0.003	–	0.18	
Quiescent Dissipation (CL/CP Device) (Per Package)	P_Q	5.0	–	0.0025	–	0.0000025	0.0025	–	0.075	mW
		10	–	0.01	–	0.000010	0.01	–	0.30	
		15	–	0.03	–	0.000023	0.03	–	0.90	
Power Dissipation**† (Dynamic plus Quiescent, C_L = 15 pF, Per Package)	P_D	5.0	P_D = (2.0 mW/MHz) f + P_Q							mW
		10	P_D = (8.0 mW/MHz) f + P_Q							
		15	P_D = (18 mW/MHz) f + P_Q							

*T_{low} = –55°C for AL Device, –40°C for CL/CP Device.
T_{high} = +125°C for AL Device, +85°C for CL/CP Device.
#Noise immunity specified for worst-case input combination.
†For dissipation at different external load capacitance (C_L) use the formula:

$$P_T(C_L) = P_D + 4 \times 10^{-3} (C_L - 15\,\text{pF})\, V_{DD}^2 f$$

where: P_T, P_D in mW (per package), C_L in pF, V_{DD} in Vdc, and f in MHz is input frequency.
**The formula given is for the typical characteristics only.

(B)

Fig. 1-19 cont'd Data sheet for the Motorola MC14011 NAND Gate. (Courtesy of Motorola, Inc.)

logic diagrams. The logic diagram shows the standard logic symbol for the NAND gate. Each of the NAND logic symbols represents four of the MOS transistors shown in the schematic.

To return to the top of the data sheet in Fig. 1-19a, three type numbers are listed, MC14011AL, MC14011CL, and MC14011CP. The basic numerical designation MC14011 is common to all three; the difference is the two-letter suffix. The first letter of the suffix indicates the quality of the device. The letter "A" identifies the device as conforming to military specifications that require more stringent device-operating parameters and greater temperature range. The letter "C" identifies the device as being of standard quality. The electrical characteristics (Fig. 1-19b) summarize the performance of each type. The last letter of the suffix indicates what kind of packaging material is used for the device. The letter "L" refers to ceramic packaging and "P" refers to plastic packaging, which is shown in the picture in Fig. 1-19a. Some specifications will differ for different types of devices, and when that is true, it will be indicated.

The information block directly below the type numbers contains a list of the salient features of the MC14011. Below this block is a second block entitled "Maximum Ratings." The term *maximum* should not be misconstrued to mean typical. The listed specifications should not be exceeded to preclude component damage or improper operation.

One item in Fig. 1-19a is a warning. Due to the fragile nature of the CMOS (McMOS in Motorola language) input capacitor, special handling is required.

To summarize Fig. 1-19a (the first page of the data sheet), notice that the following information is included:

SWITCHING CHARACTERISTICS* (C_L = 15 pF, T_A = 25°C)

Characteristic	Symbol	V_{DD}	Typ All Types	Max AL Device	Max CL/CP Device	Unit
Output Rise Time	t_r					ns
t_r = (3.0 ns/pF) C_L + 25 ns		5.0	70	175	200	
t_r = (1.5 ns/pF) C_L + 12 ns		10	35	75	110	
t_r = (1.1 ns/pF) C_L + 8 ns		15	25	55	80	
Output Fall Time	t_f					ns
t_f = (1.5 ns/pF) C_L + 47 ns		5.0	70	175	200	
t_f = (0.75 ns/pF) C_L + 24 ns		10	35	75	110	
t_f = (0.55 ns/pF) C_L + 17 ns		15	25	55	80	
Propagation Delay Time	t_{PLH},					ns
t_{PLH}, t_{PHL} = (1.8 ns/pF) C_L + 33 ns)	t_{PHL}	5.0	60	75	100	
t_{PLH}, t_{PHL} = (0.73 ns/pF) C_L + 14 ns)		10	25	50	60	
t_{PLH}, t_{PHL} = (0.60 ns/pF) C_L + 10 ns)		15	19	40	45	

*The formula given is for the typical characteristics only.

FIGURE 1 – TYPICAL OUTPUT SOURCE CHARACTERISTICS TEST CIRCUIT

All unused inputs connected to ground.

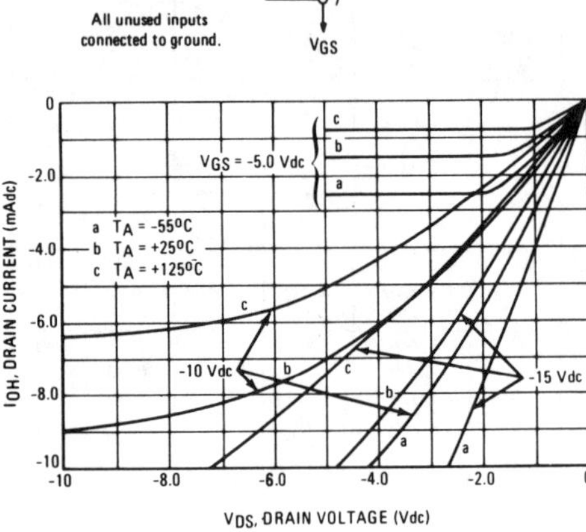

FIGURE 2 – TYPICAL OUTPUT SINK CHARACTERISTICS TEST CIRCUIT

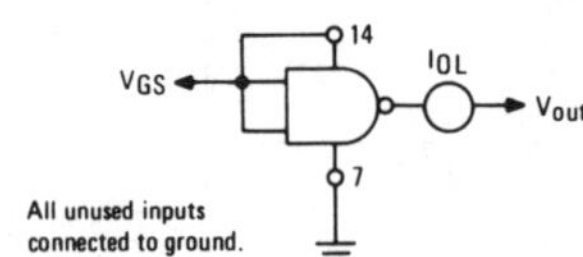

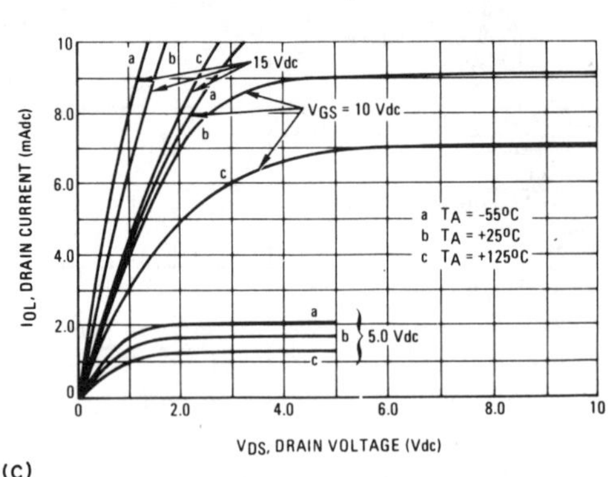

(C)

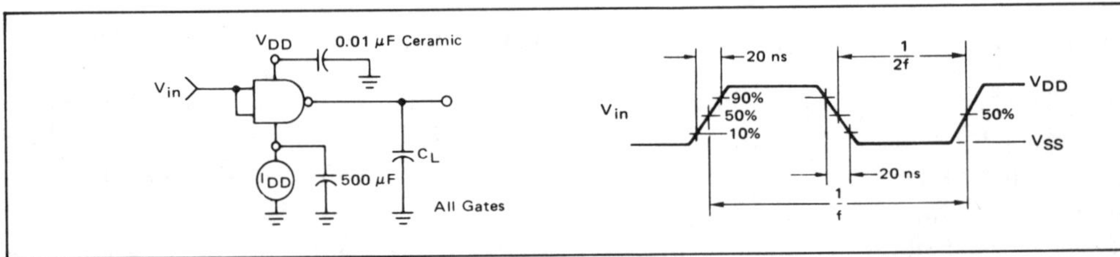

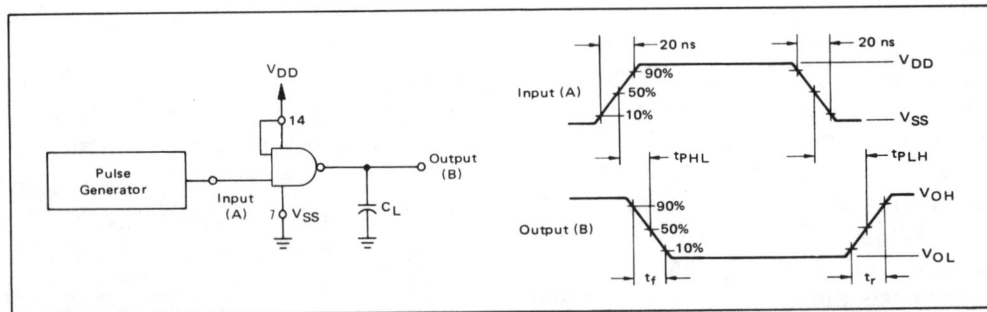

(D)

Fig. 1-19 cont'd Data sheet for the Motorola MC14011 NAND Gate. (Courtesy of Motorola, Inc.)

1. IC function, NAND gate
2. IC type number, MC14011AL, CL, and CP
3. Package type, quad two-input NAND gate
4. Family—McMOS (CMOS)
5. Typical use—low power/high noise immunity
6. Supply voltage range—3.0 Vdc to 16 or 18 Vdc
7. Fan-out, 50 CMOS fan-in units
8. Replacement, CD4011A
9. Maximum ratings
10. Handling warning
11. Schematic
12. Logic diagram

This is an impressive amount of information.

Figures 1-19b and 1-19c contain detailed operational specifications. Notice that the table of electrical characteristics is divided into three temperature ranges: low, high, and 25°C; and three supply voltages, 5.0 Vdc, 10 Vdc, and 15 Vdc. Observe also that AL devices are described separately from the CL and CP units when the characteristics differ.

The table in Fig. 1-19c describes the switching speed of the MC14011. Output rise time, output fall time, and propagation delay time are listed for supply voltages of 5.0 Vdc, 10 Vdc, and 15 Vdc. AL and CL/CP devices are treated separately.

The two graphs in Fig. 1-19c show the source current and sink current capabilities of the MC14011 for different supply voltages and temperatures. Figure 1-19d shows test circuits and waveforms for power dissipation and switching time. For critical circuit designs, this information can mean the difference between correct and erratic operation.

From the foregoing, the value of the data sheet should be evident. Data sheets are usually available at parts houses free of charge in single sheet quantities. Most manufacturers publish collections of data sheets in the form of books. These books also contain IC family and general design information that can be invaluable. The digital designer will do well to buy some of these references.

Learning Digital Technology

The concepts of digital technology can be learned by traditional study techniques. However, if one learns by reading about it and never actually manipulates the circuits, he will not have the "feel" for digital technology that comes through experience. This may seem unimportant, but such is not the case. Designs on paper can never be trusted until reduced to hardware. Timing, stray capacitance, circuit interactions, power supply glitches, and a myriad of other things can result in circuit action that is different from that predicted by a textbook.

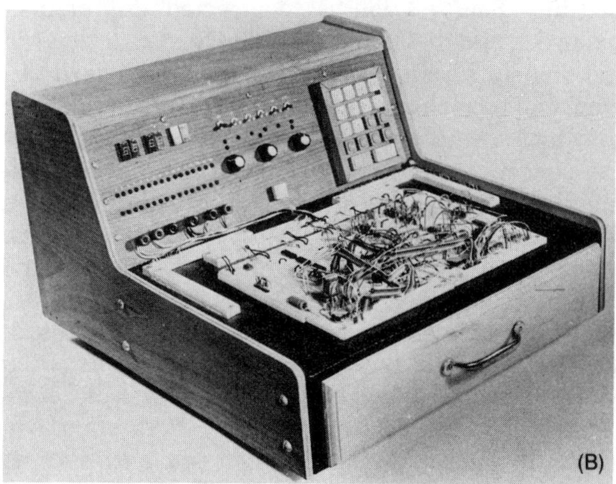

Fig. 1-20 Experimenter's breadboards: (a) Melton Special and (b) Smitty.

To explain every possibility for all these circuits would be impossible. Some of them will be covered in the form of construction guidelines. The student will be well advised to add practical experience to his textbook study. Experience is provided for in this volume in the form of experiments that cover the concepts described by the text.

Even though integrated circuits are relatively inexpensive, the number of ICs required to construct the circuits shown in this text would add up to a significant sum of money. In addition, to design each circuit using printed circuit board techniques would be very time consuming. An alternative to these processes is the use of a breadboard. A *breadboard* is a device that uses plug-in sockets to permit construction of circuits in a temporary fashion. The circuit components can be removed after the experiment is completed and reused with other circuits. Two different breadboards are shown in Fig. 1-20. Either of these units can be used for constructing the circuits in this book. Alternatively, a commercial unit can be purchased. Kits and factory constructed breadboards are available.

The Melton Special is a practical breadboard with a number of useful features. Plug-in circuit strips are used that will accept up to 32 14 dual-in-line integrated circuit packages (DIPs). It contains a 5-V power supply and a variable supply that provides from 0 to 15 V. LED state indicators are used for monitoring of circuit action. A triggering circuit is included that provides for single or multiple trigger pulses. The Melton Special is designed for use with TTL IC but will accept other families as well. Plans and construction details for this breadboard are provided in Appendix A.

As a part of the experiments in this volume, various circuits used in a second breadboard, dubbed the "Smitty," will be constructed. Rather than build the breadboard in its entirety at this point in the student's training, the various breadboard circuits will be included in several chapters of the book to coincide with the section of the text that teaches the concepts taught by the circuit. Also, constructing the circuits in this way will permit the student to understand the circuits more thoroughly as he builds them. If the student prefers to construct the Melton Special, the circuits will simply become lab exercises and will be built using that breadboard. The Smitty has a few more features than does the Melton Special. A larger number of ICs will be accommodated by the plug-in strips. Both standard-sized DIP[8] packages and the larger packages such as clocks, microprocessors, and the like, are provided for. A 16-button keyboard that provides conditioned[9] outputs is incorporated on the unit. Thirty-two LED state indicators are provided, 16 to monitor the keyboard, and 16 for use as state indicators in circuits. Six seven-segment LED readouts are mounted on the panel for use as needed. Single and variable frequency trigger circuits are included. Three +5-V regulated power supplies, one −5-V regulated supply, and both positive and negative 12-V regulated supplies are included. As can be seen, this breadboard is more versatile than the Melton Special, but the cost is increased proportionately.

In the following section, the construction of the basic breadboard will be covered. Power supply consideration for digital circuits will be included. If the student prefers to construct the Melton Special, he should refer to Appendix A for construction details. Once the Melton Special is completed, the student should return to the power supply discussion. The breadboard will be used in the next chapter to investigate digital IC gating circuits. Experiments in that chapter will demonstrate the IC family characteristics described previously.

Construction of the "Smitty" Breadboard

The following breadboard construction details are suggestions. If the student prefers, he can alter the design to utilize the materials he has available. There are some design criteria that must be followed if proper operation is to be assured. These requirements will be noted within the discussion of the breadboard circuits.

This first construction section covers the mechanical construction of the breadboard and design considerations for power supplies used with IC circuits. Other portions of the breadboard circuitry will be presented within the chapters of the book that discuss each type of circuit.

The heart of any breadboard is the breadboard socket. Several manufacturers market socket strips that are usable in breadboards. The socket is a strip of plug-in connections spaced to accept the pins of a standard DIP IC package. These strips vary in length from ones providing only a few connections to others that provide 48 or more connections. The sockets usually provide 4 or 5 connections for each IC pin. The connections in the row are paralleled so that a wire or component placed in one of these other holes will be connected to the IC pin that is plugged into the end hole. The ICs, wires, and components can be plugged in, changed, and removed as desired to create different circuits. An additional feature of some breadboard sockets is bus strips. These bus strips are as long as the breadboard socket and all of the connections are in parallel. Bus strips are used to route power and ground for connection to all the circuits constructed on the breadboard. This is not only convenient, but it also reduces power supply noise and crosstalk problems.

The Smitty is constructed around a 12- by 17-in. steel chassis. A 3- by 15-in. opening must be cut in

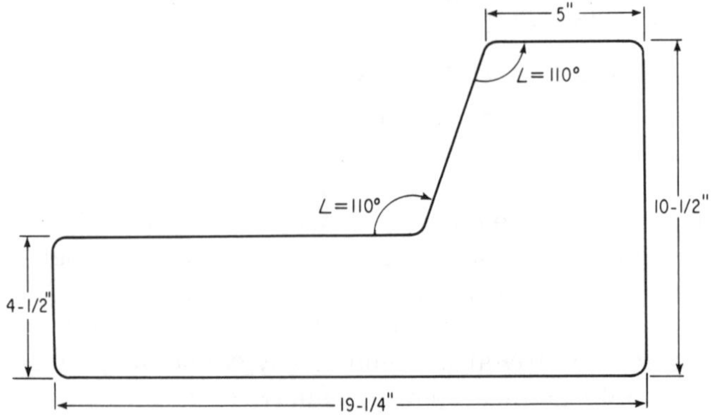

Fig. 1-21 Template for the wooden sides of the Smitty.

[8]DIP refers to Dual-In-Line Package.
[9]Conditioning is discussed in Chap. 3.

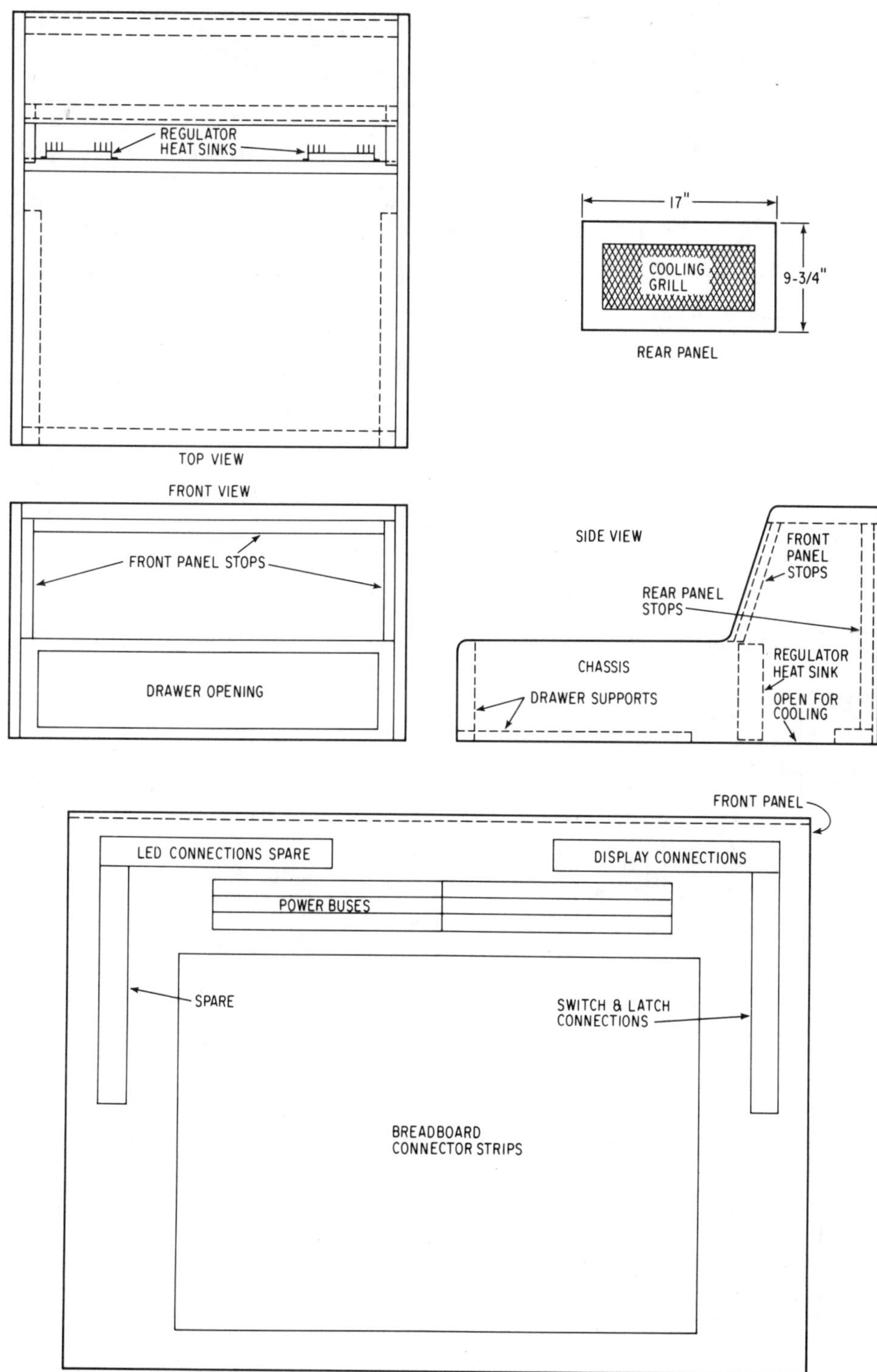

Fig. 1-22 Assembly drawing of the Smitty.

the front of the chassis if a drawer is desired. Regulator heat sinks are mounted on the rear of the chassis. Furniture grade walnut veneer plywood is cut as shown in Fig. 1-21 to use for the sides of the unit. As shown, a top, front, and back panels also are required. These parts are assembled according to the plans shown in Fig. 1-22. Figure 1-23 is a template to be used in drilling the front panel. It is recommended that all of the front panel holes be drilled at this time even though most of them will not be used until later. Figure 1-23 is designed for all of the items that will be included on the front panel

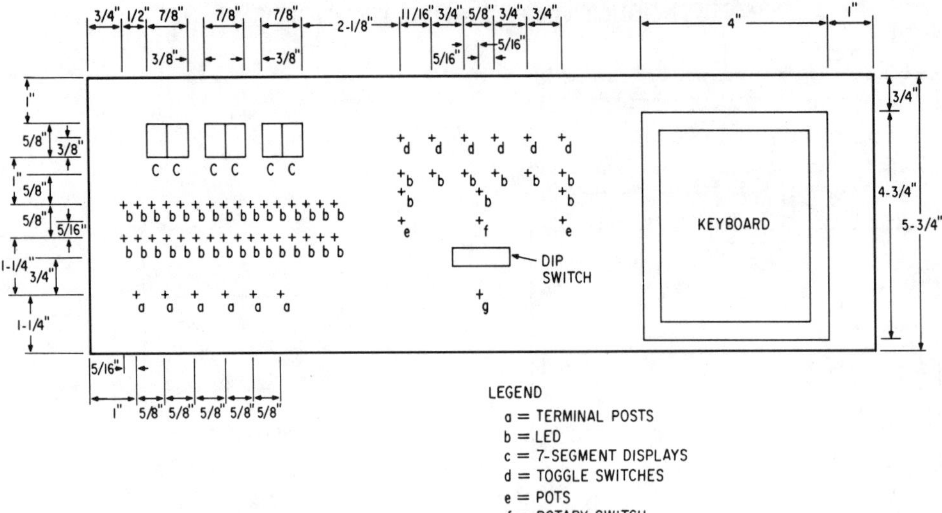

Fig. 1-23 Front panel layout of the Smitty.

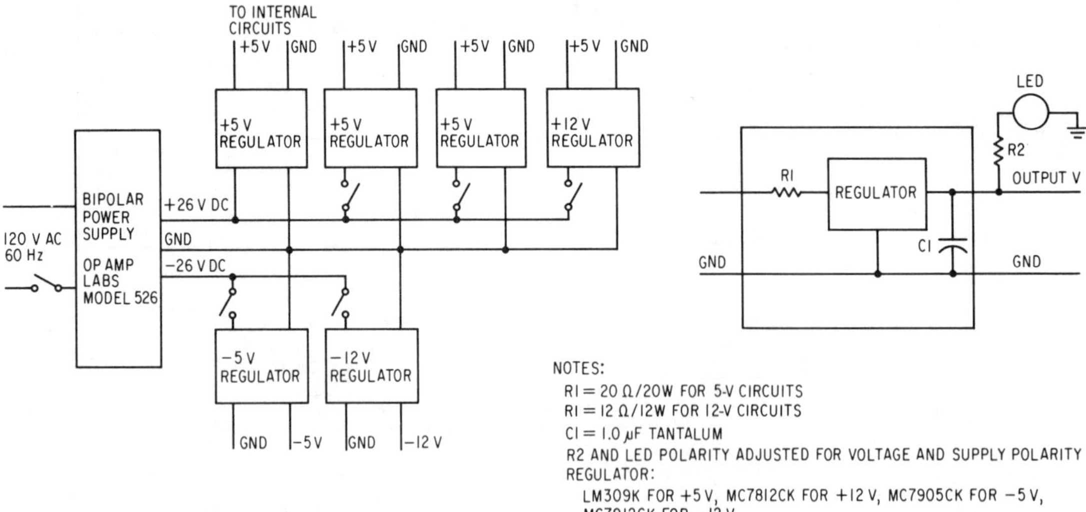

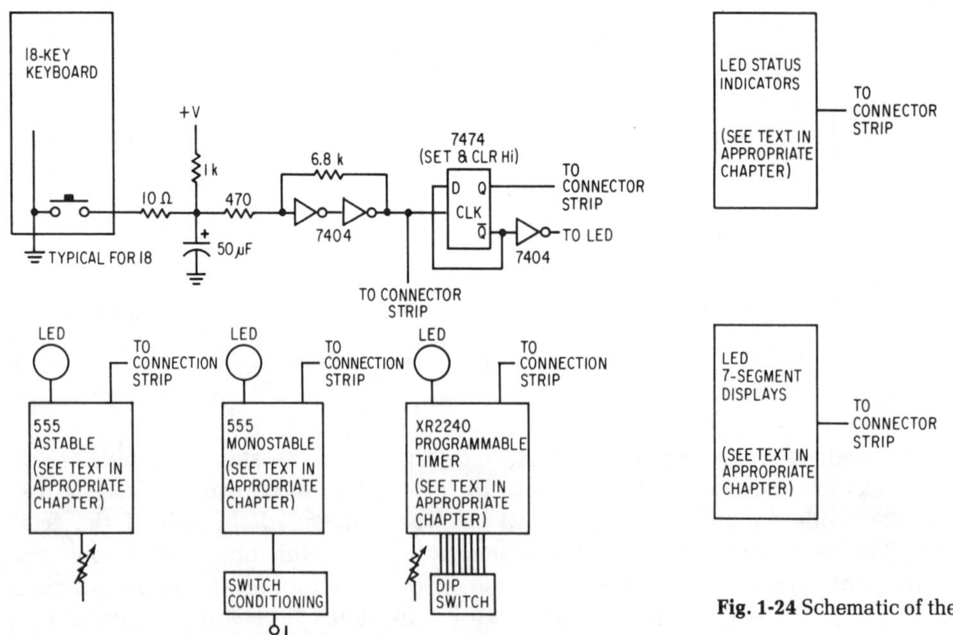

Fig. 1-24 Schematic of the basic Smitty.

when the breadboard is completed. If the student chooses to leave out some circuits or add others, he may find redesign of the front panel in order.

After the breadboard housing is completed, the sockets can be attached to the surface of the chassis. They can be either screwed to the chassis, or double-sided tape can be used to stick them to it. For the Smitty, sockets made by Continental Specialties Corporation and Radio Shack were used. Two varieties were included. The 300 series sockets fit standard 14-pin ICs and the 600 series were used to provide for microprocessors, memory, and other large package chips. Two single-strip sockets (half of a 300 series socket) were placed near the front panel to be used for connecting to the LEDs, timers, keyboard switches, and displays that form the internal circuitry of the breadboard. A schematic of the Smitty is provided in Fig. 1-24.

Power Supply Considerations

Regulations and filtering are important considerations for power supplies that are used with integrated circuits. Power supply regulation has become a simple matter with the advent of IC regulator packages. One of these units can be connected to the output of a filtered power supply, and the voltage will be held to the required value within the tolerances specified by the manufacturer. A gating power transistor can be controlled by the regulator IC if larger amounts of current are needed. The size and sophistication required in a power supply is determined by the size and nature of circuits it will power. TTL ICs require a well-regulated + 5-V supply. TTL also requires a considerable amount of current as speed and complexity of circuits increase.

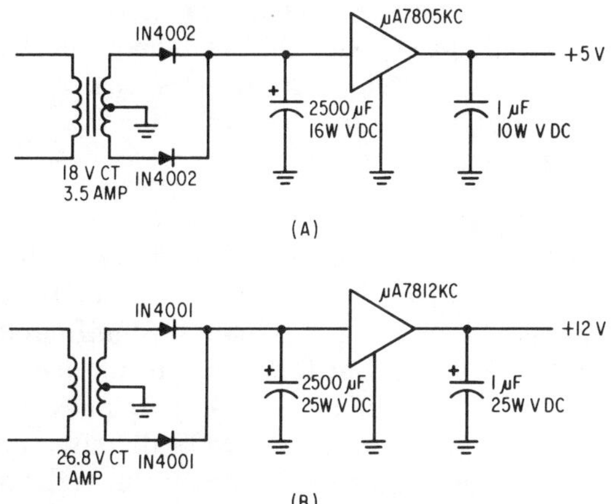

Fig. 1-25 Power supply circuits for (a) TTL logic and (b) CMOS logic.

Figure 1-25a shows a typical supply for use with TTL ICs. The supply shown in Fig. 1-25b is adequate for CMOS ICs. Combined supplies can be constructed by using a transformer that supplies adequate voltage for the highest voltage family desired. The regulators will supply voltage for the lower voltage family with the penalty of increased heat dissipation. The Smitty uses this principle. While this is not the best design procedure, it represented a valuable option for the Smitty since a ± 26 V at 5-A power supply was available in the junk box.

Supply leads from the power supply to the sockets must be large enough to display low impedance. This will assist in reducing current spikes and is especially important with TTL circuits. Despiking and decoupling capacitors also are needed. The 1-uF capacitor at the regulator output should be of tantalum construction and should be connected very close to the regulator output. Other despiking capacitors will be used within the circuits. The criteria governing the use of these capacitors will be covered in later chapters.

Summary

This chapter has provided a general overview of digital electronics. The student may be dismayed to find that there are so many options available to him. He may be concerned about the number of IC families and how to choose the right one for the job. He may be concerned about the complexity of digital circuitry and his ability to design functional devices using digital ICs. To allay these fears, however, the student has only to take a lesson from digital circuitry itself. The most complex circuit is nothing more than a combination of circuits that are either on or off, low or high. In most cases, the student can build circuits with nothing more than ICs and hookup wire. Where other components are needed, the criteria for their use is clearly defined. Compared with discrete analog circuits, digital IC design is simple and, more importantly, fun.

From this general chapter, the text will lead the student into learning experiences with specific circuits and ICs. A building block approach will be used that will permit each successive chapter to grow out of those preceding it. Gates, the simplest circuits, will be discussed next. The book will then progress into more complex circuits. The student will want to note that the circuits discussed in the advanced chapters are nothing more than combinations of the simple gate circuits, which will be discussed in Chap. 2.

TWO

Logic Gates

Introduction to Logic

Several terms used in Chap. 1 require further explanation. The term *digital* was applied to the discussion of integrated circuits on several occasions. ICs also were referred to as *logic* circuits. Are the two terms synonymous? In some cases, they are. In other cases, they are not. *Digital* refers to expressing quantities in discrete increments. For instance, the volume of a sound could be rated according to a scale that is composed of the numbers 0, 1, 2, 3, 4, 5. On the other hand, the sound volume could be measured in decibels referred to 1 mW (dBm) and could assume any value within the continuum of values. Even fractional values would be permitted. The dBm measurement is referred to as an *analog measurement* since it represents a continuous system of measurement. The other system is referred to as a *digital measurement* since only those specific digits (0, 1, 2, 3, 4, 5) are permitted. All values are made to assume one of these discrete values by rounding; i.e., if the quantity is more than 1 but less than 1½ it would be assigned the value 1. If it is more than 1½ but less than 2, it would be assigned the value of 2. This discontinuous method of assigning values is the basic concept described by the term *digital*.

Digital systems can use any numbering system. The decimal and binary numbering systems are among the more common. In math, a *digit* is defined as a character used to represent a nonnegative integer smaller than the radix. The *radix* is the number that is characteristic of the numbering system; i.e., 10 for decimal systems and 2 for binary systems. Therefore, the digits in the decimal system are 0, 1, 2, 3, 4, 5, 6, 7, 8, 9. For the binary system, they are 0, 1.

It was illustrated in Chap. 1 that an electronic circuit such as the inverter is binary in nature. It has only two states, off or on. If the off state is given the digital value "0," and the on state is given the digital value "1," the device will be able to operate as a binary digital logic element.

Logic, as used here, can be defined as the rational connection between events. According to the science of logic, inferences are made based on supplied information. Therefore, if a person observes that a light bulb is lit, he infers that the switch is on. This same process is applicable to electrical operations of ICs. If an inverter's output is observed to be high, it can be inferred that the input is low. Logic, as it applies to the design and operation of digital circuits refers to interconnecting ICs so that the interaction of their binary inputs and outputs will accomplish designated jobs. These jobs range from simple tasks such as turning lights on and off to sophisticated mathematical calculations. Any of these tasks, regardless of how complex they may be, are simply combinations of individual logical operations resulting from one of two choices, on or off. Learning to design and construct logic circuits is easiest if one begins with simple circuits and then progresses to the more difficult. This book is designed to follow this procedure. Each chapter provides information that becomes the foundation for succeeding chapters. Each experiment provides construction and design skills that will train the student in the abilities he will need to progress to the advanced circuits found at the end of the book and in his own designs.

The NOT Gate

The simplest logic element is the NOT gate. The NOT gate was discussed in Chap. 1 under its alternate name, inverter. What is the significance of calling it a NOT gate here? A logic circuit is called a gate if it passes a signal. The NOT gate passes a signal, but the signal it passes is *NOT* the signal at its input. The signal at the NOT gate output is the opposite or complement of the input signal. In logic terms, if the input signal is given the term *input*, the complement of that signal will be designated as $\overline{\text{input}}$, which means "not input." The bar over the name of the signal negates the signal. Therefore, if input is a "high" or "one," $\overline{\text{input}}$ will be a "low" or "zero."

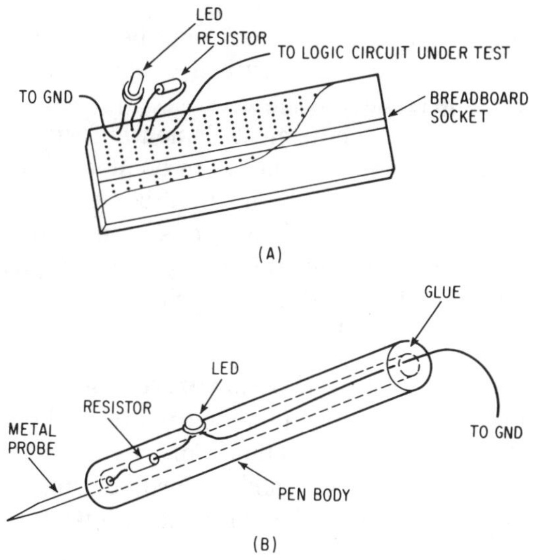

Fig. 2-1 Status indicators: (a) breadboard version and (b) probe version.

Experiment 2-1: NOT Gate Operation

This statement can be tested using the breadboard. A TTL logic block 7404[1] will be used. The TTL family is an appropriate choice for this first project since it does not require the special handling that CMOS does. Since TTL and CMOS are two of the more popular families and since the basics of logic design are fundamentally the same for all families, most of the experiments will utilize ICs from these two families. Other families will be discussed in experiments that demonstrate their unique characteristics. Special handling requirements will be presented as the experiments make them necessary.

The 7404 is a hex inverter. This means that six NOT gates are built into one 14-pin IC package. An outline drawing of the 7404 is shown in the Appendix. Notice that the logic symbol for a NOT gate is shown six times within the package. The pin connecting to the rear of the triangle is the input to the NOT gate. The pin connecting to the circle at the apex of the triangle is the output of the NOT gate. The circle on this gate and any gate means that the signal is inverted. The task of this first experiment is to monitor and record the status of the output of the gate as the input is altered between low and high logic levels. To do this, a state indicator is needed. The simplest state indicator is the Light-Emitting Diode (LED).

Figure 2-1 shows two ways of making an LED status indicator. Figure 2-1a uses the breadboard for constructing the LED indicator. All that is required is the LED and a current-limiting resistor. As shown, the resistor size is calculated by use of Ohm's law. The current required by the LED is the "I" of the formula and the number of volts that must be dropped to bring the supply voltage to the rated LED voltage is the "E" of the formula. A ¼-W resistor is most convenient for breadboard use since it has smaller leads that are easier to insert into the holes of the socket.

The logic probe shown in Figure 2-1b is handy. It can be easily moved to other test points as required. The circuit is the same as the first indicator. The student should construct one of these state indicators. Permanent state indicators will be added to the breadboard later in the chapter.

Now that the state indicator is ready, construction of the NOT gate circuit can be completed. Plug the 7404 into the breadboard. It may be necessary to squeeze the pins slightly in order to get them to match the holes in the breadboard socket. Be sure to note where pin 1 is located. The DIP package is manufactured with orientation markings. Some packages have a dot adjacent to pin 1. Most have a notch or semicircle indentation in one end of the IC. This notch is on the pin 1 end of the IC. Looking at the top of the IC, with the notch pointing away from the observer, pin 1 is the first pin on the left side of the IC. The pins then count sequentially from pin 1 to the end of the package away from the notch. On the 7404 the last pin on the left side is pin 7. Pin 8 is on the right side directly across from pin 7. The count continues along the right side (back toward the notched end) to pin 14.

Connections must be made to the IC for operating power. The 7404 requires +5 V and ground. The +5-V connection is made to pin 14. Ground is connected to pin 7. These are standard pin-outs for most TTL packages. All ICs do not follow this convention, however, and caution is advised in connecting new ICs into the circuit. Check a data sheet before connecting the power leads.

Changing states on the input of the NOT gate is easily accomplished. For a low state, the input is connected to ground. For a high state, the input is connected to +5 V.

A simple circuit such as the NOT gate may not require recording the state changes to prevent confusion. With more complicated circuits, however, recording the input and output changes will be necessary to permit understanding of the circuit action. It is recommended that the student record the NOT gate state changes so that he will become familiar with the procedure. A handy state change recording device is the *truth table*. The truth table lists all input states and combinations of input states that are

[1]The number 7404 is a generic designation. This number is preceded by a manufacturer's code such as MC for Motorola or SN for Texas Instruments. A package code follows the number.

IN	- OUT
+ 5V "1"	
GND "0"	

Fig. 2-2 NOT gate truth table.

IN	OUT
+ 5V "1"	0
GND "0"	1

Fig. 2-3 NOT gate truth table (completed).

possible. A result column records the output state for each input condition. Figure 2-2 is a truth table for the NOT gate. Only input states are listed. Change the state of pin 1, the input to inverter (NOT gate) pin 1. The following procedure will accomplish this. First, connect the + 5-V supply to pin 1. Second, connect the LED indicator to pin 2. If the LED is lit record a "1" in the truth table. If the LED is not lit, record a "0" in the truth table. Connect the input of the NOT gate to ground. Record the output state as instructed previously. Notice the truth table. Was the inverter action demonstrated? The truth table should look like Fig. 2-3. If it does not, recheck the procedures, check the breadboard wiring, or try another IC.

Another important concept in IC design can be illustrated by this experiment. Remove the wire to the input of the inverter with the probe in place at pin 2. Check the output state. Connect + 5 V to the input of the inverter. Check the output state again. The output of the inverter remains low in both cases. This indicates that an unconnected input is interpreted as a high by the inverter. This is true with all TTL gates. The unconnected input is not a true high, however, and can create noise problems. If the input is left unconnected when circuits are constructed, the inverter can display erratic operation. To prevent this, a design rule that should be followed with all circuits regardless of family is: *Provide a connection to the supply voltage or ground for all unused inputs.*

The AND Gate

The inverter (NOT gate) is functionally the simplest logic element. It is, however, rather limited. All it can do is to provide an output that is the complement (opposite) of the input. While the inverter is handy when changes of an output state are needed or when the unit is used as a buffer, it cannot provide all of the functions that may be required of a gate. Other types of gates are available for these chores. One is the AND gate.

The AND gate in its most elemental form has two inputs. In order for the output to be high, both input 1 *and* input 2 must be high. If neither input is high, the output will be low. If only one input is high, the output will be low. Only if both inputs are high at the same time will the output be high. This is the basic function of the AND gate.

Experiment 2-2: AND Gate Operation

A simple experiment on the breadboard will confirm this AND gate function. For this experiment a CMOS IC 4081 will be used. Before building the circuit, however, the following precautions for using CMOS should be noted.

1. Dissipate all static electricity charges on personnel and equipment by using a grounded wire and antistatic work surfaces.
2. If a soldering iron is used, be sure it is a three-prong grounded type, a battery operated type, or ground the tip of the iron.
3. Connect all unused inputs to ground or the positive voltage source.
4. Store CMOS ICs in conductive foam. During assembly and experimentation, place the IC on a conductive surface such as a piece of aluminum foil.
5. Place a load resistor of 1 MΩ or greater across test inputs or inputs being driven by external circuits.
6. Provide current limiting for any inputs that might force the CMOS input protection diodes into conduction.

These precautions may seem tedious. With experience, however, they will become second nature. Once the student has opportunity to experience the benefit of CMOS circuitry, he will gladly accept these few requirements.

The 4081 IC[2] should be plugged into the breadboard socket. Power connections are the same as were used for the 7404, ground to pin 7 and the positive voltage to pin 14. The positive 5 V used with the 7404 also will power the 4081. However, 5 V is not the optimum supply voltage. CMOS will operate on voltages from 3 V to 18 V. The most appropriate voltage is in the vicinity of 12 V. At 3 V, CMOS is slow, has reduced drive capability, and has reduced noise immunity. At 18 V the power dissipation is too

[2]The number 4081 is generic for CMOS ICs. As with TTL, a manufacturer's code such as MC or CD will precede the generic number. Motorola also adds a 1 to the number, e.g., MC14081. A suffix will follow the generic number.

great and the speed increase above that for a 12-V supply does not warrant the extra power.

Appendix C includes an outline drawing of the 4081. Note that there are four two-input AND gates in this one package. It is usually referred to as a *quad AND gate*. For this experiment, the state indicator described for use with the TTL family must be altered slightly. TTL circuits have a supply voltage of 5 V. For CMOS, the supply is 12 V. As described earlier, the LED indicator draws a set amount of current. The Dialight 9202, for example, draws 20 mA. The LED also has a normal operating voltage, which is 2.0 V for the 9202. The TTL dropping resistor must reduce the voltage adequately to bring the voltage level to the required value. For TTL the resistor is calculated by the following formula:

$$R = E/I\ \Omega$$

where

R is resistance of the dropping resistor in ohms
E is the amount of voltage to be dropped (for TTL, 5 V − 2 V = 3 V to be dropped)
I is the amount of current that the LED draws

Substituting the values gives

$$R = 3.0/0.02 = 150\ \Omega$$

For CMOS, the voltage drop is 12 V minus the voltage required by the LED. For the 9202 the formula for the dropping resistor to be used with a 12-V CMOS supply is

$$R = E/I$$
$$R = 10/0.02 = 500\ \Omega$$

CMOS does not have adequate drive to provide for the LED current drain and retain consistent circuit action. Figure 2-4 shows a circuit that can be used for both TTL and CMOS without loading the circuits and interfering with circuit action.

LEDs are available that will operate at TTL voltage without a dropping resistor. The Dialight 9183 is an example. Some LEDs are manufactured with built-in dropping resistors. The Xciton XC-ZZ is designed for use with 12-V systems, and the XC-556 is a 5-V model. The use of these LEDs is convenient since current-limiting resistors add to circuit complexity, size, and cost.

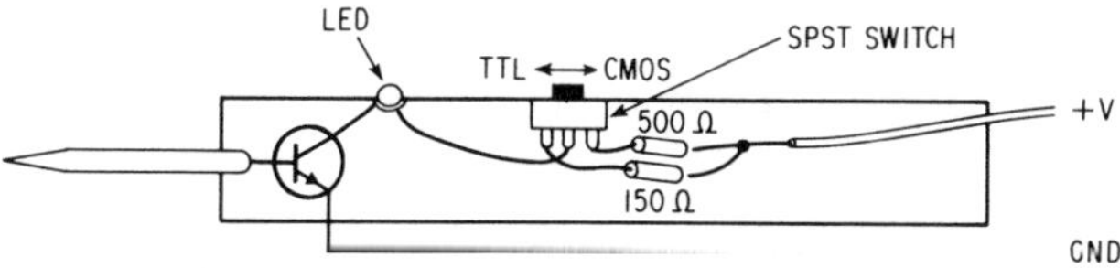

Fig. 2-4 Improved logic probe.

IN 1	IN 2	OUT
Hi (1)	Hi (1)	
Hi (1)	Lo (0)	
Lo (0)	Hi (1)	
Lo (0)	Lo (0)	

Fig. 2-5 AND gate truth table.

Construct the state indicator. Figure 2-4 shows the required modifications to the logic probe that will permit its use with both TTL and CMOS circuits. Connect the state indicator to pin 3 of the 4081. The two inputs, pins 1 and 2, can now be connected for various combinations of input levels, while the state indicator is observed to determine the output state. A truth table will be helpful for recording the input and output states and ensuring thorough systematic analysis of the AND gate function. The truth table for the AND gate will require two input columns and one output column, as shown in Fig. 2-5. Vary the input conditions and record the output states. Which input conditions resulted in a high output? As to be expected, the AND gate is high only when both inputs are high.

Multiple Input AND Gates

Some circuits may require more than two inputs on an AND gate. These circuits can be accommodated. The output of two two-input AND gates can be connected to the inputs of a third two-input AND gate. This circuit will expand the AND gate to a four-input AND gate. If an AND gate is connected to each of these four inputs, an eight-input AND gate will result. The expansion could continue in this same fashion until a very large input capability is created.

Experiment 2-3: AND Gate Input Multiplication

Construct the circuit shown in Fig. 2-6a. Only one 4081 is required to build the circuit since the 4081 contains four two-input AND gates. Since the circuit has five inputs, complete the truth table shown in Fig. 2-6b,[3] which has five input columns and one output column. Vary the five inputs. Be sure that all input signal variations are used. Record the output state for every step of the experiment. Analyze the output column. The output should be high only when all five inputs are high. The result will reflect the operation of a five-input AND gate.

AND gates are available with multiple inputs. These gates are packaged within standard DIP modules and differ from two-input gates only in number of inputs per gate and elements per pack-

[3]The number of possible combinations for input states is 2^n where n = number of gate inputs.

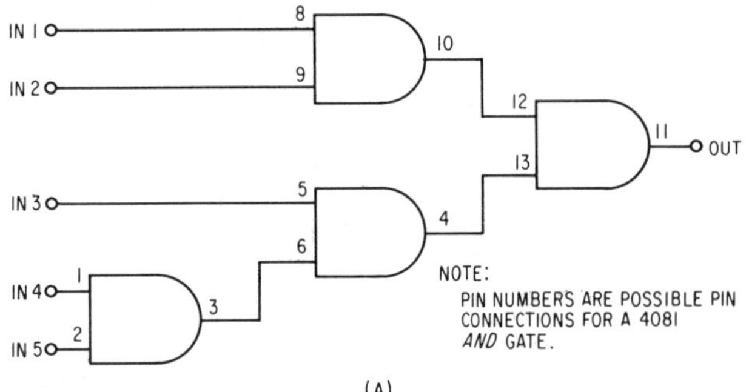

Five-Input AND Gate

IN 1	IN 2	IN 3	IN 4	IN 5	OUT
0	0	0	0	0	
0	0	0	0	1	
0	0	0	1	0	
0	0	0	1	1	
0	0	1	0	0	
0	0	1	0	1	
0	0	1	1	0	
0	0	1	1	1	
0	1	0	0	0	
0	1	0	0	1	
0	1	0	1	0	
0	1	0	1	1	
0	1	1	0	0	
0	1	1	0	1	
0	1	1	1	0	
0	1	1	1	1	
1	0	0	0	0	
1	0	0	0	1	
1	0	0	1	0	
1	0	0	1	1	
1	0	1	0	0	
1	0	1	0	1	
1	0	1	1	0	
1	0	1	1	1	
1	1	0	0	0	
1	1	0	0	1	
1	1	0	1	0	
1	1	0	1	1	
1	1	1	0	0	
1	1	1	0	1	
1	1	1	1	0	
1	1	1	1	1	

(B)

Fig. 2-6 Multiple input AND gate using a 4081: (a) functional diagram and (b) truth table.

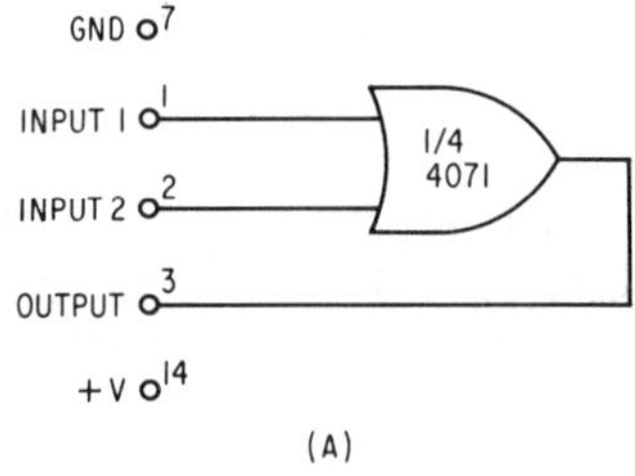

Two-Input OR Gate

IN 1	IN 2	OUT
1	1	
1	0	
0	1	
0	0	

(B)

Fig. 2-7 OR gate circuit: (a) functional drawing and (b) truth table.

age. In the CMOS family, the 4073 has three three-input AND gates built into one package. The 4082 is a dual four-input AND gate. Other variations are available but are not as common as those already indicated. Appendix C includes outline drawings for several multiple input AND gates.

The OR Gate

The CMOS IC 4071 contains four OR gates (see Appendix C). Notice the differences between the AND gate and OR gate symbols. The AND gate has a rounded front and a straight back. The OR gate has a pointed front and a concave back. When these symbols are seen on a schematic or logic diagram, the distinctive symbol ensures that the type of gate represented will be apparent even if the IC number is not given.

Experiment 2-4: The OR Gate

The OR gate functions in accordance with its name. If either one input *or* the other is high, the output will be high. The output also will be high when both inputs are high. Prove this by constructing and testing the circuit shown in Fig. 2-7a. Connect the LED indicator to pin 3. A two-input, one-output truth table, which is shown in Fig. 2-7b, will be required. After the truth table is completed, analyze the output column. The number of high output states of the OR gate is greater than the number produced by the AND gate. Indeed, only one input combination results in a low output for the OR gate. Only when both inputs are low will the output be low.

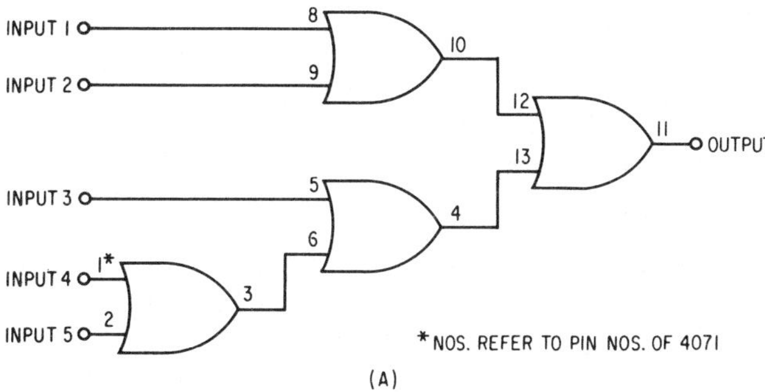

Five-Input OR Gate

IN 1	IN 2	IN 3	IN 4	IN 5	OUT
0	0	0	0	0	
0	0	0	0	1	
0	0	0	1	0	
0	0	0	1	1	
0	0	1	0	0	
0	0	1	0	1	
0	0	1	1	0	
0	0	1	1	1	
0	1	0	0	0	
0	1	0	0	1	
0	1	0	1	0	
0	1	0	1	1	
0	1	1	0	0	
0	1	1	0	1	
0	1	1	1	0	
0	1	1	1	1	
1	0	0	0	0	
1	0	0	0	1	
1	0	0	1	0	
1	0	0	1	1	
1	0	1	0	0	
1	0	1	0	1	
1	0	1	1	0	
1	0	1	1	1	
1	1	0	0	0	
1	1	0	0	1	
1	1	0	1	0	
1	1	0	1	1	
1	1	1	0	0	
1	1	1	0	1	
1	1	1	1	0	
1	1	1	1	1	

(B)

Fig. 2-8 Multiple input OR gate using a 4071: (a) functional drawing and (b) truth table.

Experiment 2-5: OR Gate Input Multiplication

OR gates can be expanded. Construct the circuit shown in Fig. 2-8a. Five inputs are available in this circuit as was true of the comparable AND gate circuit. Test the circuit and fill in the truth table shown in Fig. 2-8b. Again, the only time the output is low is when no input is high.

Insert a 4072 into the breadboard. Monitor the state of pin 1. Vary the input signal on pins 2, 3, 4, and 5. Again, the OR function is apparent. When any or all of the inputs are high, the output will be high. The 4072 is a dual, four-input OR gate CMOS IC. The 4075 is a triple, three-input OR gate IC. The other families of ICs have similar multiple input devices. Use of these multiple input gates greatly reduces the complexity of the circuitry.

DeMorgan's Theorem and the Basic Gates

In the preceding discussion of logic function, a positive logic convention was used. According to this convention, a device receives its name based on the positive voltage input conditions. Thus, a two-input AND gate has a positive output when input 1 *and* input 2 are both positive. The two-input OR gate has a positive output when either input 1 *or* input 2 *or* both inputs are positive. This method of referring to a logic device is standard. Most data sheets and manuals will use the positive logic definition when describing a logic element. However, the designer of digital circuitry who limits himself to a positive logic definition will fail to achieve maximum economy of circuits within his designs.

It is possible to redefine a logic element using a negative logic convention. According to this convention, a logic element's function is based on the grounded input conditions.

A negative logic AND gate will therefore be a device that has a zero output when all of its inputs are zero. A negative logic OR gate will have a zero output when any of the inputs are zero. Truth tables for two-input negative logic AND gates and OR gates are shown in Fig. 2-9. Do these truth tables look familiar? Compare them with the truth tables for

Negative Logic AND Gate

IN 1	IN 2	OUT
1	1	1
0	1	1
1	0	1
0	0	0

(A)

Negative Logic OR Gate

IN 1	IN 2	OUT
1	1	1
0	1	0
1	0	0
0	0	0

(B)

Fig. 2-9 Negative logic gate truth tables: (a) AND gate and (b) OR gate.

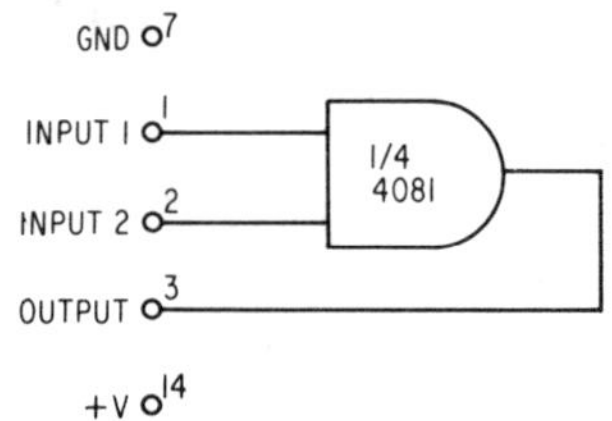

Fig. 2-10 Positive-AND gate/negative-OR gate.

their two-input positive logic counterparts. The positive logic AND gate and the negative logic OR gate have the same truth tables. The same is true for the positive logic OR gate and the negative logic AND gate. This complementary circuit action is defined by *DeMorgan's theorem*. DeMorgan's theorem is a part of Boolean algebra. The theorem describes the complementary nature of gates in the form of a Boolean formula. (Boolean algebra will not be used in this volume.) It will be more useful in this study to observe the practical result of the theorem. The breadboard will provide a convenient medium for this observation.

Experiment 2-6: DeMorgan's Theorem

Reconstruct the positive logic AND gate circuit shown in Fig. 2-10. Repeat the experiment to refresh your memory of AND gate operation. Notice that the LED is lit only when both inputs are positive and is extinguished for all other input conditions. Now conduct an experiment to validate the negative logic OR gate truth table. Vary the input conditions as required in the truth table. Notice that the output is zero and that the LED is extinguished when either or both of the inputs are negative. If both inputs are positive, the output is positive. From these findings, it is apparent that the function performed by a logic element is determined by the logic convention in use.

DeMorgan's theorem is a very powerful tool for logic designers. It will, in many cases, result in a reduction of the number of ICs that are needed for a particular circuit. As additional logic devices are described, the full benefit of this theorem will become evident.

State Definitions

The terms "high" and "low" were avoided in this discussion. The reason is that these are relative terms that can be assigned by the designer. In this volume, high refers to the more positive signal while low is the zero or more negative signal. This assignment is at the discretion of the designer; however. If he chooses to refer to the positive signal as "zero" or "low" and to the nonpositive signal as "high," that is his prerogative.

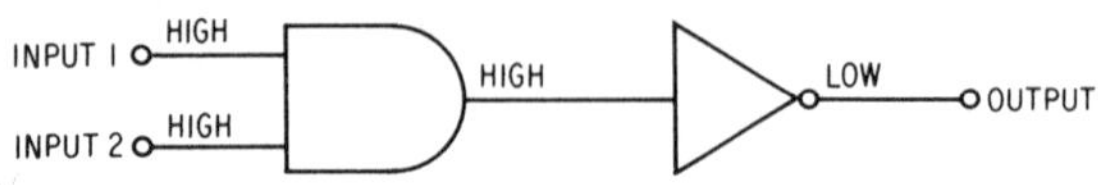

Fig. 2-11 AND gate/NOT gate circuit.

The Negated Functions

Adding a NOT gate to the output of an AND gate negates the normal AND gate action. Figure 2-11 shows the signal status at the several points of such a circuit. Notice that both inputs to the AND gate are high. As demonstrated in the previous sections, the output of the AND gate will be high. The NOT gate will now invert the high signal and provide a low signal at its output. The NOT gate has therefore negated the AND gate output by converting it to its complement. This combined circuit functions as a NOT-AND gate, which is usually referred to by the shortened term, *NAND gate*. Figure 2-12 shows the outline drawing of the NAND gate. Notice that the drawing is simply an AND gate with a circle at its output. The circle signifies that the signal is inverted. The 7400 is a quad, two-input NAND gate (see Appendix C).

Experiment 2-7: The NAND Gate

Insert a 7400 in the breadboard. Ground pin 7 and apply + 5 V at pin 14. Inputs to the first gate are on pins 1 and 2. The output of this gate is on pin 3. Connect the state indicator to pin 3. Fill in the truth table shown in Fig. 2-13 by varying the inputs as required. The output states of the NAND gate are the complement of the output states of an AND gate when the two gates are supplied identical inputs.

The NOR Gate

A second negated function logic device is the *NOR gate*. The NOR gate functions as if an OR gate is followed by a NOT gate. The name *NOR gate* is a contraction of NOT-OR. The OR gate output signals, as identified earlier, are inverted by the NOT gate. This action is suggested by the NOR gate symbol

Fig. 2-12 NAND gate functional diagram.

Two-Input NAND Gate

IN 1	IN 2	OUT
0	0	
0	1	
1	0	
1	1	

Fig. 2-13 NAND gate truth table.

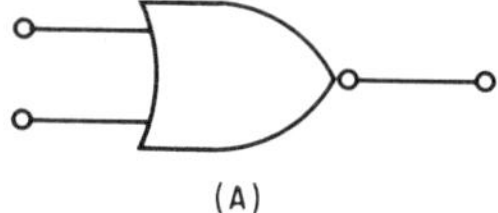

Two Input NOR Gate

IN 1	IN 2	OUT
0	0	
0	1	
1	0	
1	1	

(B)

Fig. 2-14 NOR gate: (a) functional drawing and (b) truth table.

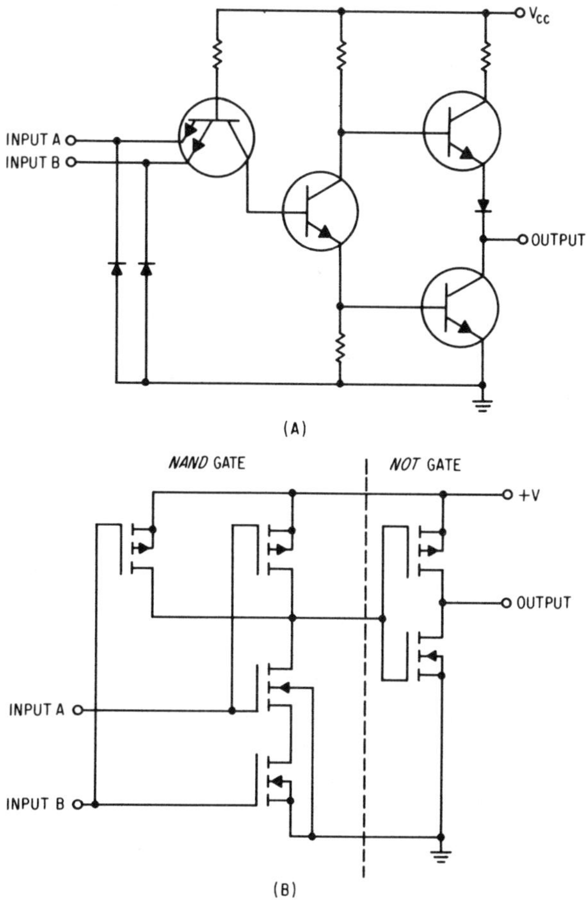

Fig. 2-15 Schematic of (a) TTL NAND gate and (b) CMOS NAND gate.

shown in Fig. 2-14a. Again, the small circle at the output of the OR gate symbol signifies inversion. Use of the breadboard in the following experiment will confirm that this is indeed the action of the NOR gate.

Experiment 2-8: The NOR Gate

Plug a 4001 into the breadboard. The 4001 is a CMOS quad, two-input NOR gate. Pin 7 is grounded and for pin 14 we shall use +12-V supply. As with the NAND gate, pin 3 is the output of gate 1 and pins 1 and 2 are input. Attach the 12-V status indicator to pin 3. Vary the input conditions and record the output states in the truth table shown in Fig. 2-14b. Compare the completed table with that obtained for the OR gate. Notice that for each input condition, the output of the NOR gate is the complement of the OR gate. The OR gate function has been negated.

Before continuing, it is important that gate structure and functioning be clarified. For simplicity, the AND gate and the NAND gate were discussed in that sequence. Functionally, the AND gate is simpler than the NAND gate, and the derivation of the name *NAND gate* suggests that the AND gate is foundational to the NAND gate; i.e., NOT-AND.

At the component level, however, it is the NAND gate that is simpler, and it is the NAND gate that is negated by a NOT gate to produce an AND gate function. The reason for this is simple. The basic transistor circuit is by nature an inverter. Figure 2-15 shows schematics of both a TTL NAND gate and a CMOS NAND gate. If reference is made to the NOT gates in Chap. 1 (Figs. 1-14 and 1-17), it will be noted that the NAND gate in both of these families is a NOT gate with an additional input. The same relationship holds for the OR and NOR gate circuitry. It also is true that NAND gate and NOR gate circuitry is the simplest circuit in both families, except for the NOT gate. NAND gates and NOR gates can be constructed in less space on the IC substrate than is true for the other gates. NAND gates and NOR gates operate with less propagation delay. For all of these reasons, the NAND gates and NOR gates along with the NOT gates represent the fundamental building blocks of digital circuitry. As more complicated devices are introduced, it should be kept in mind that each of these sophisticated circuits is constructed by interconnecting simple NAND, NOR, and NOT gates.

DeMorgan's Theorem and the Negated Function Gates

The designation NAND and NOR, as they were just used, are positive logic terms. Since these are negated function logic gates, the low outputs result from the appropriate high inputs. The input signals characterize the gates as positive logic. With the NAND gate, for example, the NAND function is demonstrated when both inputs are high. The NOR gate is demonstrated when any or all inputs are high. Due to the inversion, the output for these input conditions will be low. DeMorgan's theorem can be applied to negated logic. The NAND gate, according to the negative logic convention, will have the same truth table as a positive logic NOR gate. The negative logic NOR gate will conform to the truth table of the positive logic NAND gate.

Experiment 2-9: DeMorgan's Theorem and Negated Logic Gates

Use the breadboard to repeat Experiments 2-7 and 2-8. Write out and complete truth tables for these cases. Remember that the characteristic condition for a negative logic NAND gate is the simultaneous application of low signals to all inputs. For the negative logic NOR gate, low signals at any or all inputs is characteristic. Comparison of the truth tables in Experiments 2-7 and 2-8 (see Figs. 2-13 and 2-14b) with those you have written out for this experiment will confirm DeMorgan's theorem for negated function gates.

Exclusive Function Gates

At this point we shall introduce two additional gate circuits. These are called the *exclusive-OR gate* and the *exclusive-NOR gate*. The exclusive-OR gate provides a positive output only when one input is high. If both inputs are high or both inputs are low, the output will be low. The exclusive-NOR gate functions in the same way except the output is inverted. The symbol for these two gates is shown in Fig. 2-16.

The 7486 (TTL) and the 4070 (CMOS) are quad, exclusive-OR packages. The CMOS 4077 is a quad, exclusive-NOR package. See Appendix C for outline drawings of these ICs. Choose one of the exclusive-OR gates and construct the circuit shown in Fig. 2-17. Remember to apply appropriate voltage levels to pins 7 and 14 as required by the IC family of the chip; i.e., 5 V for TTL or 12 V for CMOS. (Note: CMOS will operate at 5 V. TTL may be damaged by 12 V.) Vary the inputs to one of the gates and record the output levels in a truth table. Notice that the output pattern for the exclusive-OR gate is different from that of the previous gates. All of the previous two input gates had three outputs of one state (high or low) and one of the other. The exclusive-OR divides the output states equally between highs and lows. Compare the exclusive-OR truth table you have written with the standard OR gate shown in Fig. 2-7b. Notice the difference in the two cases. For the exclusive-OR gate the output is low when both inputs are high, whereas the OR gate output is high for simultaneously high inputs. The exclusive-OR gate will provide a high output only if one but not both inputs are high (hence the term *exclusive*).

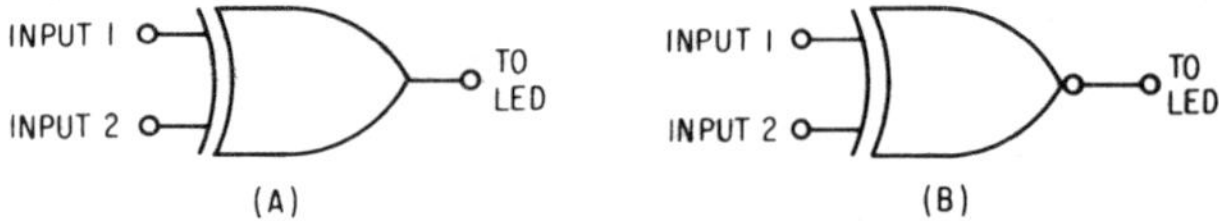

Fig. 2-16 Functional diagrams: (a) exclusive-OR gate and (b) exclusive-NOR gate.

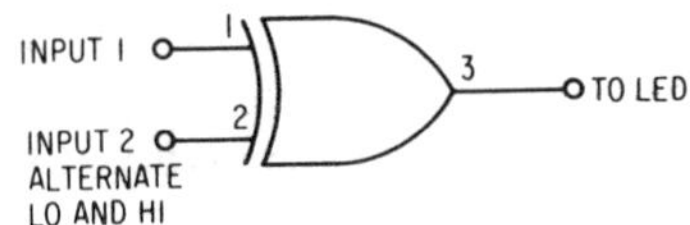

Fig. 2-17 Exclusive-OR gate circuit.

Positive Exclusive-OR

IN 1	IN 2	OUT
0	0	0
0	1	1
1	0	1
1	1	0

(A)

Negative Exclusive-OR

IN 1	IN 2	OUT
0	0	1
0	1	0
1	0	0
1	1	1

(B)

Positive Exclusive-NOR

IN 1	IN 2	OUT
0	0	1
0	1	0
1	0	0
1	1	1

(C)

Negative Exclusive-NOR

IN 1	IN 2	OUT
0	0	0
0	1	1
1	0	1
1	1	0

(D)

Fig. 2-18 Truth tables: (a) positive exclusive-OR gate, (b) negative exclusive-OR gate, (c) positive exclusive-NOR gate, and (d) negative exclusive-NOR gate.

Replace the exclusive-OR package with a CMOS 4077. The 4077 is a quad, exclusive-NOR gate package. Obtain a truth table for this gate. Compare this table with that of the exclusive-OR. Notice that the outputs are complementary, but the function is the same. The exclusive-NOR gate will provide a low output only if one input is high (not both). If both inputs are either high or low, the output will be high.

DeMorgan's Theorem and the Exclusive Function Gates

DeMorgan's theorem is operative with exclusive function gates. Truth tables for the negative exclusive-OR gate and the negative exclusive-NOR gate are shown in Fig. 2-18. Comparison of these truth tables with those of their positive convention counterparts show the following equalities:

1. A positive logic exclusive-OR gate has the same output as a negative logic exclusive-NOR gate.
2. A positive logic exclusive-NOR gate has the same output as a negative logic exclusive-OR gate.

Numerical Systems

One of the primary benefits of the exclusive function gates is their ability to accomplish binary calculations. To understand this capability, some background in binary numbering systems is required.

The binary numbering system operates on the base two. This means that the maximum number of digits in the system is two. These digits are defined, as indicated earlier, as being all of the integers that are less than the radix (the radix for the binary system is the number 2). The digits for this system are therefore 0 and 1. The binary system is not as familiar to most people as the decimal system. Understanding the similarities between the two systems may therefore assist in using the binary system. Addition, for instance, consists of adding the value of the digits within the numbering system until the value of the radix is reached. At this point, a digit is carried to the next column and addition continues. In the decimal system, each of the first column numbers has the value of one. The digits 0 through 9 are available in this column. The second column digits have values of ten each. When addition goes beyond 9, a digit must be added to the second column. This is called a *carry*.

Binary addition follows this same procedure. The only difference is that in the binary system there are only two digits rather than ten as in the decimal system. Therefore, when addition passes the digit 1, a carry is required to the next column. The resulting number is 10. This binary number has the same appearance as the decimal number 10. We do *not* refer to this number as "ten" but as "one zero." Its actual value is equivalent to the decimal number 2. The value of each digit "1" in each of 16 binary columns is shown in Table 2-1. Notice that each successive column carries the value of the successive power of two. To obtain the decimal equivalent of a binary number, the decimal values of the 1s in the binary number are added. For example, the binary number 11101101 is converted to its decimal equivalent by assigning values to the binary digits from right to left, as shown in Table 2-2.

A decimal number also can be converted to its binary equivalent. The process requires successive division by 2. To convert the decimal number 237 to binary, the procedure shown in Table 2-3 can be used. Divide the decimal number by 2. When a 1 remains, a 1 is placed in the binary number column;

Table 2-1 Decimal equivalent of digits in a binary number.

No.	*Binary Digit**	*Decimal Equivalent*
1	1	1
2	1	2
3	1	4
4	1	8
5	1	16
6	1	32
7	1	64
8	1	128
9	1	256
10	1	512
11	1	1,024
12	1	2,048
13	1	4,096
14	1	8,192
15	1	16,384
16	1	32,768

*Beginning at the least significant digit.

Table 2-2 Binary and decimal number equivalents.

Binary Column Number	*Binary Number*		*Decimal Value*
1	1	=	1
2	0	=	0
3	1	=	4
4	1	=	8
5	0	=	0
6	1	=	32
7	1	=	64
8	1	=	128
	Decimal Total		=237

Table 2-3 Converting from decimal to binary.

Division Number	*Decimal Number*	*Remainder*	*Binary Number*
1	237 ÷ 2 = 118 +	remainder = 1	1†
2	118 ÷ 2 = 59 +	remainder = 0	0
3	59 ÷ 2 = 29 +	remainder = 1	1
4	29 ÷ 2 = 14 +	remainder = 1	1
5	14 ÷ 2 = 7 +	remainder = 0	0
6	7 ÷ 2 = 3 +	remainder = 1	1
7	3 ÷ 2 = 1 +	remainder = 1	1
8	1 ÷ 2 = 0 +	remainder = 1	1‡
	*		

*Continue dividing until quotient is 0.
†Least significant digit.
‡Most significant digit.

Table 2-4 Binary addition and multiplication.

+	0	1
0	0	1
1	1	10 (↑ carry)

×	0	1
0	0	0
1	0	1

otherwise, a 0 is placed in the binary column. The binary equivalent of the decimal number 237 is 11101101.

The rules for binary addition and multiplication are shown graphically in Table 2-4. For addition, the rules are as follows:

1. 0 + 0 = 0
2. 1 + 0 = 1
3. 0 + 1 = 1
4. 1 + 1 = 10 (binary 0 with a carry to the next column)

For multiplication, the rules are as follows:

1. 0 × 0 = 0
2. 1 × 0 = 0
3. 0 × 1 = 0
4. 1 × 1 = 1

Here is an example of binary addition:

```
 1  1      (carry)
  1001  =  number 1
+ 1101  =  number 2
 -----
 10110
```

The validity of this binary addition process can be checked by converting the binary numbers to decimal and comparing the decimal sums with the value of the binary sums. The steps to check the above binary addition are as follows:

1. Convert number 1.

```
1 = 1
0 = 0
0 = 0
1 = 8
    -
    9
```

2. Convert number 2.

```
1 =  1
0 =  0
1 =  4
1 =  8
    --
    13
```

3. Convert binary sum.

```
0 =  0
1 =  2
1 =  4
0 =  0
1 = 16
    --
    22 (binary sum)
```

4. Add decimal numbers.

```
 9
13
--
22 (decimal sum)
```

Multiplication is performed as follows:

```
     1001
  ×  1101
  -------
     1001
    10010
   1001
  -------
  1110101
```

The following steps show the decimal equivalent for the above multiplication:

1. Convert the numbers.

```
1001 =  9
1101 = 13
```

2. Multiply decimal equivalents.

```
   9
× 13
----
  27
  9
----
 117
```

3. Convert binary result.

```
1 =  1
0 =  0
1 =  4
0 =  0
1 = 16
1 = 32
1 = 64
    ---
    117
```

The equivalent binary result of 117 in the previous equation equals the decimal result 117.

In binary subtraction, the procedure is altered in that instead of a carry when the digit 1 is exceeded, a *borrow* is required when the subtrahend exceeds the minuend. For instance, if binary one is subtracted from binary two, the numbers have the following form:

```
  10 (minuend)
–  1 (subtrahend)
```

The 1 of the subtrahend cannot be subtracted from the 0 of the minuend, so the 1 of column two that has a value of decimal 2 is borrowed. The completed problem is

```
 10       0 10
– 1   =  –   1
         -----
            01
```

Subtraction of larger binary numbers follows the same procedure.

```
  1101
– 1001
------
  0100
```

Table 2-5 Binary subtraction.

−	0	1 ← ②
0	0	1
1	1	0

↑ ①

① = minuend
② = subtrahend

The following steps prove this binary subtraction with its decimal equivalents:

1. Convert binary subtraction to decimal subtraction.

```
  1101  =  13
- 1001  = − 9
  0100  =   4 (decimal result)
```

2. Convert binary results.

```
0 = 0
0 = 0
1 = 4
0 = 0
    4 (binary result)
```

Note that the equivalent binary result of 4 equals decimal result 4. The rules for binary subtraction are shown in Table 2-5. They are as follows:

1. 0 − 0 = 0
2. 0 − 1 = 1 (requires a borrow)
3. 1 − 0 = 1
4. 1 − 1 = 0

Binary division uses the same trial-and-error process decimal division uses. No rules for division are given. Division is actually a process of multiplication and subtraction. Division of binary two by binary two takes the form

```
     1
10 ) 10
     10
```

Division of larger numbers follows the same process.

```
       1110
110 ) 1010100
       110
       1001
        110
         110
         110
```

The decimal proof for this binary division is

```
    110 =  6
1010100 = 84
   1110 = 14
 6 ) 84 = 14
```

Table 2-6 Decimal to BCD equivalents.

Decimal	*BCD*
0	0000
1	0001
2	0010
3	0011
4	0100
5	0101
6	0110
7	0111
8	1000
9	1001

Table 2-7 Octal to binary conversions.

Octal	*Binary*
0	000
1	001
2	010
3	011
4	100
5	101
6	110
7	111

Several systems are used to make the binary numbering system more functional and easier to use. One such system is the Binary Coded Decimal (BCD) system. In this system a 4-bit (digit) binary number is used to represent a decimal number. If the binary number 1111 (the largest four-digit binary number) is converted to decimal, the decimal number 15 is obtained. Therefore, since decimal digits include only numbers less than 10, all decimal digits can be represented by binary numbers four digits in length. Table 2-6 provides decimal-to-BCD equivalents. The binary numbers from 1010 to 1111 are not used in BCD. To represent larger decimal numbers, 4-bit binary numbers are placed side by side. The decimal number 2863 in BCD would be

0010 1000 0110 0011

The octal and hexadecimal systems are different from BCD. BCD represents the decimal numbering system in binary code. Octal and hexadecimal systems are different ways of representing binary. They make the binary system easier to use. A long string of 1s and 0s is hard for people to copy and remember with accuracy. Octal is named for the number 8. Since binary numbers three digits in length can represent eight decimal numbers, an octal digit can replace three binary digits. Table 2-7 lists the octal-to-binary conversions. Notice that "0" is a valid digit. In digital and computer applications, 0 is always a valid number. When specifying a memory location in a computer system, 00000000 is the code for a location just as 11111111 is. Octal uses decimal digits from 0 to 7 to represent three-digit binary numbers. Octal becomes, therefore, a shorthand for binary. For example, 011 in binary would be represented by the octal symbol 3. The value of this system is seen when manipulation of large binary numbers is required. For example, the number 1 011 001 011 100 110 is unwieldy in binary form. In octal the number would be 131346. This number is more convenient for human manipulation than the 16-bit binary number.

Hexadecimal (hex) is similar to octal. The major difference is that hex represents 16 values or a 4-bit binary number. The derivation of the name *hexadecimal* should be apparent: hex = 6, decimal = 10, hexadecimal = 16. Table 2-8 outlines the hexadecimal-to-binary relationships. Notice that the letters A through F are used to expand hex to 16 numbers. Also notice that hex is more efficient than octal in that all binary values available from a 4-bit number are accounted for. As with octal, hex is a shorthand for the manipulation of binary numbers. The binary number 1110 is represented by the letter E in hex. The number 1011 0010 1110 0110 is represented in hex as B2E6. Many computers use 16-bit words. Hex is a very convenient way to represent these codes.

It is important to realize that octal and hex are conveniences for people and are not working codes. For these codes to be used in digital systems, they must be converted to their binary equivalents. This conversion can be accomplished by the digital system, which will be demonstrated shortly.

Experiment 2-10: Binary Arithmetic

A single gate in isolation is somewhat limited. When gates are connected together, however, their capabilities can be magnified many fold. This experiment will illustrate the procedures to be followed in the design of logic circuits. Several common errors in circuit design and construction make these special words of caution in order:

1. Be sure that all IC packages are supplied with appropriate power and ground connections. Refer to data books or Appendix C for information concerning pin-outs and power supply requirements.
2. Use heavy wire or wide foil strips for power supply buses. A breadboard bus is adequate. Low impedance power and ground buses assist in reducing noise and spiking problems.
3. Use adequate despiking and decoupling capacitors. A 0.01-μF capacitor for each group of four TTL gates is usually adequate. Other families such as CMOS may not require this level of bypassing. A good maxim is: *If the circuit operation is erratic, add a capacitor.*
4. Be sure all inputs are connected appropriately. Every input should be connected to one of three points: to a solid input, to the positive supply bus, or to ground. Unused inputs, even if they are inputs to logic elements not used in the circuit, should be connected to an appropriate point.
5. Leave all unused outputs unconnected.
6. Do not load circuit outputs beyond their stated capacity.
7. Refine all circuits for the simplest and most efficient designs.

Armed with these few suggestions, design of some actual circuits can be undertaken.

It was stated previously that the exclusive-OR gate functions in accordance with the rules of binary addition. Reference to the rules of binary addition shows that adding 0s and 1s results in a sum of "one" only when the two numbers added are not alike. If two 0s are added, a zero sum results. If two 1s are added, a zero sum with a carry results. The exclusive-OR truth table shows that the output of the gate is "one" only if the two inputs are different. If the inputs are the same, whether high or low, the output is zero. This exclusive-OR function is analogous to binary addition if a carry capability is provided.

Construct the circuit shown in Fig. 2-19a. This circuit is called a *half adder*. It will provide an output that is the binary sum of the two inputs. Vary the inputs as required by the truth table shown in Fig. 2-19b and record the output states of the two gates. Notice that the rules of binary addition are fulfilled. The AND gate provides for the carry function when both inputs are high; both outputs are zero when both inputs are low, and only the exclusive-OR gate output is high when one but not both of the inputs is high.

Table 2-8 Hexadecimal to binary relationships.

Hexadecimal	*Binary*
0	0000
1	0001
2	0010
3	0011
4	0100
5	0101
6	0110
7	0111
8	1000
9	1001
A	1010
B	1011
C	1100
D	1101
E	1110
F	1111

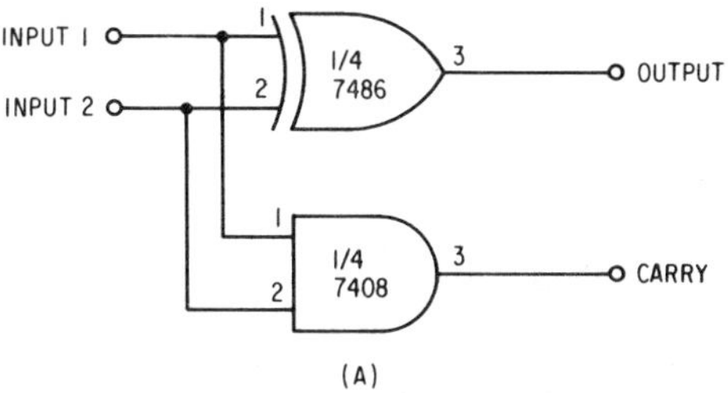

Half Adder

IN 1	IN 2	OUT	CARRY
0	0		
0	1		
1	0		
1	1		

(B)

Fig. 2-19 Half adder: (a) functional diagram and (b) truth table.

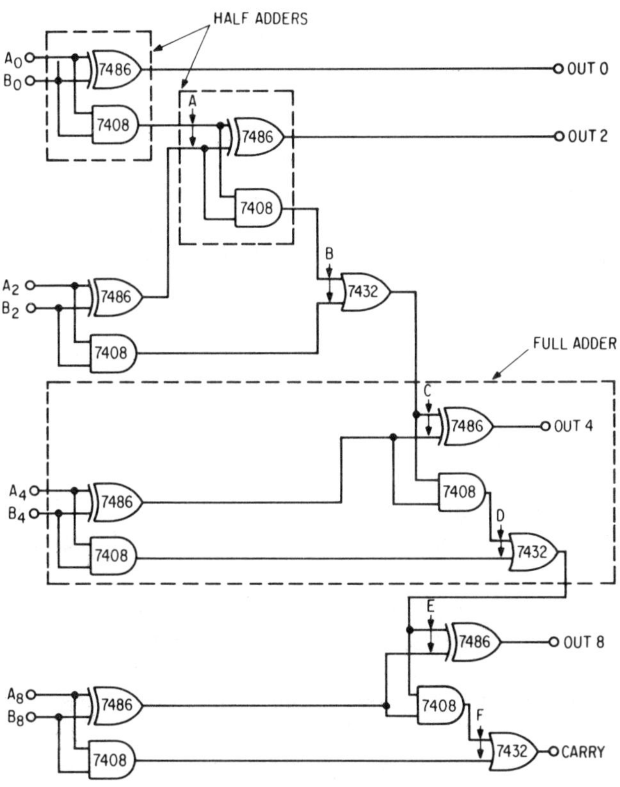

Fig. 2-20 Full adders.

Construct the circuit shown in Fig. 2-20. It should be apparent that this is simply an expansion of the half adder. This circuit is labeled in Fig. 2-20 and is called a *full adder*. The difference between the half and full adder is that the full adder is designed to permit introduction of a carry from a previous circuit. The circuit shown in Fig. 2-20 consists of four full adders that are connected in parallel. This circuit will add a 4-bit binary number introduced at the inputs marked "A" with a 4-bit binary number introduced at the inputs marked "B." To test the circuit, connect the "A" inputs for the binary number 1010.[4] (A "1" is obtained by connecting the input to the positive supply. A "0" requires the input to be grounded.) Connect the "B" inputs for the binary number 1101. The output state indicators will indicate the binary sum of these two numbers. According to binary addition, the output should be

Binary	*Decimal Equivalent*
1010	10
+ 1101	+ 13
10111	23

Add other binary numbers using the circuit. The state indicator can be used to check the state of intermediate outputs such as those marked A, B, C, D, E, and F. Trace the carries and determine the route by which they affect the output. Notice that in each case the rules of binary addition are adhered to.

Table 2-9 Multiple input gates.

Family	*Number*	*Type of Gate*
RTL	99540C	Dual, four-input AND
RTL	918	Dual, three-input NOR
RTL	919	Dual, four-input NOR
TTL	7410	Triple, three-input NAND
TTL	7420	Dual, four-input NAND
TTL	7430	Single, eight-input NAND
CMOS	4023	Triple, three-input NAND
CMOS	4025	Triple, three-input NOR
CMOS	4073	Triple, three-input AND
CMOS	4075	Triple, three-input OR
CMOS	4002	Dual, four-input NOR
CMOS	4012	Dual, four-input NOR
CMOS	4072	Dual, four-input OR
CMOS	4082	Dual, four-input AND
CMOS	4068	Single, eight-input NAND
CMOS	4078	Single, eight-input NOR
ECL	10110	Dual, three-input, three-output OR
ECL	10106	Triple, three-input NOR
ECL	10109	Dual, multiple input OR/NOR
ECL	10118	Dual, multiple input OR/AND

Multiple Input Gates

Many circuits can be simplified if multiple input gates are used. Multiple input gates reduce the need for two input gates to be connected in tandem so that the number of inputs can be increased. Table 2-9 lists various multiple input gates from some popular

[4]The 4-bit binary numbers should be entered with the least significant bit at the input marked A_0/B_0 and the most significant bit at A_8/B_8; i.e., A_0 = 0, A_2 = 1, A_4 = 0, A_8 = 1. The output also will follow the binary values; i.e., MSB = Carry, LSB = Out_0.

Fig. 2-21 LED resistor stack used in constructing the Smitty. Resistors are enclosed between two pieces of perforated board.

IC families. These special IC packages represent the most convenient method of multiplying inputs. Other ways do exist for the provision of additional inputs. Interconnecting gates with two inputs can provide an equivalent circuit of many inputs (see Figs. 2-6a and 2-8a). This technique has its drawbacks. Each gate delays the circuit a small amount. Cascading several gates can introduce an intolerable amount of delay. Power dissipation also can be a problem when using this technique.

Another method of increasing inputs is called *strobing*. Strobing, in this instance, refers to selectively removing power from the entire package. An external gate, for example, could be used to remove or apply ground to a package. This procedure would effectively increase some circuit inputs by one per gate. The limitations are obvious. The package ground is common to all gates within the package. If one gate is powered down, all will be. Therefore, the technique must be used with caution. Strobing can be effectively used only in circuits that employ all package gates in ways that will not be impacted by the procedure.

Gate expanders are available in some older IC families such as RTL. These devices permit expansion of inputs by simply paralleling outputs. While the devices were useful for RTL, later families have discontinued their use.

Open collector logic employs ICs that are designed with open collector outputs. This facility permits the designer to use the output as he desires. One use for open collector logic is to assemble several gates with their outputs in parallel. A pull-up resistor is required for activation of the resulting multiple input gate. The resistor must be grounded through one or more of the inputs to provide a zero output. Thus, open collector logic circuits function as multiple input NOR gates. In fact, open collector logic circuits are referred to as the "wired" or "implied" NOR.[5] Open collector logic also can be used to drive relays, LEDs, or other devices at voltages higher than typical IC logic voltages.

There are problems with open collector logic. The load resistor reduces the speed of the circuit. Also, noise problems are increased and troubleshooting is made more difficult.

Tri-state logic provides a form of multiplexing outputs. Numerous outputs can use the same input line. Tri-state functions also apply to inputs. Tri-state logic is often used in bus oriented systems. The uniqueness of tri-state logic is that it provides a third state. The inputs and outputs can assume a low state, a high state, or a "third state." When in this third state, the input or output appears to be disconnected from the circuits. Inputs do not load down circuits and outputs do not drive circuits. This permits numerous circuits to be connected to the same signal bus and be selectively turned on to drive the bus or to receive from it. The mechanism for placing tri-

[5]In CMOS, the wired OR function is normally accomplished by transmission gates and similar circuits due to the nature of the CMOS inverter.

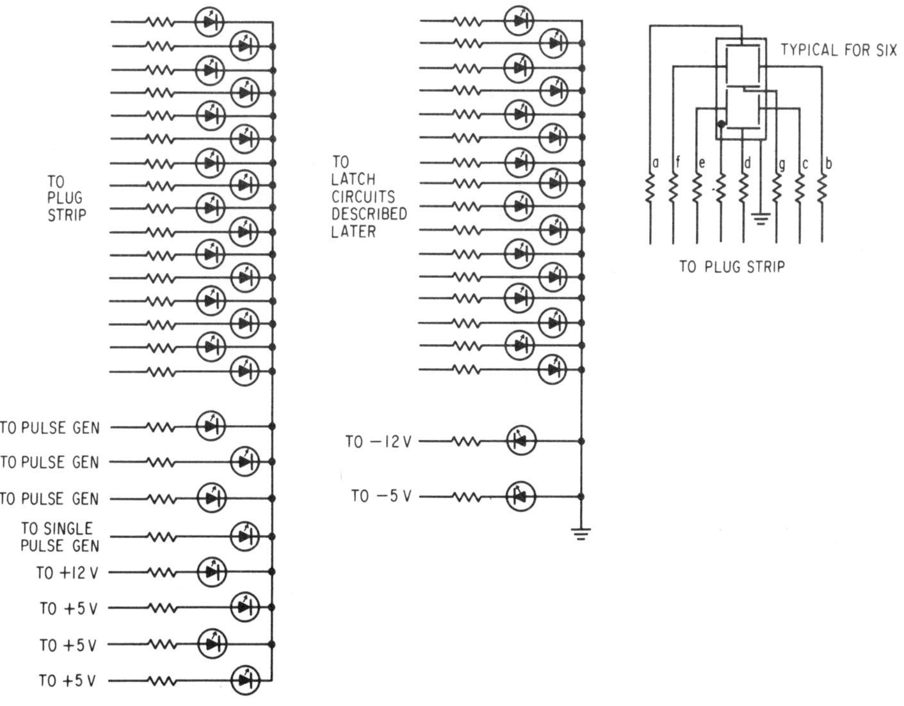

Fig. 2-22 LEDs and indicators for the Smitty.

state devices into the active mode is an "enable" or "control" pin-out. An increasing use of tri-state logic is being observed within microcomputer and microprocessor circuitry. Processors often drive several peripheral devices. A number of these devices can use the same signal bus without interference as long as the processor enables only one at a time. Some processors require input and output data to flow over the same buses at different times. Tri-state logic makes this possible. Tri-state logic will be dealt with in more detail later.

Displays for the Smitty

Constructing state indicators for each experiment is inefficient. Having only one logic probe is not satisfactory. The Smitty and the Melton Special breadboards incorporate multiple, built-in LED state indicators. The Smitty adds six seven-segment LED displays as well. As was indicated earlier, most LEDs require a current-limiting resistor. Refer to page 24 for the formulas relating to these resistors. LED displays require resistors for each segment. The Smitty, with 41 state indicators and six seven-segment displays with decimal points requires 89 resistors. If the student has access to printed circuit fabrication materials, the resistors can be placed on the PC board. The author's version of the Smitty employs perforated board to hold the resistors, as shown in Fig. 2-21. The value of the resistors used in the Smitty is a compromise that permits use of the LEDs with 5-V TTL circuits or with 12-V CMOS circuits.[6]

The Smitty uses 16 LED state indicators to monitor the status of latches on 16 push switches. In a later project, these switches and the IC latching circuits will be constructed. The bottom row of LEDs on the panel of the Smitty will indicate if the latch is high or low. A second row of 16 LED state indicators is provided on the Smitty front panel for use in general monitoring tasks. These LED inputs appear at the plug strip located to the left of the chassis near the panel. The display inputs connect to the right-hand terminal strip. Connections between circuits and the LEDs are accomplished by plugging wires between the selected LED input pins and the desired point in the circuit. LED state monitors also are provided for each power supply and timing circuit in the Smitty.

It is recommended that all state indicators be connected to the resistors at this time. The displays also should be connected to the resistors. As shown in Fig. 2-22, the cathodes of the LED state indicators are wired in common. The anode connects to the resistor. From the resistors connected to the top 16 LEDs, wires are run to the left-hand plug-in strip. The lower 16 LEDs will be connected to push button

[6]This practice provides less voltage than optimum in 5-V circuits and more than optimum in 12-V circuits, but no problems have been noted to date.

circuits in a later project. The power supply LEDs also can be connected (see Fig. 2-22).

The six displays can be soldered directly or standard DIP sockets can be used with the displays plugged into them. It would be simpler to use a standard BCD to seven-segment driver IC such as the 7447 TTL circuit to drive the display. This method of driving the display would require only four input lines per display. This method was not chosen in constructing the Smitty. Maximum flexibility and available parts dictated direct interconnection. This method of connecting the displays provides for their use with hexadecimal code. By having all segments available, a makeshift hexadecimal display can be made. Figure 2-23 shows the 16 numerals and letters devised for use with these displays. The BCD to seven-segment drivers can still be used by constructing them on the breadboard.

Logical Processes of Logic Design

Any useful digital logic circuit design is the product of logical planning. This planning process will normally follow a sequence of several design decisions.

1. *Decide on Design Goals* The basic question is "What function must this circuit accomplish?" Depending on the complexity of that function, a single circuit goal or a broader circuit goal with several subordinate circuit goals may be required. For example, a simple circuit might be described with the following circuit goal: This circuit will provide a low logic level only when both of its inputs are at a logic level high. As you may have deduced, this goal statement defines a NAND gate. A more complex circuit might require a number of circuit goal statements. For example: This circuit will turn the house lights on 1 hour after dark. It is apparent that this statement contains several different functions. To make the described circuit more manageable for the designer, several supporting circuits should be identified. *Main Goal:* (Subgoal 3) lights on/ (Subgoal 2) 1 hour/ (Subgoal 1) after dark.

   Subgoal 1: This circuit will provide a high output when the sun goes down. (This might be a photocell circuit with logic level output.)

   Subgoal 2: This circuit will provide a high output 1 hour after its input goes high. (This is a timer circuit.)

   Subgoal 3: This circuit will turn on the house lights when its input goes high. (This is an interface circuit that probably would activate an SCR or a relay via logic level signals.)

   More complex circuits might require additional subgoals and may even require subfunctions for each subgoal. The more specific functions are made during this planning phase, the simpler the actual circuit design will be.

2. *Decide on Design Approach* The question to be answered under this decision is "Should I design a new circuit or should I use an available design?" Again, the complexity of the circuit will determine how the decision question is answered. For simpler tasks, only one decision will be needed. For complex tasks, the appropriate decision will be required for each subgoal and each subfunction. This method of circuit design lends itself to a design concept called *modularity*. If each subfunction circuit performs a single task, it can be used as a module in the other circuits. As the designer's file of single task circuits increases, his design work will become module interconnection. He will design new circuits only for those tasks that require circuits not in his design file. Modularity will increasingly speed up the design process and reduce design errors as the design file increases in size.

3. *Decide on Design Refinement* The question to be answered here is "Can the design be simplified?" In answering this question, several other questions must be answered:

   a. Is this circuit doing more than the original goal required? Adding capability to a system may enhance its flexibility, but it also may decrease its cost effectiveness. If superfluous circuitry is discovered, it should be removed but not discarded.

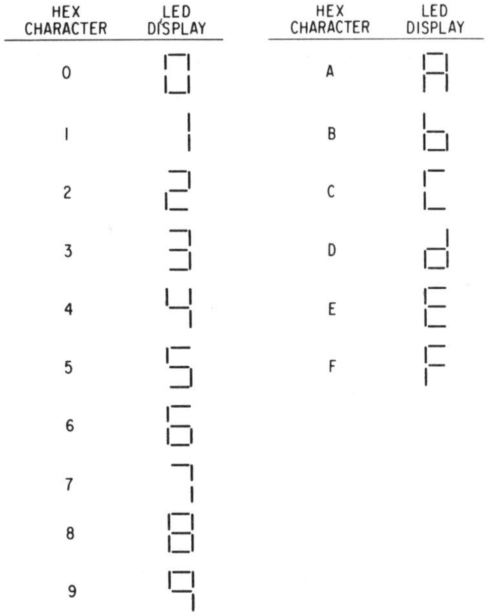

HEX CHARACTER	LED DISPLAY	HEX CHARACTER	LED DISPLAY
0		A	
1		B	
2		C	
3		D	
4		E	
5		F	
6			
7			
8			
9			

Fig. 2-23 Pseudo hex display using seven-segment displays.

Rather, it should be recorded and included in the designer's design file.

b. Can the number of circuit elements be reduced by application of DeMorgan's theorem? If inverters must be used to interface circuits, DeMorgan's theorem may simplify the circuit design.

c. Can several circuit functions be accomplished by special purpose logic chips? The simplest logic design uses gates only. Indeed, all logic functions can be accomplished with gates. However, the tedium, cost, and size of design using standard gate packages would be prohibitive. As will be seen in succeeding chapters, special purpose logic chips are available that perform functions that would require numerous gates. Small-Scale Integration (SSI) and Medium-Scale Integration (MSI) include flip-flops, timers, and similar circuits. Large-Scale Integration (LSI) provides the power and resources of large memory arrays and microprocessors. These units can replace hundreds and even thousands of gates. The decision to use these special circuits must include cost considerations. Cost of sockets, PC boards, and development of PC board layout are all involved in this design decision.

Experiment 2-11: The Design Process and Number Conversions

This experiment will use the design decisions just described to construct a circuit.

Step 1: Define the Goals of the Circuit The major goal of the circuit is to display the equivalent decimal sum of two four-digit binary numbers. The following subgoals are required to accomplish the major goal:

1. *Subgoal 1:* This circuit will provide a BCD output that is the sum of two binary numbers.

    a. Subfunction 1-1: This module will add two binary numbers and provide a binary output.

    b. Subfunction 1-2: This module will convert a binary input to a BCD output.

2. *Subgoal 2:* This circuit will light the appropriate segments of seven-segment displays to show a decimal equivalent of a BCD input.

    a. Subfunction 2-1: This module will convert a BCD input to a decimal output.

    b. Subfunction 2-2: This module will provide the conversion and interface required to drive seven-segment LED displays.

Table 2-10 Binary, BCD, and decimal equivalents.

Binary	*Add*	*BCD*		*Decimal Equivalent*
00000		0000†	0000‡	0
00001		0000	0001	1
00010		0000	0010	2
00011		0000	0011	3
00100		0000	0100	4
00101		0000	0101	5
00110		0000	0110	6
00111		0000	0111	7
01000		0000	1000	8
01001		0000	1001	9
01010	0110	0001	0000	10
01011	0110	0001	0001	11
01100	0110	0001	0010	12
01101	0110	0001	0011	13
01110	0110	0001	0100	14
01111	0110	0001	0101	15
10000	0110	0001	0110	16
10001	0110	0001	0111	17
10010	0110	0001	1000	18
10011	0110	0001	1001	19

†Tens weighting.
‡Ones weighting.

Step 2: Decide on Design Approach In this example, the preliminary design will use only the gate circuits that have been introduced to the student. Also, the design file will be limited to the circuits that the student has constructed. In the refinement step, some alternate circuit solutions will be suggested.

Subfunction 1-1 calls for a circuit that adds two binary numbers. Such a circuit was constructed in Experiment 2-10. The schematic of it is shown in Fig. 2-20. Since the circuit has been built previously, it can be used intact in this experiment. Subfunction 1-2 requires a circuit to convert binary to BCD. The design file does not contain a circuit that will accomplish this task. Circuit design is therefore called for but should be preceded by a logical analysis of the problem. In this case binary code must be converted to BCD. How do these codes differ? BCD and binary are the same up to the number 1001 (9 in decimal). At this point, binary continues to alter its code in a standard binary fashion. BCD, however, initiates a carry to indicate the number ten, and the other four digits are returned to zero. The process continues until the BCD digits are 1001 again. Another carry is initiated. This process is continued for each decade. The important point for this experiment is that BCD and binary numbers are six counts apart for numbers between 9 and 20, not inclusive.[7] A BCD adder is needed to make the conversion. A BCD adder operates under the same rules as binary addition. The one difference is that the BCD adder adds six to the sum of the first binary adder when the numbers

[7]To keep the number of ICs reasonable, this experiment will limit the input to numbers which will provide a sum of decimal 19 or less.

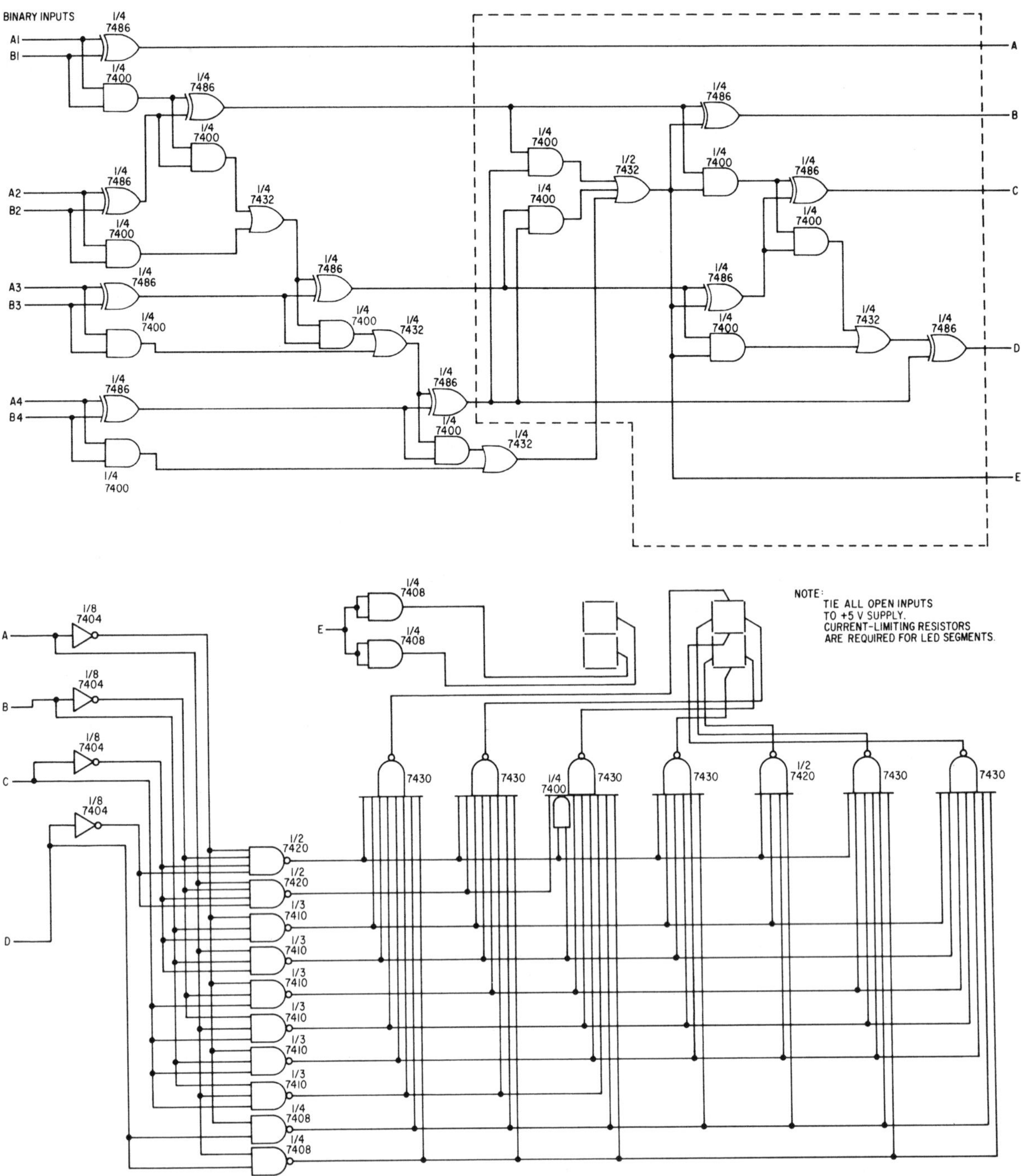

Fig. 2-24 Binary coded decimal schematic diagram.

are greater than 1001. Table 2-10 shows the binary and BCD codes for numbers between 0 and 19. For the numbers between 10 through 19, add a binary 0110 (decimal 6) to the binary number and compare the result with the BCD column. The agreement of the two numbers indicates that adding six to these numbers will provide a proper BCD output. The circuit that will convert binary to BCD must therefore add six to each number over 9 and less than 20. The circuit shown within the dotted lines in Fig. 2-24 will accomplish this task. Careful examination of the circuit shows that 0110 will be added to the binary adder output if the two's and four's columns or the two's and eight's columns are simultaneously hi. Table 2-10 shows the numbers between 0 and 19. Notice that all numbers over 1001 have the simultaneous highs required for the addition of 0110 to take place. Also notice that no number below 1001 will result in the addition. This circuit, therefore, meets the conditions required by subfunction 1-2. To expand the adder to handle numbers larger than 19 requires an additional BCD adder for each decade.

Subfunction 2-1 requires conversion from BCD to a one-of-ten or decimal output. Four inverters and ten multiple input gates are used to produce one-of-ten outputs from the four BCD inputs. The carry is used to drive a separate LED display. This can be done since no number over 19 will be permitted in this experiment. The important design concept for this circuit is that inputs to the gates are chosen so that all inputs are high for a particular output only when the unique combination of 1s and 0s representing that number are placed on the inputs. For many of the decimal outputs, three inputs suffice. For others, four inputs are required and for two, only two inputs are needed. Multiple input gates were used in this design for all digits. Inverters are used to provide highs to the gate inputs when required. Subfunction 2-2 provides for a decimal to seven-segment LED driver circuit. One two-input gate and seven multiple input gates are used to decode the decimal output. Each multiple input gate drives one LED segment through a current-limiting resistor. While the multiple input gates are shown as positive logic NAND gates, their actions are more easily understood as negative logic NOR gates. A negative logic NOR gate provides a high output when any or all inputs are low. When one of the decimal input gates has all inputs simultaneously high, it provides a low output. This low output is applied to a negative logic NOR gate, which goes high and lights the appropriate LED segment.

Step 3. Decide on Design Refinement Question a: Is the design doing more than the original goal required? The answer is no. The circuit accomplished exactly what the goal required.

Question b: Can the number of circuit elements be reduced by application of DeMorgan's theorem? Again, the answer is no. Usually DeMorgan's theorem will be indicated if numerous inverters are required. The only inverters used in this circuit are those at the input of the BCD-to-decimal converter, and this use of inverters is a special application that cannot be avoided using DeMorgan's theorem.

Question c: Can several circuit functions be accomplished using special purpose logic chips? The answer is yes. While these special purpose chips have not been covered, their function can be identified by reference to data sheets. The circuit can be simplified by using several special purpose logic blocks.

The 7483 is an example. This is a full adder built into a single 16-pin DIP. This one chip will replace the entire adder circuit of our design. It also can be used for the BCD adder. Only the "add 0110" circuit must be retained. Thus seven exclusive-OR gates, seven dual input AND gates, and three dual input OR gates will be replaced by one logic block. Two of these blocks replace twelve exclusive-OR gates, ten AND gates, and four OR gates. Use of a 74185A in place of the BCD adder will negate the need for the

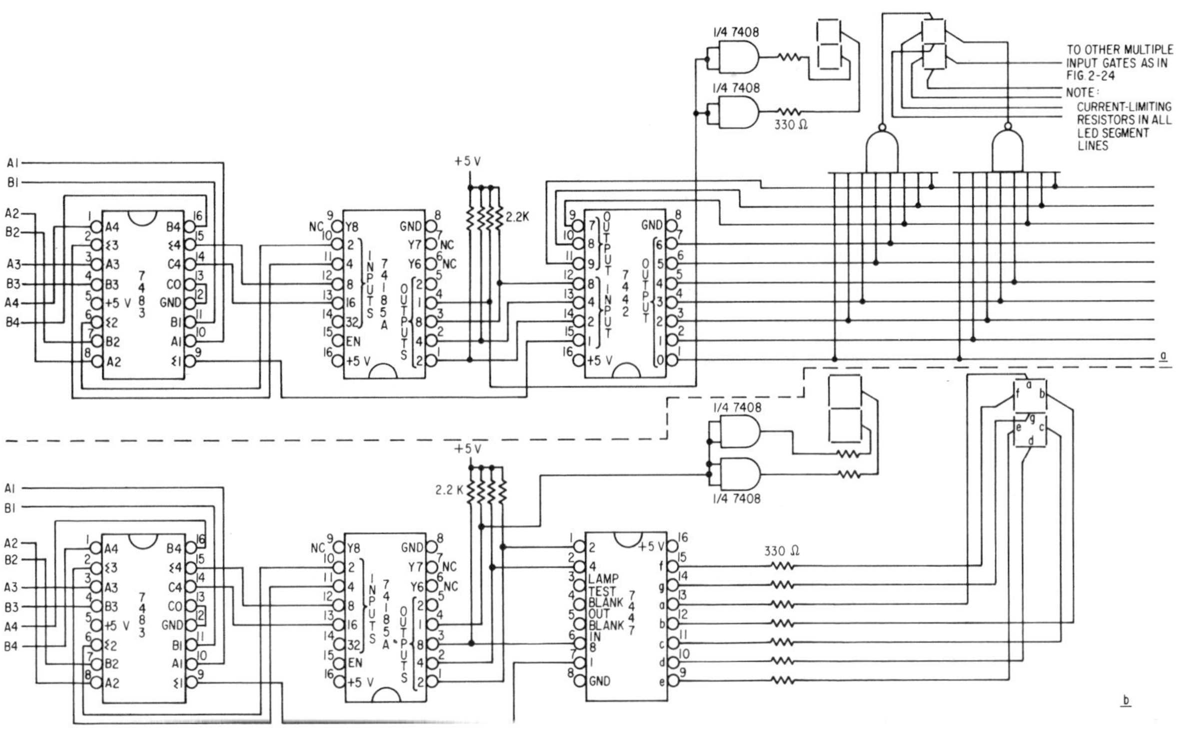

Fig. 2-25 Binary coded decimal schematic diagram (simplified designs).

Table 2-11 Numbers for addition.

IN A	*IN B*	*Display*
0001	0001	— —
0010	0001	— —
0011	1000	— —
0100	1001	— —
0101	0011	— —
0111	0111	— —
1000	1000	— —
1001	1001	— —
1011	1000	— —
1100	0110	— —
1101	0110	— —
1111	0010	— —
1010	0010	— —
1000	0010	— —
1010	1001	— —

addition of 1001. The 74185A is a binary-to-BCD converter.

The 7442 logic block also can be used to simplify the design. The 7442 is a BCD-to-decimal decoder. This one 16-pin DIP will replace four inverters and the ten multiple input gates used in the original design.

This is not the best choice, however. If the conversion to decimal is bypassed, the 7447 offers considerable advantages. The 7447 is a BCD to seven-segment driver. This chip will replace the four inverters, the ten gates of decimal converter, and the seven eight-input gates and one two-input gate of the decimal to seven-segment decoder circuit. Figure 2-25 demonstrates rather dramatically the extent of simplification that is permitted by using special purpose chips.

Construct this circuit[8] and add the binary numbers listed in Table 2-11. Record the decimal digits displayed on the readout. Does this final circuit accomplish the original design goals? This same design process can be used with any design task.

---

[8] The student may wish to construct each of these circuits and then trace the logic through them to learn how each obtains the final results.

THREE

Synchronizing Circuits

Synchronous and Asynchronous Logic

The gate circuits discussed in the previous chapter are the basic building blocks of logic circuitry. Their operation is relatively simple: change the input signals to the gate and the output will immediately change. If the inputs revert to their first conditions, the output immediately returns to its former state. This simple action is adequate for many circuit functions, but it represents limitations for others. Many circuits require timing and memory capability. Simple gate circuits cannot provide these. A second type of logic called *clocked logic* is required.

Clocked logic is composed of gates, but these gates are connected in interactive circuits that provide for timing and memory. In computer circuits, for example, data may be available from several inputs and may be routed to several outputs. This data is manipulated by the computer one input or one output at a time. The only way the multiple inputs and outputs can be serviced by the computer is for the data flow to be synchronized by system timing pulses. The logic elements that are governed by system timing commands are clocked logic devices such as flip-flops, counters, and microprocessors. Differentiation can be made between simple gates and clocked logic on the basis of synchronization. Clocked logic is referred to as *synchronous logic* while nonclocked logic is referred to as *asynchronous logic*. As was stated previously, the primary functional difference between the two is timing and memory. As will be shown, clocked logic operates on the same two logic levels that gates use. The two types of logic are therefore compatible and often are combined in circuits.

The various synchronous logic devices that are available will be covered in later chapters. This chapter discusses the synchronizing or timing circuits used with clocked logic.

The Switch and Clocked Logic

A mechanical toggle switch has characteristics similar to clocked logic. It can be "clocked." Switch it one way and data flows. Switch it the other way and the flow of data stops. The switch has memory. Once it is turned on, it will remain on until it is turned off. A switch can be used as a simple system clock. A toggle switch connected to the positive voltage source will vary its output between 0 and full supply voltage each time it is switched. The switch output will be a series of square waves. The rate of the square waves will be determined by the "toggle frequency" or the frequency at which the switch is turned on and off. This output could be used for logic system timing except for one problem: a mechanical switch is noisy. When mechanical switch contacts "make," they "bounce." Rather than producing one clean closure, the switch contacts actually touch and separate several times before they settle into final contact. For slow devices such as light bulbs and power supplies, this bouncing is not noticed. Logic circuits, however, operate so rapidly that each contact "bounce" will be interpreted as a contact closure. The result will be improper circuit timing.

Suppose the Highway Department desired to count the number of cars that use a certain highway. To count the cars the department might stretch a hose across the roadway and connect the hose to an air-activated mechanical switch. As a car passes over the hose, the tires compress the hose, and air is forced down the hose activating the switch. An IC counter could be used to count the number of times the switch closes. The problem occurs when the counter records 15 or 20 switch closures for each car. The Highway Department might build a superhighway where a farm-to-market road is actually needed.

The solution to the noise switch problem is an interface circuit called a *contact conditioner*.[1] A contact conditioner is designed to produce one and only one output pulse for each switch closure. It will disregard the contact bounces. Figure 3-1 shows

[1] This circuit also is referred to as a "noiseless pushbutton circuit" or a "switch debouncer."

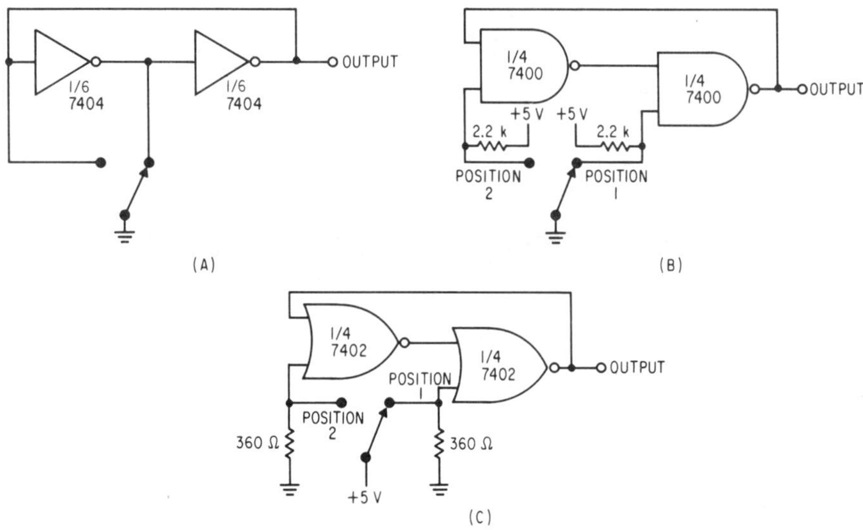

Fig. 3-1 Switch conditioner circuits: (a) dual inverter, (b) dual NAND gate, and (c) dual NOR gate.

contact conditioners constructed from several different types of logic elements. The circuit of Fig. 3-1a is the simplest. As shown, this circuit uses two inverters. They are connected so that the output of one inverter drives the input of the second inverter and the output of the second inverter drives the output line, as well as being cross-coupled to the input of the first inverter. Consider the action of this circuit if the switch is disregarded. If the output of inverter 1 is a logic high, it will present a logic high to the input of inverter 2. Inverter 2 will invert the input and drive the output line and the input of inverter 1 with a logic low signal. The circuit is now stable with each inverter being held in its present logic state by the other inverter. If the switch is now added to the circuit, it provides a means whereby the balance of the circuit can be upset. When the switch is moved to the right-hand position, it grounds out the high output of inverter 1 causing a low input for inverter 2. Inverter 2 responds by changing its output to a high that forces a low output for inverter 1. The circuit is now stable once again. Since this circuit has two stable states, it is said to be bistable and is called a *bistable multivibrator*. The reason that the circuit is useful as a contact conditioner is that it will change states at the first contact closure and will remain in this new state until the switch is returned to the original position. The circuit operates so quickly that all contact bounces, except the initial closures are disregarded.

The contact conditioners shown in Figs. 3-1b and 3-1c provide the same function as the inverter circuit. The principle benefit of these circuits is that they represent better design philosophy. Grounding the output of a gate has the potential of destroying the gate.

The conditioner of Fig. 3-1b uses two NAND gates. The 7400 quad NAND gate is the flagship of the TTL line and is used in this circuit. The circuit is essentially the inverter circuit with a second input for each gate. The two gates are again cross-coupled so that each drives one input of the other. The second input on each gate connects to the positive voltage source through current-limiting resistors. These inputs also connect to the switch that holds one or the other at ground potential. Since a NAND gate has a low output only when both inputs are high, the gate that has one of its inputs grounded by the switch will provide a high output. The other gate will have one input at the positive supply voltage level. The high output of the first gate provides the second high input for this gate and results in a low output. This is a stable condition that will remain until the switch is placed in the opposite position. When moved to position 2, the switch will remove the ground from the input of gate 1 and place it on the input of gate 2. Gate 2 output will go high providing the second high input for gate 1. Gate 1 output will go low, and the circuit will again be stable. This circuit reacts to the initial contact of the switch and ignores the switch bounces just as the previous conditioner did.

The circuit shown in Fig. 3-1c also accomplishes the switch conditioning. The only difference is the use of NOR gates and the consequent requirement for only one high input to cause the output to change. The switch in this circuit applies the positive supply voltage to an input of one or the other gate. In position 1, the switch places a high on the input of gate 1 forcing the output low. Gate 2 now has both inputs low and switches its output high. With two high inputs, gate 1 is stable and the circuit remains in this condition. When the switch is placed in position 2, the circuit assumes the second stable state. This circuit, as with the two previous, disregards the switch bounces. The output from these circuits can therefore be used to clock synchronous logic devices. Due to the nature of the TTL gate circuit, the resistors used in the NOR gate version of the circuit shown in Fig. 3-1c must be small. This configuration increases power dissipation and reduces circuit noise immunity. The NAND gate network shown in Fig. 3-1b is thus preferable.

The Set-Reset Flip-Flop

The NAND gate and NOR gate contact conditioners previously discussed have a hidden benefit. The circuit is responding to logic levels each time the switch position is altered. This makes it possible to force the circuits into their opposite states using outputs of other IC elements. This permits electronic switching to be employed with these circuits. When using electronic switching, however, the function of the circuit is not contact conditioning. The electronically switched circuit acts as a memory element. Figure 3-2a shows a variation of the NAND gate conditioning circuit. Notice that the circuit has the same interconnections between the gates as before, but the circuit has been redrawn in the form typical of a *flip-flop*. The square wave signals shown at the inputs represent electronic trigger pulses. To analyze the circuit, it will be assumed that the output marked Q will be a high logic level. Notice that this output is connected to one input of the lower NAND gate. The second input of this lower gate is held high until the trigger pulse is received. With both inputs held high, the output of the bottom NAND gate will be low. This low is applied to the top NAND gate holding the output of that gate high. The circuit is therefore stable in this condition. Notice that as long as the bottom input of the top gate is low, the output of this gate will remain high regardless of the state of the second input. The circuit will retain the states of the Q and the $\overline{Q}$ outputs until the trigger goes low at the input of the bottom gate. When this input goes low, the gate no longer has two high inputs, and the output will therefore swing positive. This will place a high on the bottom input of the top gate. A high already exists at the second input of this gate. A low output results at Q and at the top input of the bottom gate. The circuit is stable again. In this condition, the circuit will not respond to changes on the bottom trigger line. A pattern for this circuit should be apparent. The circuit will change states in response to an input change if that input did not cause the immediately preceding change. Thus, if input $\overline{S}$ causes the circuit to change or flip to the opposite state, the circuit will no longer respond to the $\overline{S}$ input. The $\overline{R}$ input must now trigger the circuit and cause it to change or flop back to the original state. This action gives the circuit its name. It is called a *set-reset flip-flop*.[2] The truth table for the circuit is shown in Fig. 3-2b. Notice that the simultaneous application of low trigger to both inputs results in a *disallowed state*, which is a condition that does not conform to the rules of logic. In this case, simultaneous low triggers force both outputs high. If one output is "Q" and the other is designed to be "not Q" ($\overline{Q}$), both outputs cannot be simultaneously high. Indeed, the two outputs must be complementary. Thus, simultaneous low triggers are not permitted on the inputs.

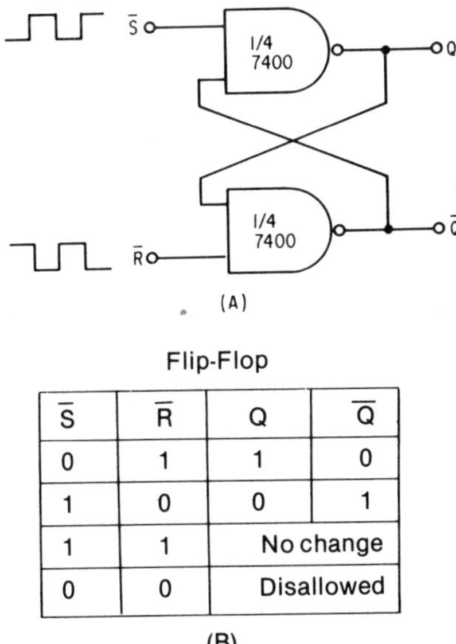

Flip-Flop

$\overline{S}$	$\overline{R}$	Q	$\overline{Q}$
0	1	1	0
1	0	0	1
1	1	No change	
0	0	Disallowed	

(B)

Fig. 3-2 R-S flip-flop using NAND gates: (a) circuit diagram and (b) truth table.

Experiment 3-1: NOR Gate R-S Flip-Flop

R-S flip-flops can be built using NOR gates. Draw a circuit that uses two NOR gates to provide (1) two complementary outputs, (2) a high at output 1 when the first high input trigger is received, (3) a high at output 2 when a second high input trigger is received, and (4) stability in either state until the next trigger is received.

When designing the circuit, remember that R-S flip-flops are always cross-coupled. Remember also that the circuit has two inputs and two outputs.

Compare your design with Fig. 3-3. The input lines of the circuit in Fig. 3-3 are reversed from the NAND gate flip-flop. The reason for this is that the *set input* is defined as *the input that when made positive will cause the Q output to go or stay positive.*

[2] The set-reset flip-flop is by convention called an *R-S flip-flop*.

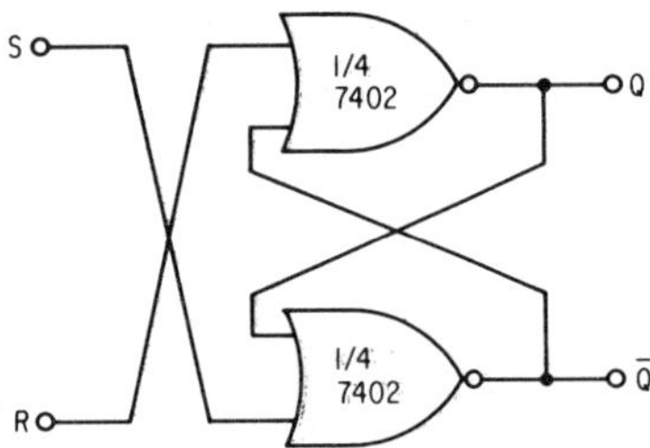

Fig. 3-3 R-S flip-flop using NOR gates.

S	R	Q	$\bar{Q}$
0	1		
1	0		
0	0		
1	1		

Fig. 3-4 Truth table for NOR gate flip-flop.

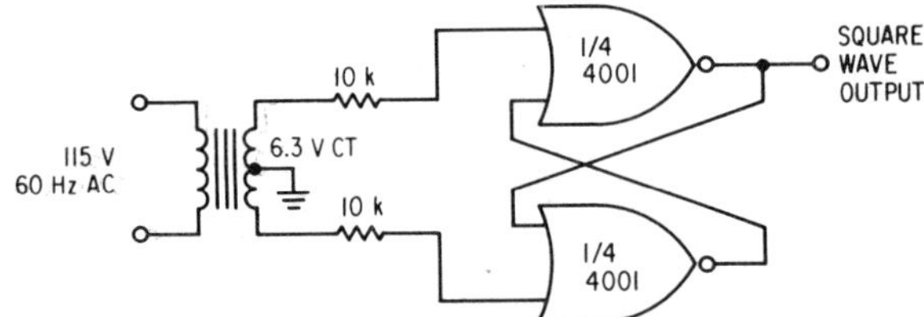

Fig. 3-5 Square wave generator/ac line conditioner circuit.

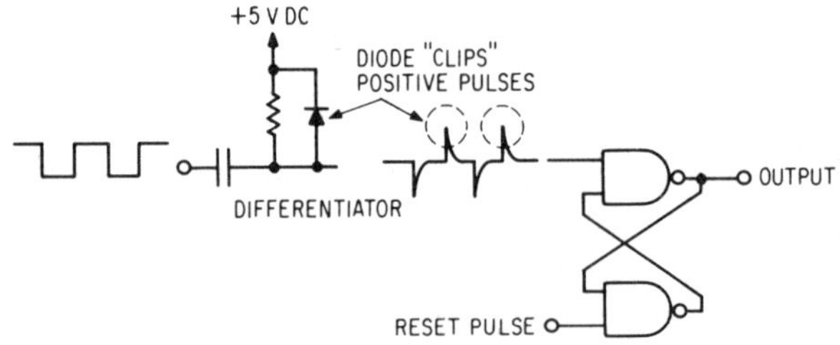

Fig. 3-6 Differentiator circuit.

Conversely, the *reset input* is defined as *the input that when made positive will cause the $\bar{Q}$ output to go or stay positive*. As can be seen, for this action to be accomplished with NOR gates, the input to the lower gate must be the S input, and the input of the upper gate must be the R input.

Construct the NOR gate R-S flip-flop. Vary the inputs as indicated and fill in the truth table shown in Fig. 3-4. Construct a NAND gate R-S flip-flop as shown in Fig. 3-2a. Verify the truth table in Fig. 3-2b. Compare the two truth tables. The two flip-flops operate in complementary fashion. The NAND gate flip-flop requires logic low triggers. Simultaneous low inputs result in a disallowed state. The NOR gate flip-flop responds to high triggers. Simultaneous highs are disallowed. Using one or the other of the two circuits, any application should be able to be accommodated.

R-S flip-flops are available as standard DIPs. The 4043B (CMOS) is a quad NOR gate flip-flop and the 4044B is a quad NAND gate R-S flip-flop. Refer to the data sheet for these units in Appendix C. Notice that the truth tables shown on the data sheets differ from those shown in Figs. 3-2b and 3-4. The reason for the discrepancy is that the 4043 and 4044 employ a tri-state, inverting buffer at the flip-flop output. When the enable line is held high, the inverted flip-flop output appears at the appropriate pin.

Pulse Generation

Clocking logic blocks with a push button is seldom satisfactory. Automatic timing circuits are needed. These circuits usually take the form of square wave generators or oscillators that are conditioned for use with ICs. The circuit shown in Fig. 3-5 demonstrates the use of a bistable multivibrator (R-S flip-flop) for conversion of a sine wave signal to a square wave signal. Circuits similar to the one shown are often used to produce a 60-Hz timing signal. Since the signal is referenced to the 60-Hz line frequency, the timing accuracy is held to close tolerance. The output of the bistable will be a square wave of equal duty cycle. If this signal is used for the master timing signal in a logic device, all of the circuits would be on for one-half of the time and off for one-half of the time. This is seldom the action desired. Most clocked circuits are designed to turn on at some specific time and stay on for a set or an indefinite period of time. The circuit will remain on until a second trigger pulse is received or the on time will be set by the components used in the circuit.

Precise timing is more easily achieved if switching coincides with the start or finish of the trigger pulses. A logic circuit can be made to respond to edge triggering through the use of a resistor/capacitor input circuit. The resistor/capacitor arrangement shown in Fig. 3-6 is called a *differentiator*. The components are chosen to have a short time constant relative to the input pulse. The time constant must be chosen to provide a pulse that is long enough to activate the flip-flop but short enough to prevent a disallowed state when the reset pulse arrives. The result will be an output from the RC combination that is a positive spike at the leading edge of the input pulse. Notice that the pulse is rising from zero to the positive supply level at its leading edge. A negative spike is generated at the trailing edge of the input pulse where the level falls back to zero. The R-S flip-flop responds to the negative pulse by changing its output states. The positive pulse is disregarded

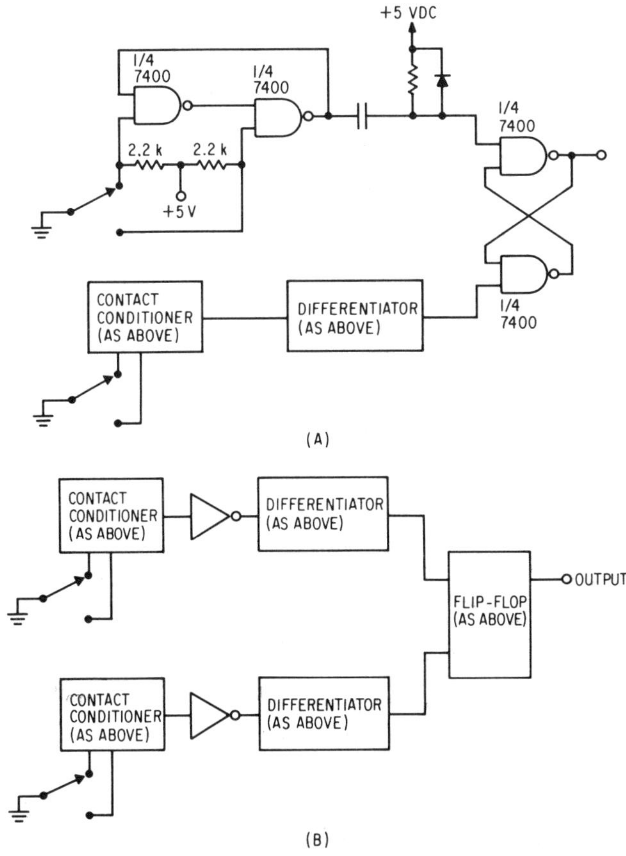

Fig. 3-7 (a) Edge triggering of a flip-flop and (b) edge-triggered flip-flop with inverters added.

by the flip-flop. Using edge triggering, the circuit will alter its states only in response to the falling edge of an input pulse.

The circuit can easily be made to respond to a rising pulse edge by using an inverter prior to the differentiation.

Experiment 3-2: Edge Triggering

Construct the circuit shown in Fig. 3-7a. The circuit uses contact conditioners and differentiators to drive an R-S flip-flop. Check the LEDs to determine whether the Q or the $\overline{Q}$ output is high. Depress the appropriate switch to set or reset the flip-flop. Hold the switch in the depressed condition for a second before releasing. Did the state change occur when the switch was depressed or released? Add inverters as shown in Fig. 3-7b. Repeat the experiment. When did the circuit change states this time? The experiment confirms that the R-S flip-flop is being activated by the falling edge of the input trigger.

The Half-Monostable Multivibrator

The bistable multivibrator requires both a set and a reset pulse. If a reset pulse is not provided, the flip-flop will be placed in a condition with $\overline{Q}$ high by

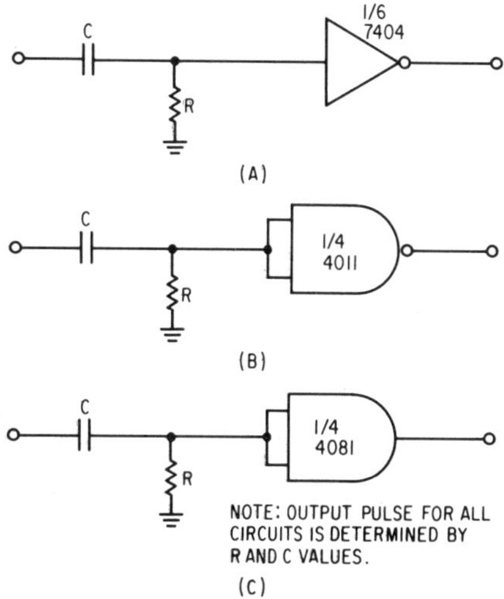

Fig. 3-8 Half-monostable: (a) using an inverter, (b) using a NAND gate, and (c) using an AND gate.

the set trigger, and it will remain in that condition. A circuit using an R-S flip-flop must therefore provide for both pulses. This is not always possible.

One solution to the bistable problem is the monostable. The *monostable*, as its name suggests, has only one stable state. The circuit is triggered into a temporary state. This state is held for a length of time determined by the circuit components. After this temporary state is completed, the circuit returns to the stable state and remains there until another trigger pulse is received.

The simplest monostable uses a single gate and is called a *half-monostable*. The half-monostable relies on RC time constants for its operation. Figure 3-8 shows several forms of the half-monostable. In each case, the temporary state is determined by the input RC network. The basic circuit action is as follows:

1. A trigger pulse is received at the input.
2. The output changes state.
3. The capacitor charges through the resistor until a level is reached that causes the gate to revert to the original state.
4. The trigger pulse is removed.

The half-monostable is limited in several respects. The input pulse must be of longer duration than the output pulse. The trailing edge of the output pulse is not well defined. This is due to the capacitor discharging from one input level to the other. As the capacitor charge changes, the gate is slowly biased into the second state. Rather than a sharp, well-defined transition, the half-monostable produces a ramp waveform at the trailing edge of the output pulse. The half-monostable must not be retriggered until the RC network has returned to its original state. These limitations reduce the utility of the half-monostable, but because of its simple and inexpensive construction it is used in numerous applications.

Schmitt Triggers

The sloppy output of the half-monostable can be improved. The use of "B" series TTL devices improves pulse shape. The use of *Schmitt trigger* circuitry also will improve half-monostable operation.

The Schmitt trigger is a circuit that uses feedback to speed switch time. Figure 3-9 shows a Schmitt trigger constructed from a noninverting buffer. The feedback resistor (Rf) and the input resistor (Ri) form a unique voltage divider circuit. The circuit is unique because it couples a portion of the output back to the input and because it changes function in the process of gate activation. A standard CMOS gate circuit changes output states at an input level that is approximately one-half of the supply voltage. As the input level rises, the gate remains unchanged until that "magic" level is reached that causes the gate to switch to the opposite state. The gate will remain in this state until the input level exceeds this critical value on the low side. Then the gate reverts to the original state. The problem is that as the point of change is approached, the gate circuitry has difficulty determining which state is appropriate. The result is a transition that is not as rapid as it should be. What is needed is a way to delay the switch point until the input level is past the halfway point. The Schmitt trigger does that and more. As the input to the Schmitt trigger increases toward the midpoint of the supply voltage, the input and feedback resistors act as a voltage divider that reduces the voltage applied to the gate input. By the time the input signal reaches a level sufficient to bias the gate on, it is far above the halfway point. When the gate output now begins to swing positive, a portion of the output level is fed back through Rf to the gate input. The gate rises in level more quickly and the pulse transition from low to high is made much more quickly. The action is reversed as the input level moves back toward the low level. Figure 3-10 is a typical transfer characteristic curve for the Schmitt trigger. Notice the steep rise time and the presence of two curves. The state change to a high output takes place above the midpoint and the change to a low output occurs below the midpoint. The intervening space between the two curves is called a *hysteresis* or *dead band*. The voltage difference of the two curves is equal to (Ri/Rf) × V.

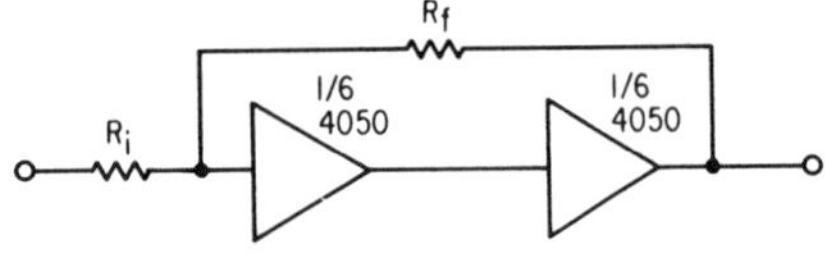

Fig. 3-9 Schmitt trigger.

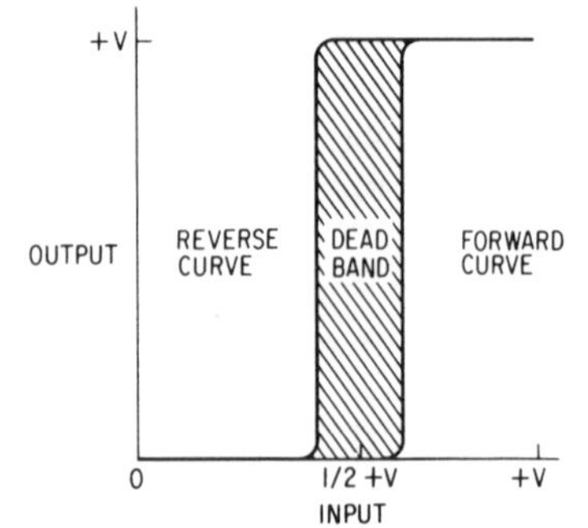

Fig. 3-10 Transfer characteristic of Schmitt trigger.

Schmitt triggers can be used to improve the contact-conditioning circuits presented earlier. The "snap" action of the Schmitt trigger represents a more acceptable trigger pulse for logic devices that are designed for high-speed operation. Figure 3-11 shows one-sixth of a 4584 hex Schmitt trigger being used as a contact conditioner. Notice the symbol being used for the Schmitt trigger. The small "hysteresis curve" within the gate symbol identifies the gate as a Schmitt trigger device.

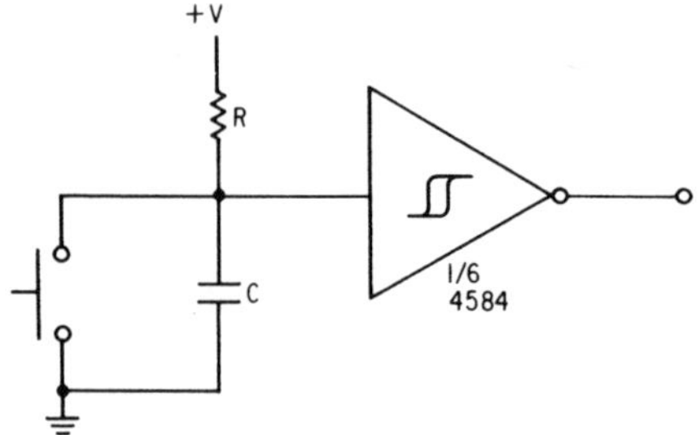

Fig. 3-11 Schmitt trigger contact conditioner.

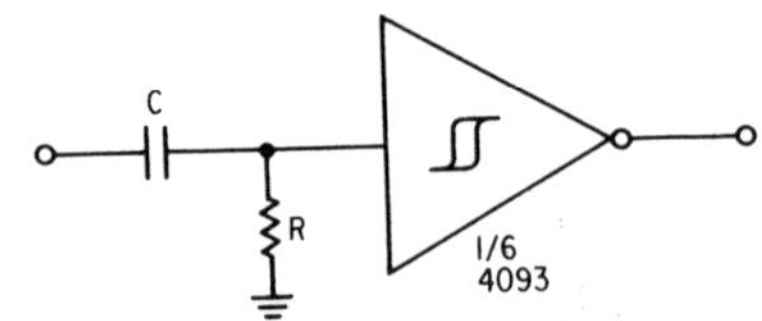

Fig. 3-12 Schmitt trigger monostable circuit.

Figure 3-12 shows the Schmitt trigger 4093 used as a half-monostable. Due to the snap action of the Schmitt trigger, the transition time of the output of the half-monostable is significantly improved.

Experiment 3-3: The Schmitt Trigger and the Half-Monostable

Construct the Schmitt trigger circuit shown in Fig. 3-13. (Remember to connect all unused inputs to appropriate points.) Use the variable resistor to vary

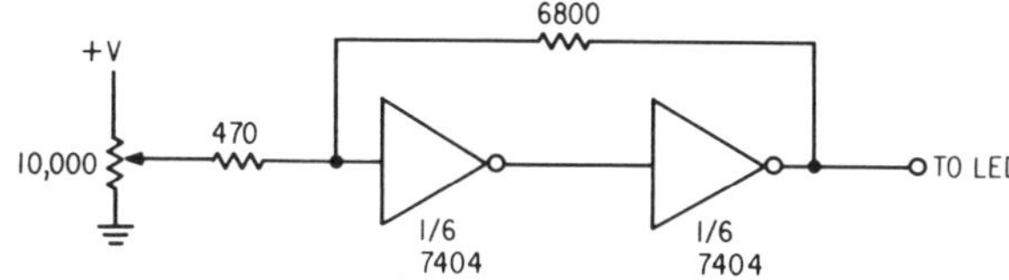

Fig. 3-13 Schmitt trigger test circuit.

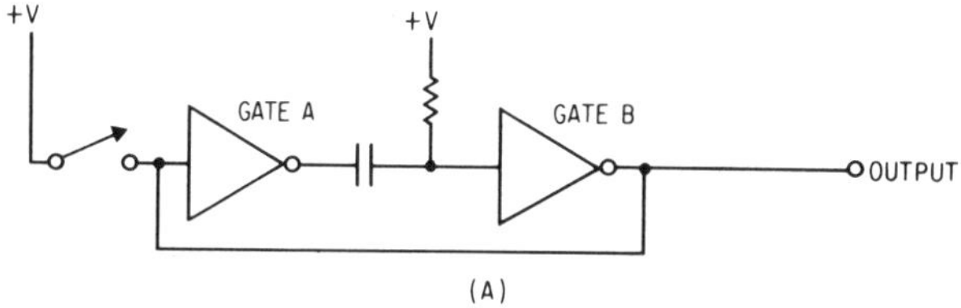

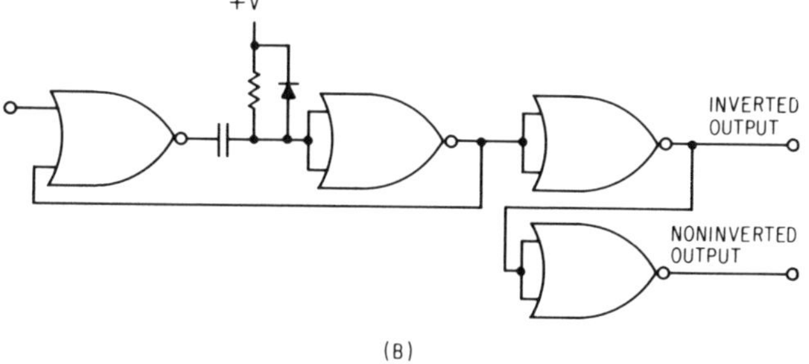

Fig. 3-14 Full monostable circuits: (a) inverter contact conditioner and (b) NOR gate monostable with complementary outputs.

the input voltage to the trigger. Observe the LED indicator and note the voltages at which it turns on and off. The intervening voltage is the hysterisis or dead band of the circuit.

The Monostable Multivibrator

Another way to improve the half-monostable is to add the other half. The *full monostable* uses two gates instead of one. The circuits bear many resemblances to the dual inverter conditioning circuits discussed earlier. The primary difference between the two devices is the RC circuit incorporated into the monostable design. Indeed, the circuit shown in Fig. 3-14a is a contact conditioner. The circuit also assists in the explanation of the monostable circuit action.

Suppose gate "A" has a high output. Gate "B" will therefore have a low output and a high input. Both ends of the capacitor are at a high logic level. The capacitor is neither charged nor charging. This is the stable state of the monostable. If the switch is now depressed, the output of gate A will be brought low. Gate B will flip to a high output that will hold gate A output low when the switch is released. The capacitor now has a voltage difference across it and begins to charge through the resistor. After a period of time determined by the RC time constant, the capacitor will charge to a level adequate to force the output of gate B low. At this time, gate A output will go high and the monostable will be stable again.

The circuit shown in Fig. 3-14b is a monostable constructed from CMOS NOR gates. The circuit action is similar to that of the prior monostable with the added benefit of the dual inputs on the first gate. The diode has been added to assist the diodes in the gate in discharging the capacitor. The circuit may look wasteful with two gate inputs connected in parallel at several places. Actually, the entire circuit is built with one 4001 quad NOR gate. Also, the paralleled inputs improve speed and drive capability. The first inverting buffer gate is used to square up the output pulse from the monostable and provide a complementary output. The second gate is used to invert the complementary output. If complementary output will work in the circuit, the last gate need not be used.

This last monostable can be triggered electronically. The output pulse length will be independent of the input trigger pulse length if the input trigger is shorter than the output pulse. If the input trigger remains in the trigger state for a period of time longer than the length of the output pulse, the monostable will retrigger and the output pulse will continue. The monostable does not lend itself to retriggering. In this case, the retriggering is not entirely satisfactory. Retriggering will only occur at the end of an output pulse. Thus, for the circuit to be retriggered, the trigger pulse must appear after the monostable has timed out. Even after the monostable has timed out, a short period is required for recovery. If this recovery period is ignored, the output pulse will be shorter than expected. If the second trigger pulse arrives before the circuit times out, it will be ignored.

Experiment 3-4: The Monostable Multivibrator

In some circuits, it may be advantageous to construct a monostable from individual gates. More commonly, however, integrated monostables will be used. Several monostables are available as standard packages. TTL monostables include the 74121, 74122, and the 74123. CMOS monostables include the 4047 and the 4528. These monostable packages offer a variety of capabilities. Among the more versatile of these "prefab" monostables is the 74122, which is shown in Fig. 3-15a.

Construct the circuit shown in Fig. 3-15a. Advance the variable resistor to maximum value. Slowly depress and release the switch. Notice the output conditions as indicated by the LED state indicators. Did the output change in response to a low-to-high transition or to a high-to-low transition? Note the length of time that the LED connected to the Q output remains lit. Depress the switch rapidly several times in a row. Did the Q output remain high longer than

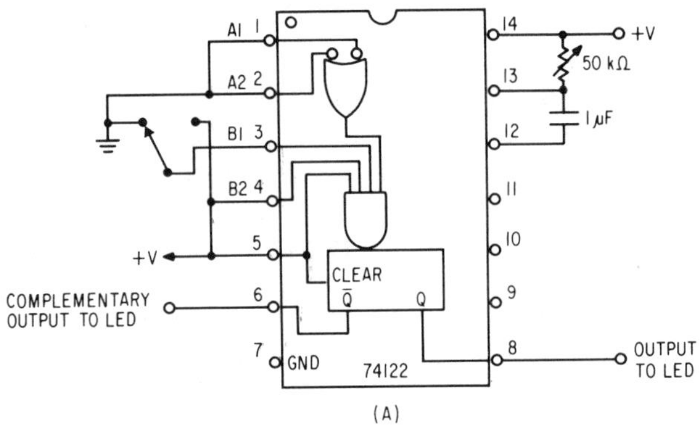

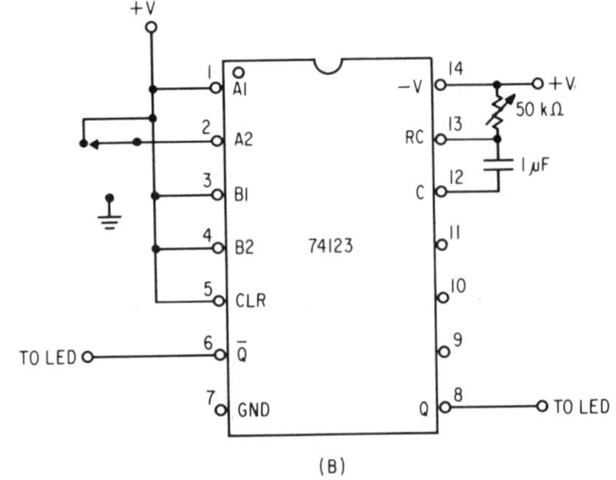

Fig. 3-15 TTL monostable: (a) 74122 wired for monostable operation and (b) 74123 wired for monostable operation.

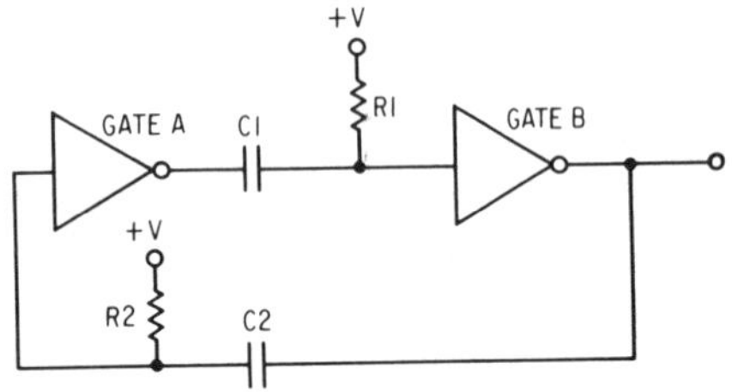

Fig. 3-16 Astable multivibrator.

before? This illustrates the retriggerability of the 74122. This is a less limited retriggerability in that the total output time begins again with each new trigger even if the current unstable cycle has not been completed.

Construct the circuit shown in Fig. 3-15b. This circuit is the same as the prior one except that the input responds to a high-to-low transition. Depress the switch and determine that the unit activates on the high-to-low transition of the trigger. Rapid operation of the switch will confirm that retriggerability also is possible in this configuration. Reduce the variable resistor to a lower resistance and note the length of the output pulse. As the resistance is decreased, the length of the output pulse will be observed to shorten. If the RC time constant is made short enough, the output pulse will occur too quickly for the LED to respond.

The Astable Multivibrator

The bistable multivibrator had two stable states. The monostable has only one. A third type of multivibrator has no stable states. This circuit is called the *astable multivibrator*. The astable differs from the other multivibrators in a significant way. The astable requires no triggering. It is *free running*, which means that the device produces a continuous stream of pulses without outside direction to do so. The astable provides for true system synchronization. Pulses can be generated by the astable and used to clock all other circuits in the system. The astable is prevalent in simple as well as complex circuits.

The circuit of a simple astable is shown in Fig. 3-16. The circuit is essentially the basic monostable with an additional RC circuit. The second RC circuit is inserted in the feedback line between the output of gate 2 and the input of gate 1. This second RC network prevents the multivibrator from having a stable state. If the circuit is in the condition where gate A has a high output, gate B will have a low output. With a low output for gate B, capacitor C2 begins to charge through R2. When C2 charges sufficiently, it will force gate A to change state causing gate A output to go low. Gate B output will now be forced high by the negative voltage of C1, and C1 will be charging through R1. When C1 charges sufficiently, it will cause gate B to change states. C2 has been charging during the high output state of gate B. When gate B goes low, C2 places a negative voltage on gate A, forcing the output positive. C2 begins to charge through R2 again. This process will continue as long as power is applied. There is no stable state.

Figure 3-17 shows an astable constructed from NOR gates. The inputs are tied together so that the stages act as simple inverters. In this circuit, only one RC circuit is used. If gate A is in the condition that provides a high output, gate B will have a low output. R1 and R2 couple a feedback voltage to capacitor C, which begins to charge and forces the input of gate B low. In a length of time determined by the time constant of the RC network, C reaches a level low enough to force gate B to begin to change

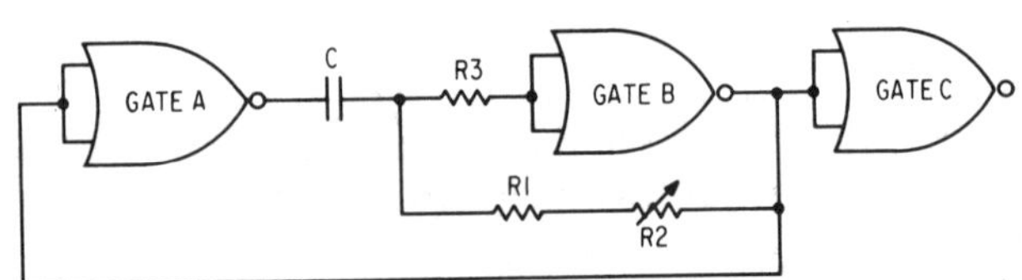

Fig. 3-17 NOR gate astable circuit.

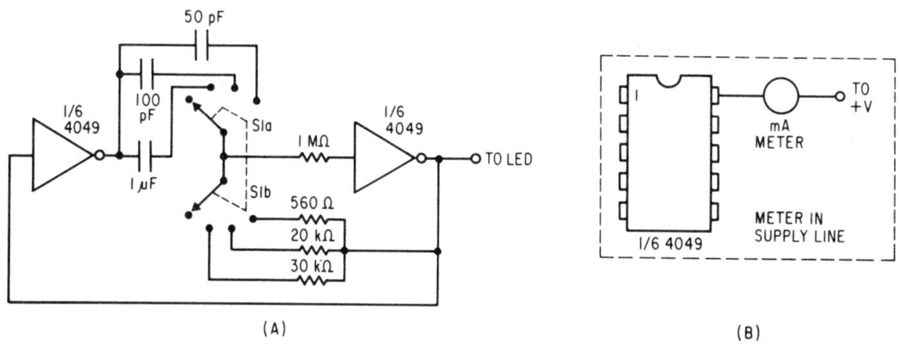

Fig. 3-18 Current test circuit.

state. When it does, gate A is forced to change state, and its output swings low. The charge of C adds to the input level of gate A and causes gate B to rapidly complete its state change. Gate A now has a low output and the resistors charge C in the positive direction. When C becomes positive enough, it forces gate B to begin reverting to a low output. This forces gate A to its alternate state. With gate A output high, positively charged C adds to gate A output and forces a rapid completion of gate B's state change. The initial condition has been restored and the process continues.

Several refinements are used in this circuit. First, R2 is made variable. It can be used to vary the frequency of the astable's operation. R1 sets the maximum frequency for the circuit. R3 is used to prevent the input protection diodes of gate A from interfering with the circuit action. Without this resistor, the operating frequency of the astable would be influenced by the supply voltage. R3 should be at least ten times the combined resistance of R1 and R2. R3 will not enter into the circuit timing calculation.

Gate C is used as an inverting buffer. This extra gate provides increased drive for the astable while ensuring that the output pulse will have rapid rise and fall times and a square output pulse.

Experiment 3-5: CMOS Current vs. Frequency

The current demands of CMOS circuits increase as frequency increases. Higher frequency astables require construction using higher current devices

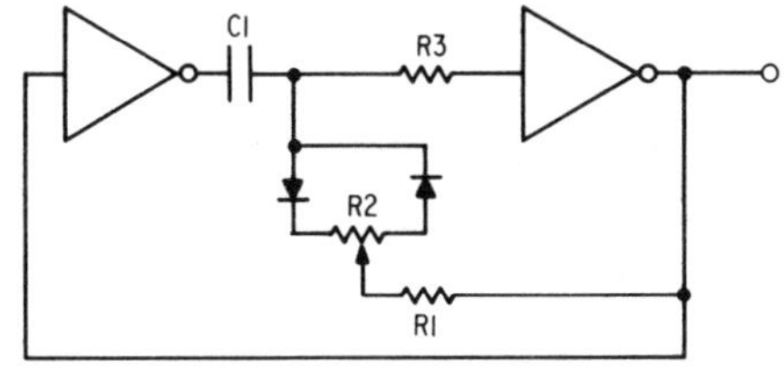

Fig. 3-19 Circuit used to improve symmetry of output pulse.

such as the 4049 or the 4502. Construct the circuit shown in Fig. 3-18. Wire a milliammeter in the positive supply line, as shown in Fig. 3-18b. Place switch S1 in each of the four positions. Record the current drawn by the package at each position. The current increases as frequency increases. At high frequencies, CMOS will draw as much current as TTL. Notice that idle current as measured at switch position 1 is almost negligible.

Symmetry

The astable circuits previously discussed have no adjustment for signal symmetry. Obtaining 50-50 symmetry without additional circuits is very difficult. The use of a frequency divider to obtain 50-50 symmetry will be covered in the following chapter. A modification to the basic astable that will achieve a fair measure of symmetry is shown in Fig. 3-19. The primary change to the circuit is the addition of diodes D1 and D2. These diodes select the portion of the variable resistor that will be incorporated into the circuit during positive and negative charging cycles. By varying the resistor (R2), the duty cycle of the astable can be altered. If a 50-50 duty cycle is desired, the resistor can be adjusted until the output on and off times are equal. The frequency of the astable will be set by R1 and R2 in combination with C1.

This circuit also can be used to produce a nonsymmetrical pulse train. By adjusting R2 more toward one direction than the other, the duty cycle will be skewed. A typical pulse train might be that shown in Fig. 3-20, with short pulses occurring at relatively long intervals.

The IC Timer

While astable and monostable circuits can be constructed using standard gates, the production of timer ICs has reduced the popularity of "home brew" multivibrators. One of the first (and still popular) IC timers is the NE555, which is manufactured by Signetics. The 555 was designed to be a versatile timer with both monostable and astable

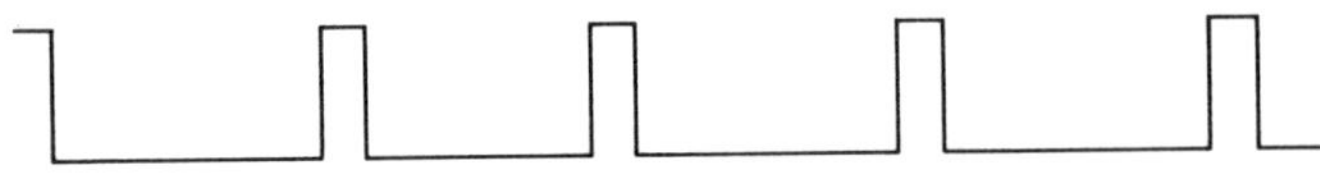

Fig. 3-20 Typical pulse train.

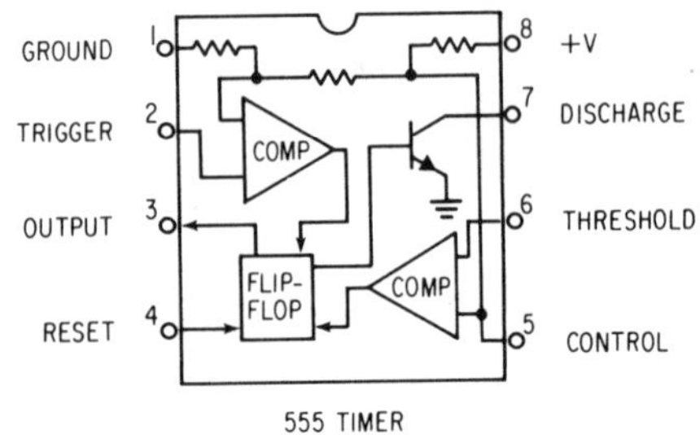

Fig. 3-21 Functional outline drawing of the NE555.

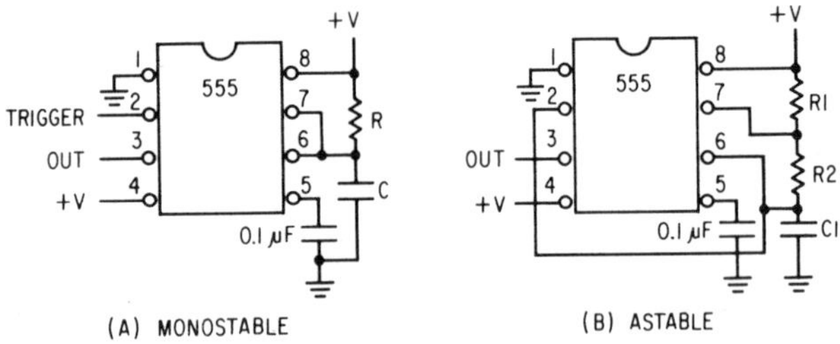

Fig. 3-22 The 555: (a) used as a monostable and (b) used as an astable.

capability and a large frequency range. The 555 has six operational connections, which are shown in Fig. 3-21. These connections and various external components are interconnected to obtain the desired mode of operation. Figure 3-22a shows the 555 connected for operation as a monostable. A differentiator is required when the trigger pulse remains low longer than the duration of the output pulse. Since the input trigger is level sensitive without the differentiator, the monostable would erroneously retrigger at the end of the output pulse.

The circuit action of the 555 is based on two voltage comparators. Internal to the 555 is a voltage divider that divides the input voltage into three equal parts. The negative side of the upper comparator is connected internally to the divider at a point that is two-thirds of the supply voltage. The negative side of the lower comparator is connected to the divider at a point that is one-third of the supply voltage. When a low trigger is applied to the trigger input, the output goes high and the capacitor C1 begins to charge. C1 will charge to a level equal to two-thirds of the supply voltage in a length of time determined by the RC time constant. When C1 reaches this level, the upper comparator will reset the flip-flop. This is the stable state of the monostable, and the circuit will not change until another trigger pulse is received.

Astable operation of the 555 is similar to monostable operation except that the trigger is produced by the circuit. Figure 3-22b illustrates the astable circuit of the 555. In the case of the astable, two resistors are used in the RC circuit, and the threshold is connected with the trigger rather than with the discharge line. In operation, the flip-flop changes state each time the charge on C1 rises above two-thirds of the supply or falls below one-third of the supply. This process is continuous since there is no stable state.

Symmetry of the 555 astable output is determined by the ratio of R1 to R2. If R2 is made large with respect to R1, symmetry is improved. Achieving perfect symmetry, however, requires the addition of a flip-flop at the output of the timer. (R1 should not be less than 1K and the combined resistance of R1 and R2 must not exceed 20 M for a 15-V supply and less for lower voltages.) The frequency of the 555 can range from microseconds to hours. The low frequency operation is limited only by the leakage of the timing capacitor.[3] For shorter times, standard TTL and CMOS multivibrator chips should be used.

Programmable Timers

The 555 is among the most popular timers, but other timers have been developed. The programmable timers are particularly interesting. The block diagram of the binary XR-2240 Programmable Timer/Counter (manufactured by Exar Integrated Systems) is shown in Fig. 3-23. Notice that the first part of the circuit is very similar to the 555. The comparators and flip-flops are the timer circuitry for the 2240. An RC network is attached to pin 13 and the timing cycle follows the 555 pattern of charging and discharging the capacitor while comparing the capacitor voltage level to the level of the internal voltage divider. The timer circuit operates as an astable at all times even though the 2240 as a whole can be operated as either an astable or a monostable. There are several apsects of the 2240 timer section that are different from the 555. First, the timer output is asymmetrical. Since this output is counted by the counter section of the 2240, asymmetry causes no problem. Also notice that there are no separate threshold, trigger, and discharge pin-outs. These are internally interconnected and are brought out to the RC pin-out. This results in a loss of symmetry control. The modulation pin-out (pin 12) connects to the upper comparator via the voltage divider. This permits external synchronization or time-base period control.

[3]The 555 data sheet contains formulas and charts that will assist the designer in choosing RC values.

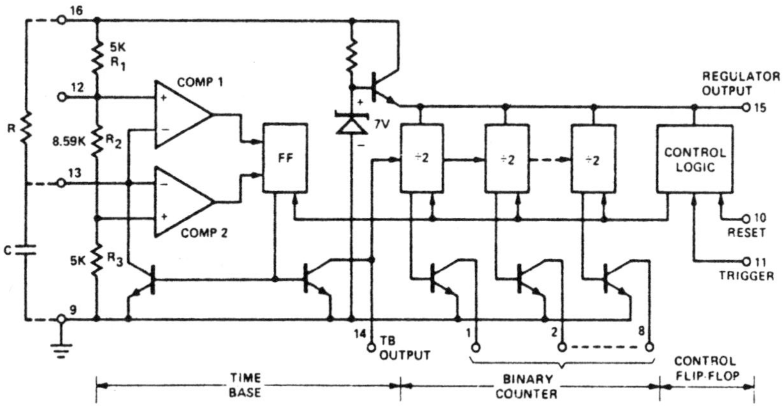

Fig. 3-23 Block diagram of XR-2240 Programmable Timer/Counter. (Courtesy of Exar Integrated Systems, Inc.)

One other aspect of the 2240 that is different from the 555 is the trigger input. The trigger is used to reset the counter and initiate the astable operation. This is required since the 2240 assumes the reset state (astable not running) when first powered up. A trigger is required to start the astable unless the trigger pin-out (pin 11) is connected to the regulator pin-out (pin 15). If this connection is made, the timing cycle will begin automatically.

The counter portion of the 2240 is designed to divide the astable output in binary fashion. Eight divide-by-2 counters are included in the 2240 and can be programmed for any time division from one astable time period to a length equal to 255 astable periods. Figure 3-24 shows the connections for monostable and astable operation. Programming the counter section is accomplished by placing jumpers

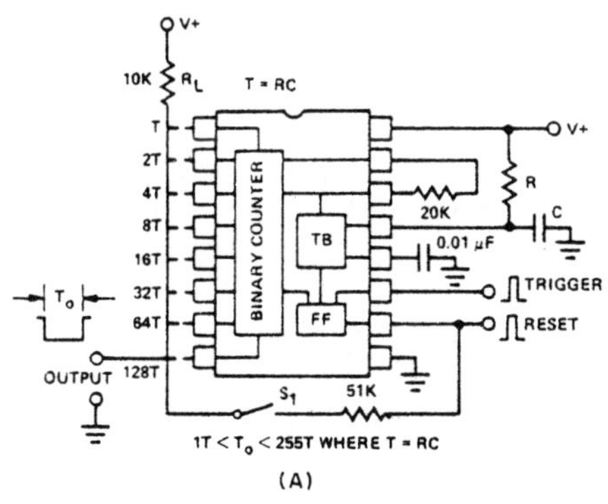

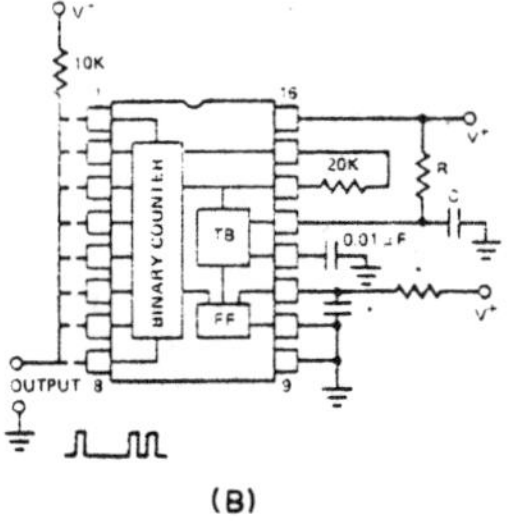

Fig. 3-24 The XR-2240: (a) wired for monostable operation and (b) wired for astable operation.

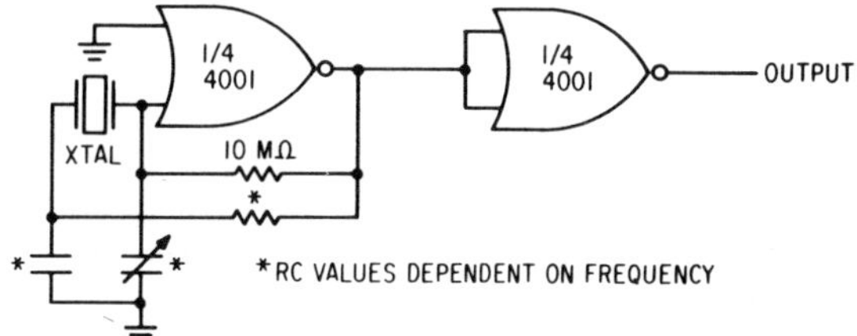

Fig. 3-25 The 4001 used as a crystal oscillator.

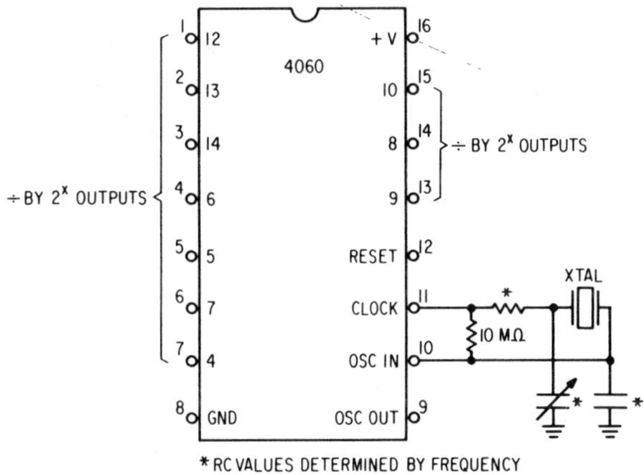

Fig. 3-26 The 4060 used as a binary ripple counter.

or making switches between the appropriate counter pin-outs and the output line, as shown in Fig. 3-24. Remember that the counter programming follows a binary pattern.

Other programmable timers include the 2250, 8240, 8250, and 8260. All of these timers are basically the same except that the counting circuit is designed for different timing schemes. The XR-2240 and the 8240 use binary counting. The XR-2250 and the 8250 count according to a BCD scheme. The 8260 provides counting in seconds, minutes, and hours. These counters can be interconnected for time delays up to several years in length. Refer to the data sheets for additional design information.

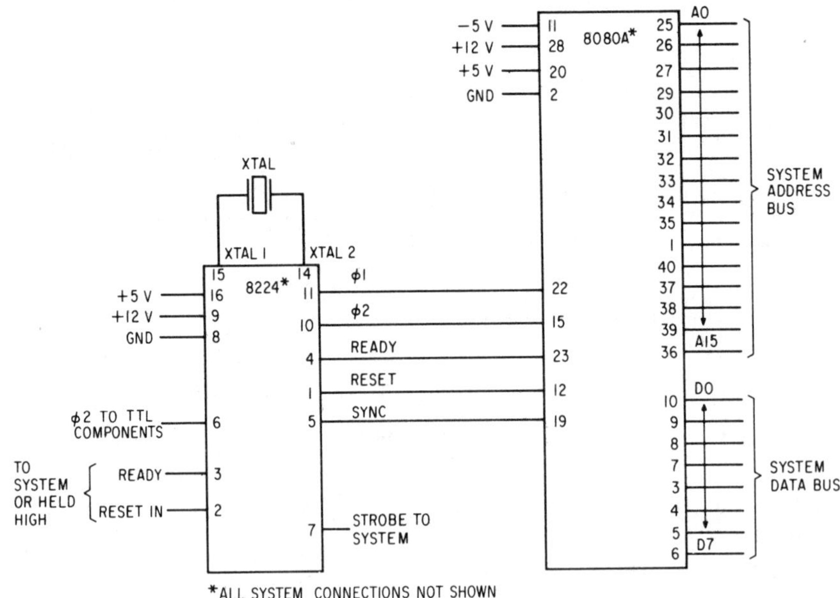

Fig. 3-27 An 8224/8080 typical circuit.

Crystal-Controlled Timers

The timers considered up to this point have depended on RC networks for determination of the frequency of their operation. Many applications require greater accuracy than simple RC circuits can provide. A bistable multivibrator that was synchronized by the power line frequency was discussed earlier. This circuit will provide a relatively stable and accurate output frequency. However, the 60-Hz signal may not be the preferred choice for all uses. To provide for a more versatile frequency standard, piezoelectric crystals may be used. Due to the piezoelectric effect, which says that a piece of crystalline material will resonate at a definite frequency as determined by its size and manufacture, a crystal can be used to lock an oscillator to a specific frequency. Figure 3-25 shows the circuit of a crystal-controlled oscillator made from 4001 CMOS NOR

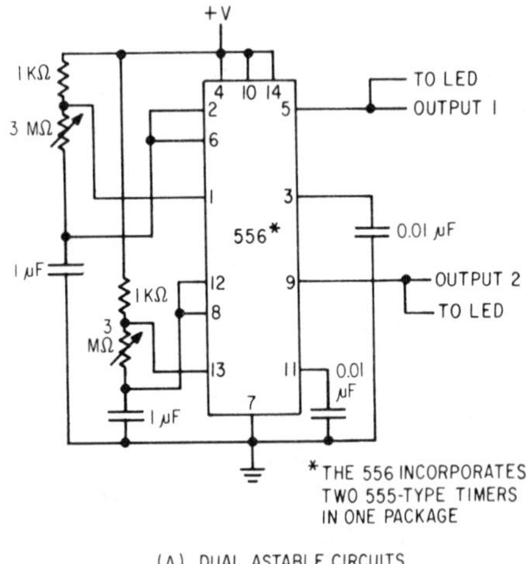

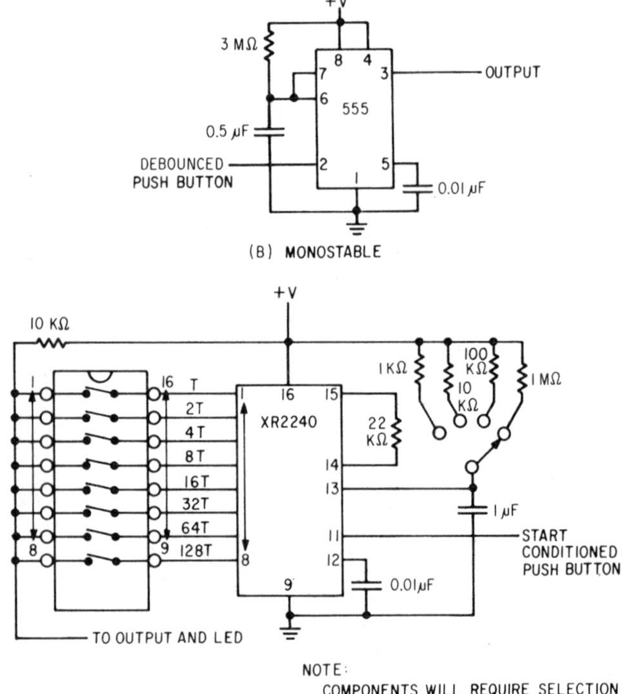

Fig. 3-28 Smitty timer circuits.

gates. The first gate is biased into its active region and oscillates at the crystal frequency. The second gate acts as a buffer. If a TTL load is to be driven, both inputs must be connected in parallel to provide adequate drive.

Figure 3-26 shows a crystal oscillator that uses the 4060 binary ripple counter. This unit has a built-in oscillator that is divided by the counter circuitry. The crystal is used with the built-in oscillator to ensure an accurate and stable frequency. The outputs provide divisions by powers of two; i.e., $2^4 = 16$, $2^5 = 32$, and so on. No outputs are provided for division by 2, 4, 8, and 2048.

The Clock Generator and Driver

Special timing circuits are required for microprocessors. The 8224 that is used with the 8080A microprocessor is an example. The 8224 provides for the specific needs of the 8080A. It is crystal controlled using crystals of various frequencies up to approximately 11 MHz. The frequency is chosen by the designer according to the speed at which he wishes to run his processor. (Microprocessor design will be covered in Chaps. 6 and 7.) Overtone crystals can be used with the 8224 if a tuned tank circuit is connected to pin 13.

The 8224 provides two outputs for the 8080A at the pin-outs labeled "Φ1" and "Φ2" (pins 10 and 11). Figure 3-27 shows the 8224 in a typical 8080A circuit. The reset, ready, sync, and status strobe interface with the microprocessor system to ensure that system timing is coordinated for the microprocessor operations. Among the functions provided for are power-up reset for the 8080A, ready synchronizing flip-flop, advanced status strobe, and oscillator output for the timing of external circuits.

Experiment 3-6: Timers for the Smitty

The Smitty breadboard includes three built-in timers. The 555s are used for a variable frequency astable and a variable pulse width monostable. An XR-2240 is used as a switch-programmable astable. The circuits of the timers are shown in Fig. 3-28.

The astable uses the circuit described earlier for the 555. With the components shown, the output frequency is variable from approximately 1 Hz to 5 kHz. Symmetry suffers toward the higher end of the frequency range.

The monostable circuit is similar to that discussed earlier. The pulse width is variable from 1 μsec to 100 μsec.

The XR-2240 uses the astable circuit previously discussed. A rotary switch is used to select RC components to provide oscillator frequencies of 1 Hz, 10 Hz, 100 Hz, and 1 kHz. An eight-switch DIP switch is used to program the counter section. The range on the XR-2240 is 1 Hz to 1 pulse every 128 sec.

FOUR

Clocked Logic

Clocked Logic

In Chap. 3, it was found that the major differences between synchronous and asynchronous logic were two things: timing and memory. This chapter deals with the logic blocks that depend on timing or clocking for their operation. Included among these devices are clocked flip-flops, shift registers, and counting circuits. These clocked logic circuits are at the heart of all sophisticated logic designs. Clocked or synchronized logic is required for complicated tasks to be accomplished in the step-by-step fashion characteristic of computer and microprocessor systems. Indeed, clocked logic brings a whole new dimension and power to the gating circuits in Chap. 2. The maximum benefits of synchronous and asynchronous logic are realized only when these two elemental logic forms are combined into synergistic[1] systems that accomplish complex tasks one logical decision at a time.

Clocks

The use of the word *clock* in this instance is similar to the use of the word *trigger* in the previous chapter. The difference between trigger and clock has to do primarily with repetition. A *trigger* is a signal that initiates a logic action. This signal may be repeated at a fixed rate, at random intervals, or it may occur only once. The usual connotation associated with trigger is a nonrepetitive signal.

A *clock*, on the other hand, conveys the idea of a repetitious signal. A clock is usually thought of as a timing signal that repeats at some predetermined rate. A clock signal is often a string of trigger pulses that may be either square waves or pulses derived from analog signals. In either case, the clock pulses occur in a repeated sequence rather than randomly as is true of trigger signals. In general, this is the distinction implied by the use of trigger and clock signals in this volume.

There are two types of clocking that are used in logic circuits. The first is *level clocking* and the second is *edge clocking*. In level-clocking circuits, the trigger results from the level or status of the clock signal. Depending on the circuit, a clock signal will be interpreted as a trigger based on whether the clock is low or high. As will be seen, the clocked logic block is either caused to initiate its logic function by the clock or it is enabled to respond in some way to a separate data input. Level-clocking circuits have some restrictions. The circuit remains clocked for the duration of the clocking pulse. If the clock does not revert to its original state prior to the completion of a logic clock cycle or if the input data changes more than once during the clocking pulse, internal, erroneous circuit action may result. This operational idiosyncrasy requires that the input data change *only once, immediately after the clocking pulse assumes its clocking level.*

Edge clocking does not suffer from the same problem. Edge-clocked logic interprets only the change from one level to the other as a valid clock pulse. This circuit is not level dependent. It responds to the transition only. If it responds to a negative-going trigger, it is referred to as *negative-edge clocking*. If the trigger results from a rising trigger, it is called a *positive-edge clock*. With edge clocking, the data can be changed at any time since it will be assimilated by the circuit only when the appropriate clock edge is received. As determined by the application, appropriate timing circuits such as those described in Chap. 3 can serve adequately for generation of clock pulses.

Level-Clocked Flip-Flop

The R-S flip-flop discussed in Chap. 3 is a form of memory element. It does not, however, include provisions for timing capability. Therefore, when the trigger is applied to the appropriate input, the output immediately changes. This characteristic of the R-S flip-flop limits its use. Without some method of clocking, the circuit cannot be made to hold an output while the input changes.

[1]Refers to technological loop in which one logic element supports and works cooperatively with other logic elements to form a sophisticated device capable of actions impossible without synergism.

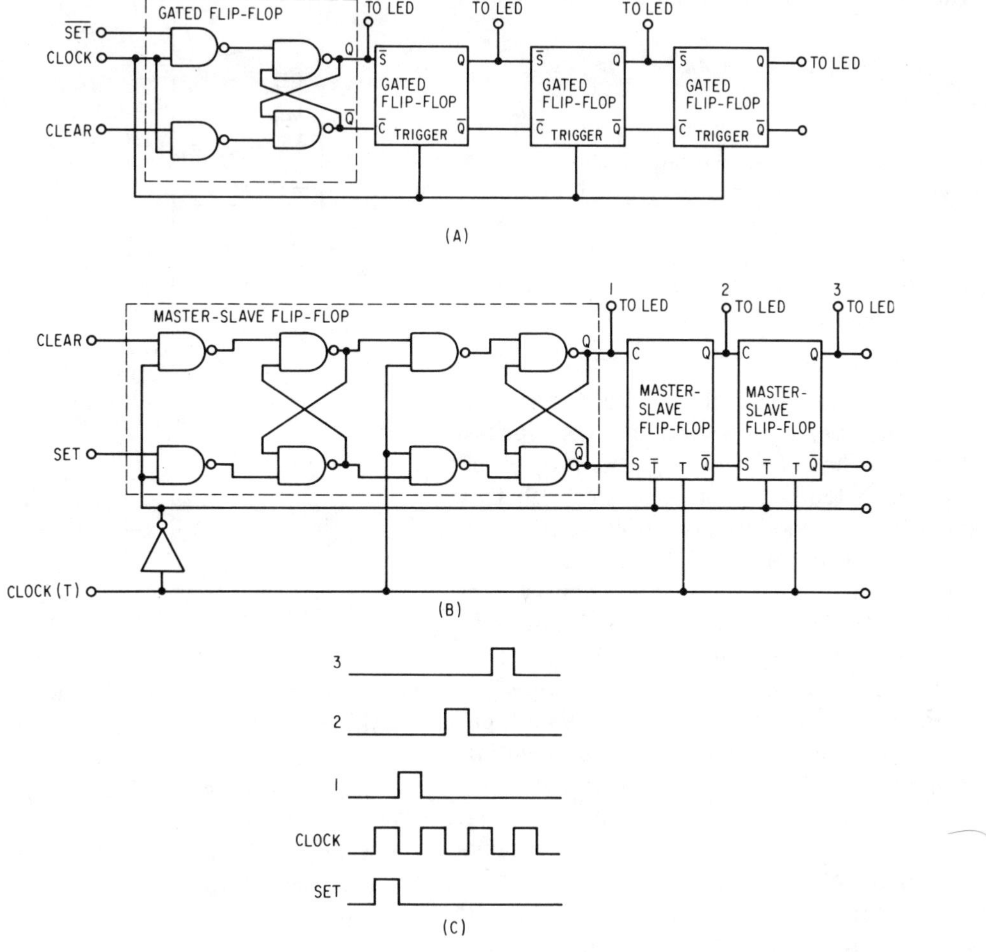

Fig. 4-1 Circuits for serial-to-parallel conversion: (a) proposed, (b) corrected, and (c) pulse diagram for corrected circuit.

The following design problem illustrates the inadequacy of nonclocked R-S flip-flops. Suppose it is desired to convert serial data into parallel data. *Serial* means data bits in a string, one following the other (in time, on a single line). *Parallel* means data bits assembled into groups (on separate lines) that move in a side-by-side fashion. Most microprocessors accept parallel data. Many peripherals such as tape decks and teletypes use serial data. To convert one form of data to another, a string of R-S flip-flops with gating circuits might be used. The circuit shown in Fig. 4-1a might be prepared to convert the serial-to-parallel data. A clock pulse placed on the gate input will enable the flip-flop. If the serial data string presents a high to flip-flop 1, a high output will result. This flip-flop will now trigger the next flip-flop, and it the next, and so on. All of these changes will occur during one clock pulse and will pass one piece of data through all of the flip-flops. This is not the action intended, however. The action needed is for one piece of data at a time to be shifted down the flip-flop chain so that the serial data string will be available as parallel data at each flip-flop output.

To obtain the desired circuit action, twice as many flip-flops must be used. An inverter also must be inserted in the clock line feeding the gates of succeeding flip-flops (see Fig. 4-1b). The result is that when one flip-flop is enabled, the next flip-flop is disabled. When the clock of the first flip-flop returns to its original state, the complement is applied to flip-flop 2. Flip-flop 1 is now off and flip-flop 2 is on. The data has passed from the first flip-flop to the second, and at the next clock the data bit will enter the third. The result of this process is that the input data is entered into the first flip-flop at the first clock pulse. It then enters the second flip-flop when the clock goes high. At the next clock, flip-flop 1 is free to enter the next data bit, remaining latched for a high or changing for a low. At the same time, flip-flop 3 will accept data from flip-flop 2 and when the clock returns high the second time, flip-flops 2 and 4 will be loaded from flip-flops 1 and 3 respectively. This process will continue throughout the length of the flip-flop chain.

A pattern can be seen emerging in the circuit shown in Fig. 4-1b. This pattern is shown in Fig. 4-1c. Each two flip-flops form identical circuit elements.

The characteristics of these elements are as follows:

1. The first flip-flop is loaded at the clock with input data.
2. The second flip-flop is loaded by the first flip-flop when the clock signal disappears.
3. The two flip-flops will not be loaded simultaneously.

These characteristics describe the operation of a *master-slave flip-flop*.

The J-K Flip-Flop

As is true of many useful circuits, the master-slave flip-flop has been integrated. In other words, a special logic chip has been developed that combines simple gates constructed from the same substrate into circuits that function as master-slave flip-flops. These circuits present a few refinements over the basic master-slave flip-flop, but the circuitry is essentially the same. Refer to Appendix C for the functional diagram of the 7476 dual J-K flip-flop. The J and K data inputs give the device its name. The choice of the designations "J" and "K" was apparently an arbitrary one. The *J-K flip-flop* is essentially a master-slave flip-flop with some additional inputs. The two added inputs are preset and preclear. These inputs force the outputs into predetermined states regardless of the inputs or the clock. When the preclear input is brought low, the output (Q) immediately goes low regardless of the input on the J and K inputs. If the preset is brought low, Q immediately goes high. Preset and preclear are called direct inputs. They impose output conditions regardless of the input and are not dependent on clocking. Simultaneous grounding of preclear and preset results in a disallowed state with both Q and $\overline{Q}$ high and should be avoided.

J	K	SET	CLR	CLK	Q (After Clock*)	$\overline{Q}$ (After Clock*)
0	0	1	1		NC	NC
0	1	1	1		0	1
1	0	1	1		1	0
1	1	1	1		Binarily Divides	
X	X	0	1	X	1	0
X	X	1	0	X	0	1
X	X	0	0	X	Disallowed State	

*Clock signal is ⎍
X = Don't care
NC = No Change

Fig. 4-2 Truth table for the 7476 J-K flip-flop.

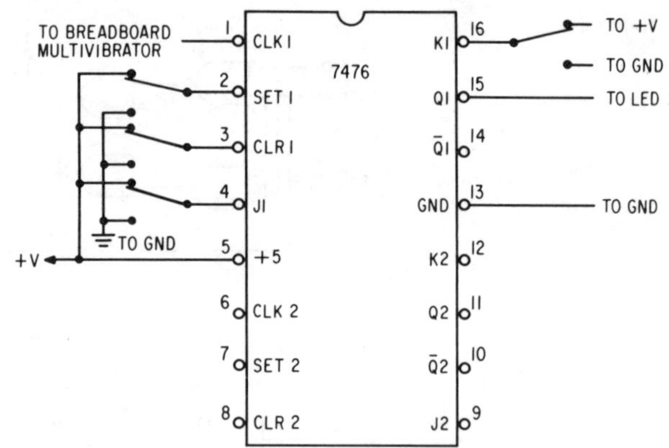

Fig. 4-3 Test circuit 7476.

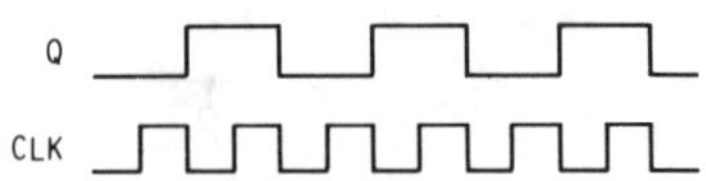

Fig. 4-4 Comparison of input/output pulse trains.

Normal operation of the J-K flip-flop sets the output in response to the data on the J-K inputs. Setting occurs only when the clock pulse arrives. The truth table shown in Fig. 4-2 describes the interrelated actions of the J-K flip-flop inputs. Notice that the output changes only at clocking and then only if the appropriate conditions are present on the J-K inputs.

Experiment 4-1: The J-K Flip-Flop

Install a 7476 dual J-K flip-flop in the breadboard. Set the multivibrator of the breadboard for a slow pulse rate (one easily counted) and connect as shown in Fig. 4-3. Vary the inputs in accordance with the truth table in Fig. 4-2. Confirm that the outputs are as indicated.

Now, connect both the J and the K inputs to + 5 V. Note the Q output pulse frequency as compared with the input pulse frequency. Figure 4-4 shows the two pulse trains aligned to illustrate their synchronism. Notice that the Q output changes each time the clock goes low. The result is an output pulse that is one-half the frequency of the input. The J-K flip-flop is functioning as a binary divider. In this configuration, it is sometimes called a "T" flip-flop.

Connect the second flip-flop in the 7476, as shown in Fig. 4-5a. Verify the output pulse train shown in Fig. 4-5b. What is the result of placing two binary dividers back to back? Several of these circuits connected in a string would function as a binary counter. In the circuit shown in Fig. 4-5a, add the binary values of the Q outputs, if each high is given a binary equivalent equal to its position as shown. Reference to the timing diagrams of Figs. 4-4

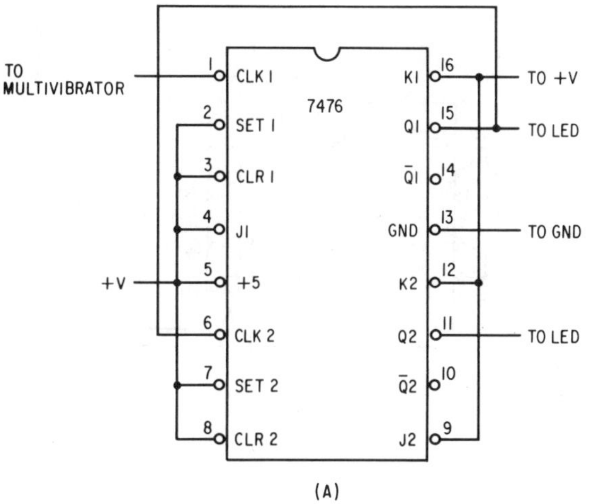

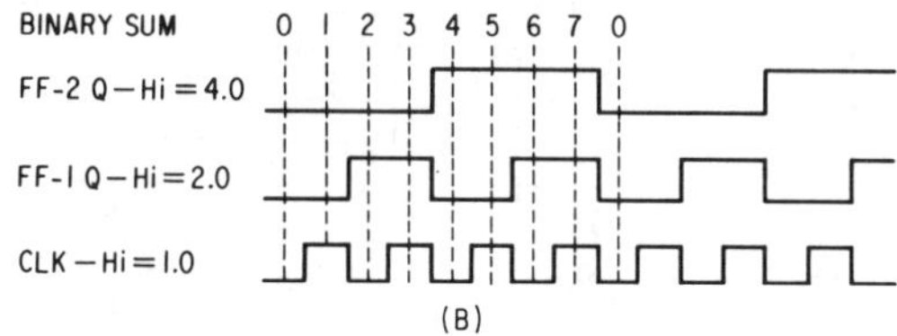

Fig. 4-5 The 7476: (a) used as a binary divider and (b) pulse train.

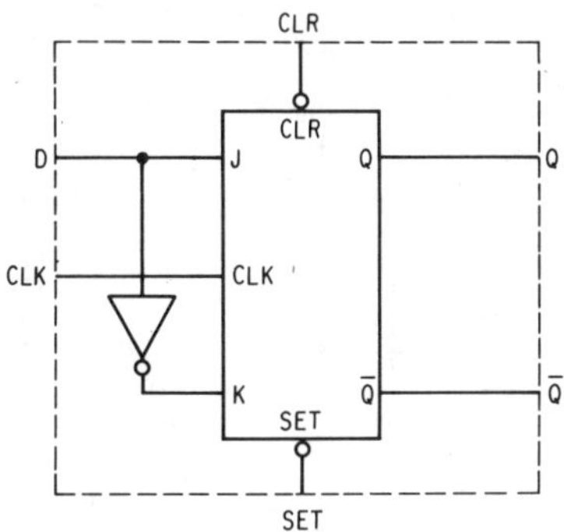

Fig. 4-6 D flip-flop functional diagram.

and 4-5b will help. The concept of counters will be expanded later in this chapter.

The D-Type Flip-Flop

The J-K flip-flop has a few idiosyncrasies that limit its usefulness in some circuits. For one thing, there are two data inputs, J and K. Unless the circuit action requires dual inputs, the additional circuitry required to "program" the J-K inputs may reduce the economy of the design. Also, the inputs to the J-K flip-flop must be changed immediately after clocking, and then they can change only once. This requirement is necessitated because the J-K flip-flop is level clocked. The nature of the clock pulse is important. Ideally, the clock should have rise and fall times of 5 μs or less. Problems result when the clock rise and fall times are very slow. If the clock moves slowly to the low level, the J-K flip-flop will be biased into its active region and may behave erratically and will be more susceptible to noise.

The type D *flip-flop* improves on most of these problems. As shown in Fig. 4-6, the type D is nothing more than a J-K with an inverter from J to K. This means that J and K will always be complementary. Also, only one input will be used and this input will be called the "D" or "data" input. The most obvious effect of this type D circuit change will be that the data states will be eliminated—the "something always happens" (i.e., highs on both inputs) and the "nothing ever happens" (i.e., lows on both inputs). The only remaining input possibilities will be the two complemented input conditions; i.e., J = 1; K = 0; and J = 0; K = 1. These two conditions will be determined by a single input such that J will be high and K low if D is high. Similarly, K will be high and J low if D is low. The D-type flip-flop will change states when the clock goes high if the D input is at a level that requires a change. Thus, if the Q output is high, it will change at clocking if the D input is low. If the Q output matches the D input, the output will not change.

Preset and clear function as direct inputs in the same way they did with the J-K flip-flop. They must both be held high for normal clocked operation of the flip-flop. Simultaneous lows on the preset and preclear inputs create a disallowed state.

Clocking of the D-type flip-flop is edge triggered. Clocking occurs only during the time of transition from a low level to a high level. This permits the input to be changed at any time, except during the brief rise time of the clock. At all other times, the flip-flop will disregard the input level.

The D-type flip-flop removes the limitations stated for the J-K flip-flop. It requires only one input; it can change the input at any time except the instant of clocking and it responds to the clock transition not its level.

Experiment 4-2: The D-Type Flip-Flop

Construct the circuit shown in Fig. 4-7a. Monitor the Q and Q outputs while varying the inputs. Complete the truth table shown in Fig. 4-7b. Notice that the number of input options have been reduced from that of the J-K flip-flop. This indicates that the improvements that the D-type flip-flop made to the J-K flip-flop were obtained at the expense of versatility. The choice of flip-flop type must be based on the circuit requirement.

Construct the circuits shown in Figs. 4-8a and 4-8b. Draw in the pulse train at each flip-flop output. A

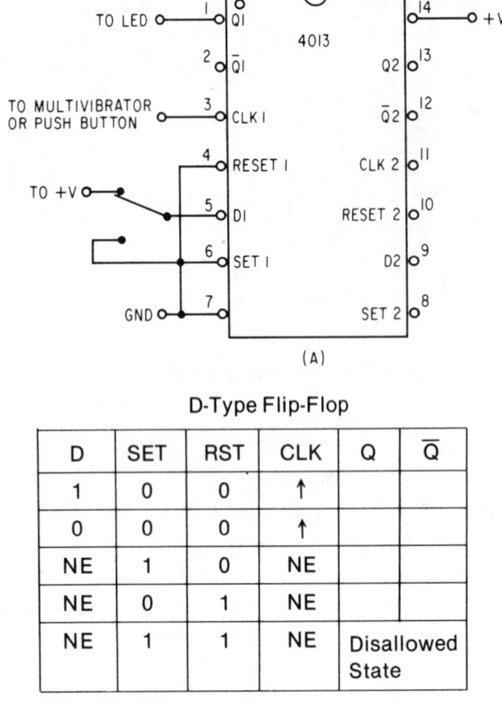

D-Type Flip-Flop

D	SET	RST	CLK	Q	$\overline{Q}$
1	0	0	↑		
0	0	0	↑		
NE	1	0	NE		
NE	0	1	NE		
NE	1	1	NE	Disallowed State	

NE = No Effect

(B)

Fig. 4-7 The (a) 4013 counter circuits and (b) truth table.

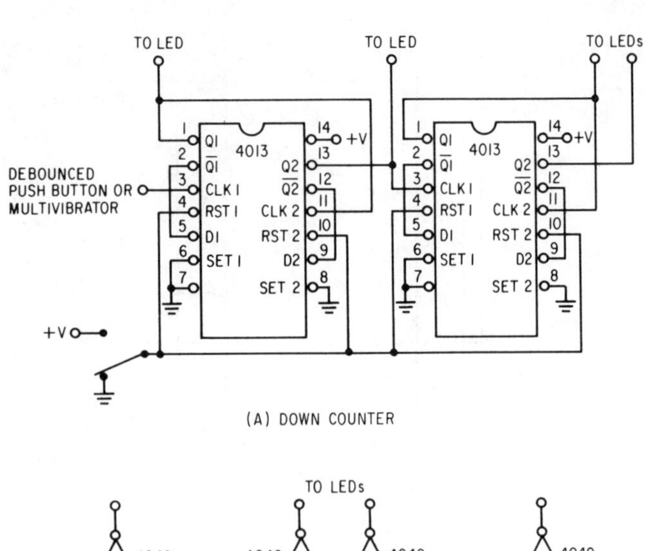

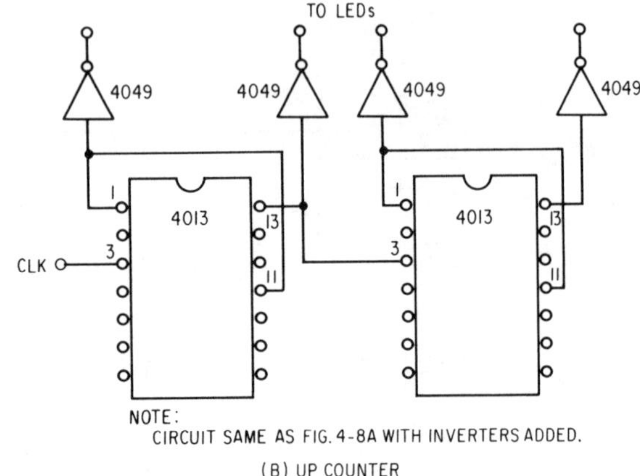

Fig. 4-8 The 4013 counter circuits: (a) down counter and (b) up counter.

debounced push button can be used for the clock if desired. This will make recording of the state changes easier as long as you maintain accurate count of the input clocking. At the input clock transition from low to high, record the binary equivalent of the flip-flop outputs giving a low the value of "0" and a high the value of "1." Be sure the flip-flops begin in the condition in which all Q outputs are low by bringing the reset inputs high briefly. The results of this experiment show the actions of a binary "up" counter and a binary "down" counter.

Ripple Counters and Propagation Delay

The counting systems constructed in the preceding experiment are examples of the simplest counters. They are called *binary ripple counters*. As discussed previously, all logic circuitry is characterized by two mutually exclusive output states: high or low. This circuit characteristic is compatible with the binary numbering system. Logic circuits can represent binary numbers without conversion being required. Therefore, the simplest devices to construct using IC logic are those that are based on the binary numbering system. It also has been demonstrated that asynchronous gating circuits operate more simply than synchronous circuits. No clocking is required, and logic events are permitted to take place immediately.

Ripple counters are transitional between asynchronous gates and fully synchronized devices. The ripple counter has outside clocking applied to the first stage only. Succeeding states are then clocked by preceding stages. Ripple counters can be constructed to provide for division of very large numbers. Each stage will increase the division by a power of two. The problem with this procedure is the time delays that result.

In circuits that operate at relatively slow rates, the ripple counter is adequate. At faster speeds or longer counts, however, the counter may fail to include successive counts due to propagation speeds. Suppose a ripple counter consisting of eight flip-flops is operating at a clocking rate of 20 MHz (see Fig. 4-9). If the counter is at count 127 and another input count is received, at clocking, the seven highs will go low one after the other as the count ripples through the counter and will finally result in bit 8 going high. The time required for this to happen will be the sum of the propagation delays caused by the flip-flops. According to the data sheet for the 7474, 17 ns is typically required for a low-to-high state change at clocking and 20 ns for a high to low state

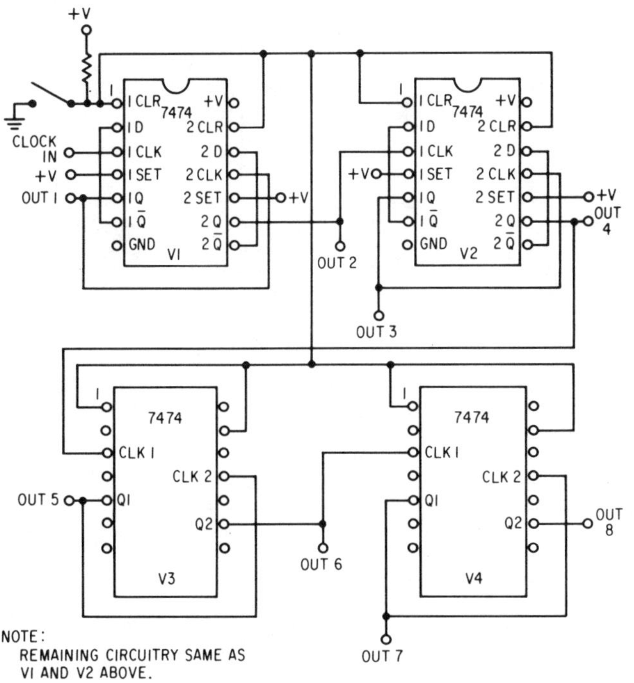

Fig. 4-9 Eight-stage 74175 ripple counter.

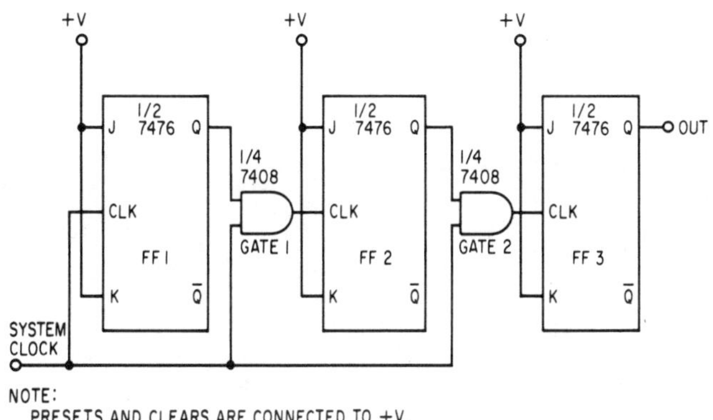

Fig. 4-10 Synchronous counter.

change. In this circuit, seven high-to-low changes and one low-to-high change take place at count 128. The total delay equals

$$\begin{array}{rcl} 7 \times 20\,\text{ns} & = & 140\,\text{ns} \\ +\ 1 \times 17\,\text{ns} & = & 17\,\text{ns} \\ \hline \text{Total} & & 157\,\text{ns} \end{array}$$

Meanwhile, the clock is clocking at 20 MHz that converts to a clocking interval of 50 ns. From this it can be seen that three additional clocks will be received in the time required for count 128 to ripple through the counter. It is essential that the counter operate more rapidly.

Synchronous Counters

The ripple counter has inherent delay. There is no way to significantly improve on this delay if the ripple counter design is retained. Some improvement would result from the use of selected components or Schottky devices. Even so, the speed will be limited.

Synchronous counters reduce the propagation delay problem. This is accomplished by clocking all of the counters with the system timing pulse rather than with the outputs of the preceding stages. Additional gating is needed so that succeeding flip-flops will be enabled by the previous stages and the clock. Figure 4-10 shows a synchronous counter constructed from J-K flip-flops that have been wired for and operated as T flip-flops.[2] All flip-flops are clocked from the system timing pulse via gates. The counting sequence begins with all flip-flops at reset so that all Q outputs are low. The first clock pulse is applied to the first flip-flop and to one side of each of the gates. The first clock pulse will cause flip-flop 1 to change states so that its Q output goes high. The clock pulse also is applied to gate 1. Since the other side of the gate is connected to the Q output of flip-flop 1, the clock will pass through to flip-flop 2 only if both the clock and the Q output are high. Prior to flip-flop 1 changing states, the Q output is low and therefore only the first flip-flop changes state.

On the second clocking pulse, the Q of flip-flop 1 is high and gate 1 will pass the clock to flip-flop 2 and the flip-flop will change state. At the same time, the clock pulse will cause flip-flop 1 to revert to its original state with its Q low. The third clock will enable only flip-flop 1 since the low Q of flip-flop 1 now disables gates 1 and 2. On the fourth clock, all three flip-flops change. Flip-flop 3 produces a high Q, and the other two revert to low Qs. All three flip-flops are clocked simultaneously. Propagation delay is limited to little more than the delay of one flip-flop.

Experiment 4-3: Synchronous Counters

Construct the counter circuit shown in Fig. 4-11a. Use LEDs to monitor each Q output and each gate output. Set the breadboard pulse generator for a slow pulse rate and use Fig. 4-11b to draw the pulse trains indicated by each LED. Remember to clear the flip-flop prior to starting the count.

Compare the waveforms of the synchronous counter with those of the ripple counter. The Q output waveforms are the same as that of the ripple counter. The difference is that the clocks are not the Q outputs of preceding stages but are actual gated system clocks that force toggling of all flip-flops in synchronism.

[2] A *T flip-flop* is a "toggle" flip-flop or a binary divider. If the J and K inputs of a J-K flip-flop are tied positive, the Q output will toggle between high and low with each clock pulse.

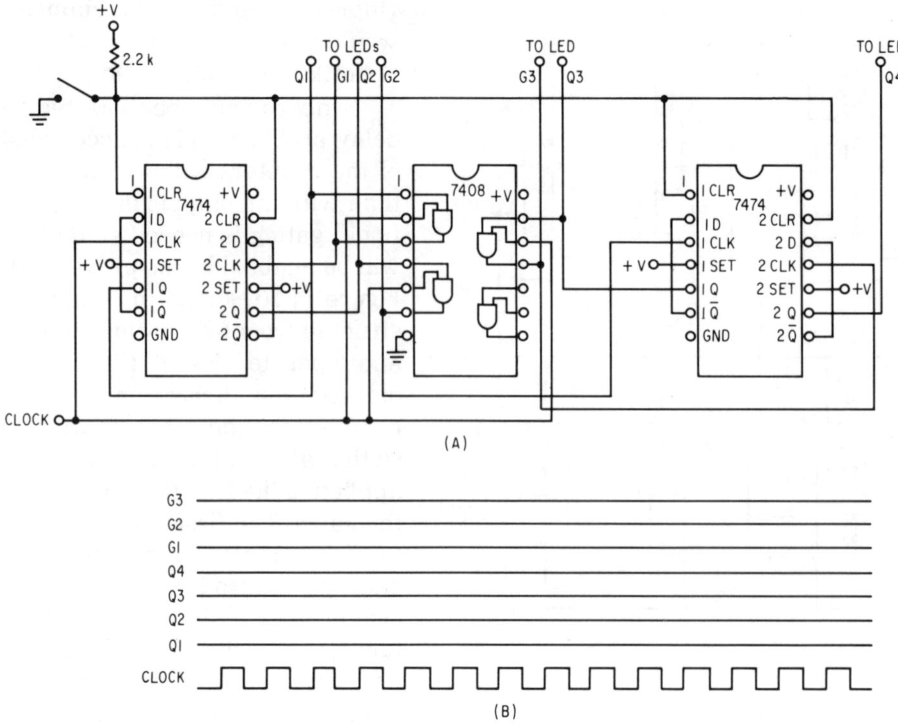

Fig. 4-11 (a) Synchronous counter circuit and (b) pulse train to be completed by reader.

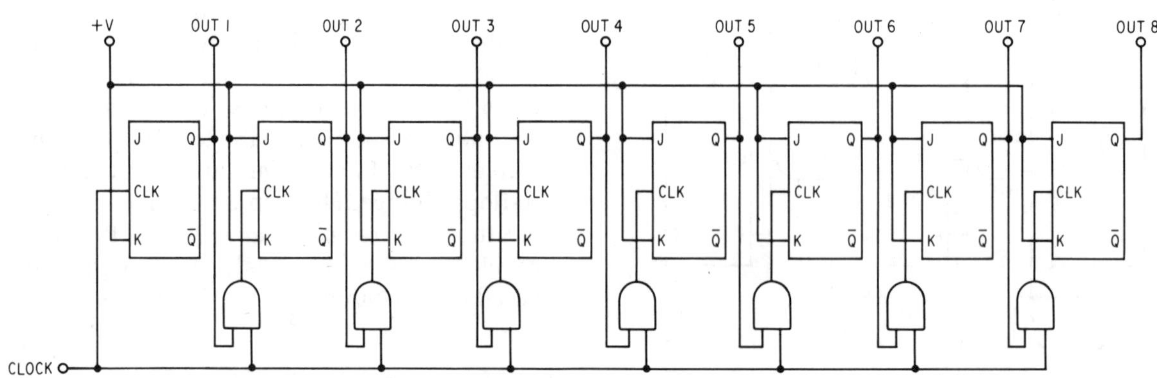

Fig. 4-12 Eight-stage synchronous counter.

If an eight-stage synchronous counter were constructed and a count of 127 were entered, the conditions would be the same as for the ripple counter described earlier. Figure 4-12 is a block diagram of such a counter. Since the counter is loaded with the number 127, all of the Q outputs except the last are high. When count 128 arrives, the clock pulse is applied to each stage at the same time. The clock will cause all of the flip-flops to change state simultaneously. All of the highs will go low and the last stage will go high. The counter will then register the count 128 and will reach that count in approximately 20 ns, rather than 157 ns that was required by the ripple counter. The benefits for high-frequency operation are apparent.

Counter Characteristics

Counters are used in many applications. In fact, counters are available in greater variety than most devices. They are fundamental to many kinds of circuits. Because the demands on counters are so varied, a number of specialized characteristics have been devised to differentiate between them.

An important counter characteristic is modulus. The *modulus* of a counter is an integer indicating the number of states through which a counter sequences during each cycle. It is usually expressed as the modulo of a counter. A modulo-4 counter, for example, would sequence through four distinct states before repeating. The *modulo* of a counter is equivalent to the number of clock pulses required to return

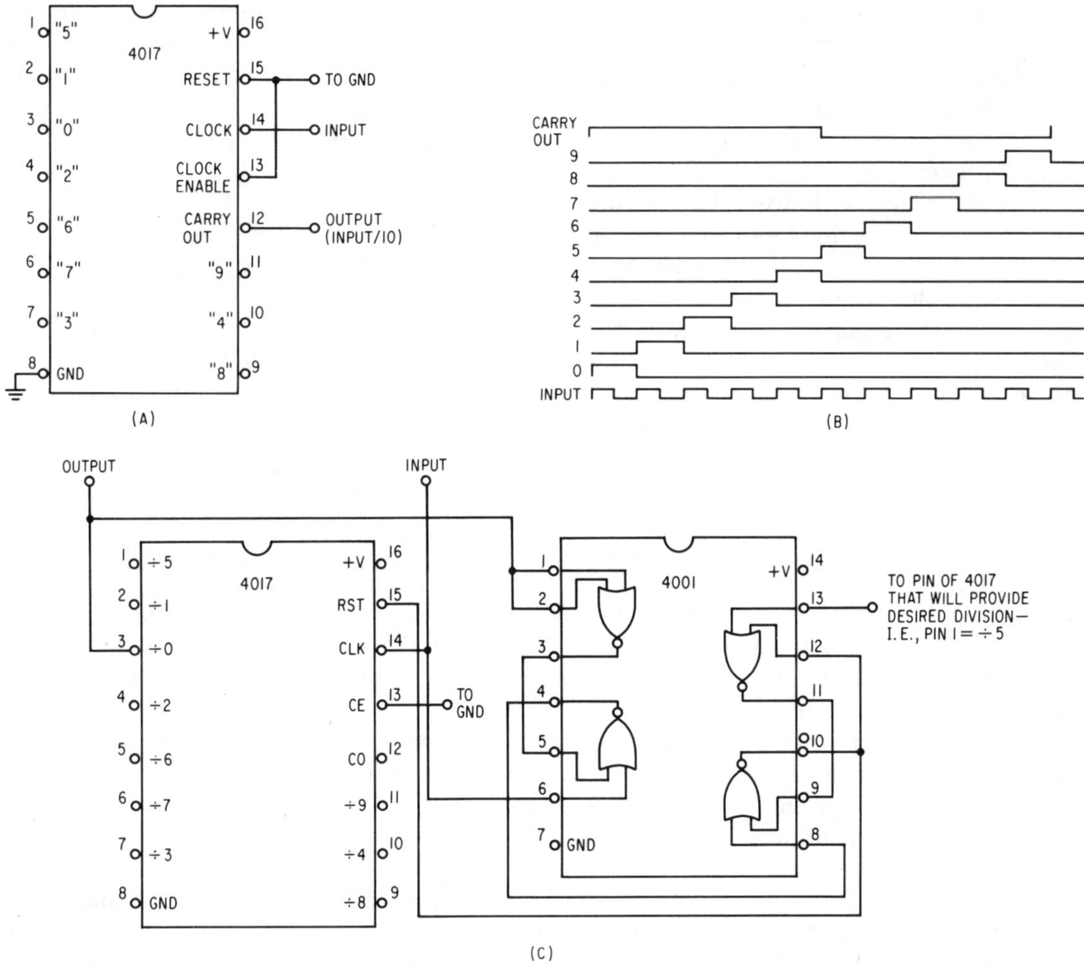

Fig. 4-13 The 4017: (a) wired for divide-by-10 operation, (b) pulse train, and (c) wired for selectable division operation.

the counter to its original state or to produce a carry. In terms of components, the modulo of a counter is the binary equivalent of the number of T flip-flops in the counter chain. A single T flip-flop will reset after two clocks. A two flip-flop chain will reset after four clock pulses. Three flip-flops will count to eight and so on. This binary number representing the number of pulses in a cycle is called the "n" of the counter as well as its modulo. A modulo-n counter refers to a variable or programmable counter. The modulo-n counter is capable of counting any number within a range of numbers. The 4017 is a synchronous decade counter that can be used with external gating to produce any count from 1 to 10. Figures 4-13a and 4-13b show the pins and pulse diagram of the 4017. For simple decade division, the 4017 can be wired as shown. For divisions other than 10, an external gate chip is required, as shown in Fig. 4-13c. The 4001 provides necessary buffering and resetting for the 4017. To obtain any desired division, pin 13 of the 4001 is connected to the appropriate 4017 pin-out, as indicated on the outline drawing (see Fig. 4-13c).

Other programmable counters operate in different ways. The 4522 and the 4526 use an interesting programming method. Both are programmed by a code that is placed on the pins marked D1, D2, D4, and D8 (see outline drawings included in Appendix C). The 4522 is a decimal counter while the 4526 is a binary counter. Except for the numbering system, the counters operate the same. The 4526, for instance, will count backward in binary code at each low-to-high clock transition. The binary number at which the counter will begin its count is determined by the binary code placed on the "D" inputs. This code can be reloaded into the counter each time the zero count is reached if zero out, pin 12, is connected to "load" pin 3. The counter will then divide continuously by the binary number placed on the D inputs. The modulo of these counters is determined by the value of the D inputs. There are other methods of determining the modulus of a counter. Some of these will be noted in the descriptions of counters provided later in this chapter.

Another important characteristic of counters relates to the value assigned to each binary digit of the counter output. This characteristic is referred to as *coding* or *weighting*.

Earlier discussions showed that counters are simply multiple binary dividers connected one after the other. This type of counter results in a binary weighting or coding of its output. According to this coding, the least significant bit has a value of "1." The second bit carries a value of "2." The third bit is weighted at "4"; the next bit at "8"; the next at "16"; and so on. This is a binary coded counter.

If the coding provides a weighting scheme that is binary until the count "9" is reached and then resets to zero with a carry, the counter is a binary coded decimal counter. Each four binary bits represent one decimal digit.

The decimal counter counts to 10 and provides an output at the tenth count. An octal counter resets at the eighth count. The hexadecimal counter provides for 16 counts. Each of these counters is coded to provide for their particular weighting scheme. Weighting does not have to follow normal binary or decimal systems. A BCD counter can arrive at count "9" using different weighting schemes. One counter could give the first bit a value of "1," the second bit a value of "1," the third bit a value of "2," and the fourth bit a value of "5." Another counter could give the first bit a value of "1," the second bit a value of "2," the third bit a value of "2," and the fourth bit a value of "4." In each case the counter weights add up to "9." As long as the counter and its associated decoder match, any coding scheme is usable.

Several characteristics that are unique to some counters and not others help determine if a particular counter will be suitable for a particular application. The direction of the counter's count is one such characteristic. One counter may begin at count zero and count up to its maximum count. Another counter will begin at its maximum count and count downward. Some counters can be programmed for either count direction. The benefit of being able to have one or the other count direction is evident in the case of a stopwatch versus a timer. A stopwatch counts the seconds that elapse while an event takes place. The length of the count is dependent on the duration of the event; i.e., a runner running a 100-yd dash. An up counter would be useful for this application. Students taking a timed test are racing the clock, however. The time period is set. They must complete the test before the count reaches zero. A down counter is the logical choice in this case.

Constructing counters of infinite length is not possible. If long counts are needed, the normal method of obtaining them is to connect several counters in series. The ability to connect counters in this way is called *cascadability*. Some counters can be cascaded, some cannot. The choice of a counter for a particular circuit must include determination of whether cascading the counters will be needed or not.

A counter that cannot be reset to zero will be unsuitable for many applications. A counter that has provision for returning it to count zero is referred to as *resettable*. If the counter can be forced into any desired state within its count sequence, the counter is said to be *presettable*. Some counters provide reset provision but no preset. Some have both. Some have neither. The application will decide which is the appropriate choice.

Designing Counter Circuits

There are several options open to the designer of counting circuits. For simple or small "n" counters, flip-flops may be the most appropriate choice. If only one binary divider is needed in a circuit, the designer may find that it is most economical to construct a flip-flop from surplus gates that are a part of chips used elsewhere in the circuit. Some very sophisticated digital equipment has been designed with many multiple gate ICs on a circuit board. The individual gates are then interconnected to produce the circuit action required. This often results in one IC chip supplying gates to four or more different circuits. Some of these circuits may even be located on other circuit boards. The chip economy of this construction is good, but the difficulty of circuit tracing and maintenance is increased.

Use of integrated circuit flip-flops is more convenient than interconnecting gates for flip-flop action. Typical TTL flip-flops include the 74107 and 7476. These two chips are both dual J-K flip-flops. They differ in the fact that the 7476 is a 16-pin DIP and provides both preset and preclear, while the 74107 uses a 14-pin DIP and deletes the preset input on both flip-flops. The 7474 is a popular D-type flip-flop that has both preset and preclear inputs. Numerous special purpose flip-flops are available. The 7470 and 7472 are examples of these. Appendix C contains pin-outs of these five TTL flip-flops as well as examples of popular flip-flops from other families.

Numerous integrated counters have been fabricated. Some are very specific as to function while others are designed for flexibility of operation. To decide which counter is best for any specific application, several characteristics of counters must be considered. The following questions represent a systematic method of deciding which counter to use:

1. What coding is required? (binary, BCD, decimal, etc.)
2. What modulo is required? (3, 4, 5, etc.)
3. What clock or toggling frequency is required? (hertz, kilohertz, megahertz)
4. What decoding is required? (1 of n, all of n, etc.)
5. Is cascadability required?
6. Is low-power operation required?
7. Is circuit board space limited? (Is a single package needed?)
8. What family is needed? (DTL, TTL, CMOS, ECL, etc.)

To illustrate how to use these questions, the following will be helpful. Suppose a divide-by-16 counter is needed to use in a CMOS circuit. What counter would be the best choice? Answering the previous questions results in the following specifications:

1. The coding is binary.
2. The modulo is 16.
3. The clock frequency is 4 MHz (determined by circuit).
4. The decoding required is 1 of 16.
5. The cascadability depends on counter chosen.
6. Low power is required (CMOS will provide low power).
7. Circuit board space is limited.
8. The family is CMOS.

The counter needed is a binary CMOS counter capable of division by 16 at a clock rate of 4 MHz. One counter that might satisfy these requirements is a four-stage binary ripple counter constructed from two 4013 dual D-type flip-flop DIPs. The question of propagation delay would need to be considered. The 4013 is capable of 10 MHz clocking at 10-V operation (only 2.5 MHz at 5 V). Propagation delay per stage is 175 ns for a maximum of 700-ns delay. A 4-MHz clock will result in a pulse once every 250 ns. The overlap of clocks and counts will create errors.

A synchronous counter built from 4013s (see Fig. 4-14a) will do the job if a separate decoder chip is supplied. Since space is limited on the circuit board, however, a solution using fewer chips would be better. The 4520 can be arranged to be used as a hexadecimal counter. The function diagram shows that the 4520 has a binary coded output. The maximum clock frequency will permit the 4520 to operate at the 4-MHz clock rate of this circuit. A 1-of-16 decoder such as the 4514 will provide the proper output (see Fig. 4-14b).

Several counters can be programmed for a variety of counting sequences. The CMOS 4018 is a divide-by-n counter. It provides a single chip counter for the division 2, 4, 6, 8, and 10. Adding a NAND gate will permit division by 3, 5, 7, and 9. The unit is cascadable. Figure 4-15 details the required connections of the 4018 for various counting schemes.

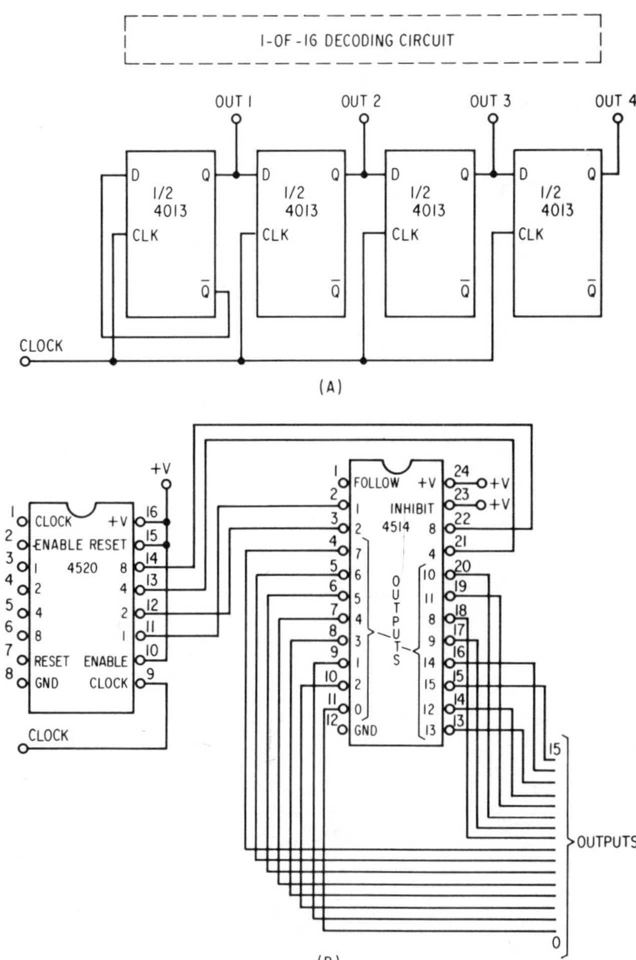

Fig. 4-14 (a) 4013 synchronous counter circuit and (b) 4520/4514 counter decoder circuit.

The 4018 is characteristic of a walking-ring (Johnson) counter. A *walking-ring counter* operates by shifting numbers down its length and back. Figure 4-16a is a walking-ring counter having a modulo of 8. Four D-type flip-flops are used. Begin with each of the outputs low. At the next clock, a high from the Q output of the last flip-flop will load a high into the first stage. The count is now 1000. Each of the next three clocks will load highs into stage 1 and will shift highs down the counter from one stage to the next. The counts progress as shown in Fig. 4-16b. When stage 4 has a high Q, a low will be loaded into stage 1. The next clocks will move the lows across the counter until all outputs are low and the sequence starts again.

The sequence of the walking-ring counter does not conform to other coding schemes that have been considered; i.e., binary, BCD, or decimal. To provide useful output, decoding is needed. This is easily ac-

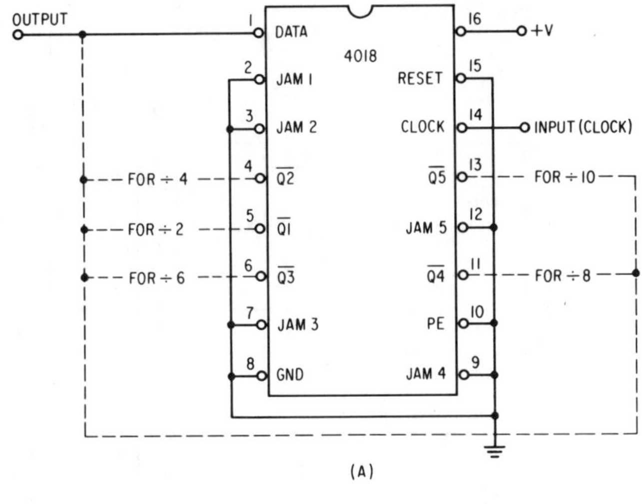

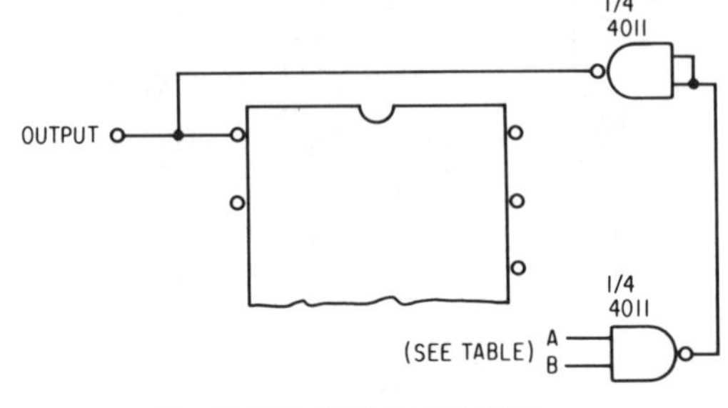

	CONNECT TO	
÷ BY	A	B
3	Pin 5	Pin 4
5	Pin 4	Pin 6
7	Pin 6	Pin 11
9	Pin 11	Pin 13

(B)

Fig. 4-15 The 4018: (a) wiring diagram for selectable even counts and (b) wiring diagram for selectable odd counts.

complished with two-input gates, as shown in Fig. 4-17. Disallowed states appear within longer walking-ring counters and must be removed via resetting, presetting, or gating. Disallowed state elimination and decoding are built into the 4018 and similar integrated counters.

Low-Modulo Counter Circuits

Counting sequences from 2 through 16 are obtainable using existing integrated circuits. Figure 4-18 shows several ways of obtaining these counting sequences. The selection shown illustrates techniques that can be used but are not intended to be all inclusive.

Cascaded Counter Circuits

For counts or division greater than 16, cascading of the basic counter circuits as just described will be required. This is simple cascading of counters. The output of one counter is used to clock the input of the next counter. Some counters will permit synchronous cascading. Several of these cascaded counter circuits are shown in Fig. 4-19.

Decade Cascadable Counters

The 4522 is an example of programmable counters that permit cascading by decades. This differs from normal cascading in that the counters are weighted according to their decimal position. As discussed earlier, a programmable counter will function according to the number placed on its D inputs. If 0011 is placed on the D inputs, the counter divides by 3. If 1000 is used, the counter divides by 8. If programmable counters are cascaded normally, the second counter would divide the count of the second. Suppose the divide-by-8 counter is followed by a divide-by-3 counter. The resulting division is 24. If, however, two 4522s are cascaded by decades (see Fig. 4-20), the resulting division will be 83. The divide-by-8 counter becomes a weighted 10 counter and the divide-by-3 counter is given units' weighting. Notice that the zero output of the weighted 10 counter is used to feed the cascade input of the units' counts, and the zero output of the units stage is connected to its own and the tens counter load inputs. The output will indicate a zero only when both counters are at zero. When this occurs, the D input will be loaded into the counters due to the zero

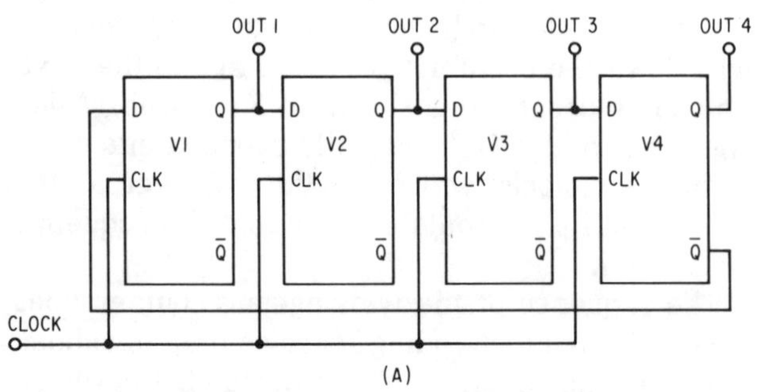

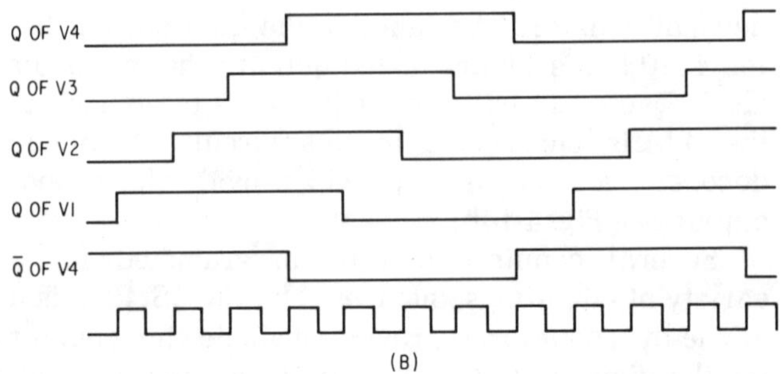

Fig. 4-16 (a) Walking-ring counter and (b) pulse train.

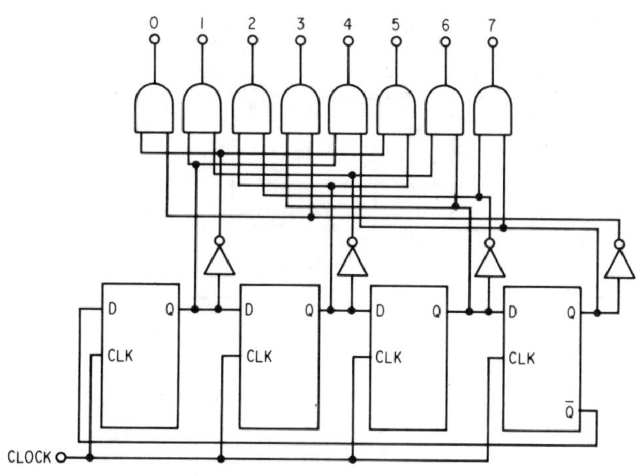

Fig. 4-17 Decoder circuit for walking-ring counter.

output from the unit's stage going high briefly at the zero count. The counter will then decrement this number each time a clock pulse is received until zero is reached.

High-Frequency Counting Circuits

A major limitation of the most popular digital IC families, TTL and CMOS, is speed. TTL suffers from the basic slowness of saturated logic. This is compensated for, to an extent, by the use of the totem pole output. While the totem pole speeds up the state change, it also increases power consumption and creates noise problems. CMOS does not utilize a totem pole output. Power consumption is low except at higher switching speeds and noise problems are minimal. The one major shortcoming is speed. Since there are no circuit elements forcing the state change, the change takes place at its own pace. That pace is comparatively slow.

RTL and DTL suffer from saturated logic and RC time constant problems. Indeed, slowness and power dissipation have contributed to the decline in popularity of these families and the elevation of TTL and CMOS.

For many applications TTL and CMOS operate at adequate speeds. But, as the speed requirements increase, these families become inadequate. Several improvements have made TTL usable for frequencies up to approximately 200 MHz. Since standard TTL is limited to approximately 35 MHz, this improvement is significant. One TTL improvement is called *high-speed TTL*. It can be recognized by an "H" within the type number; i.e., 74H00. The high-speed modification consists of adding diodes that prevent the transistors from saturating and adding amplifiers to the output stage to increase

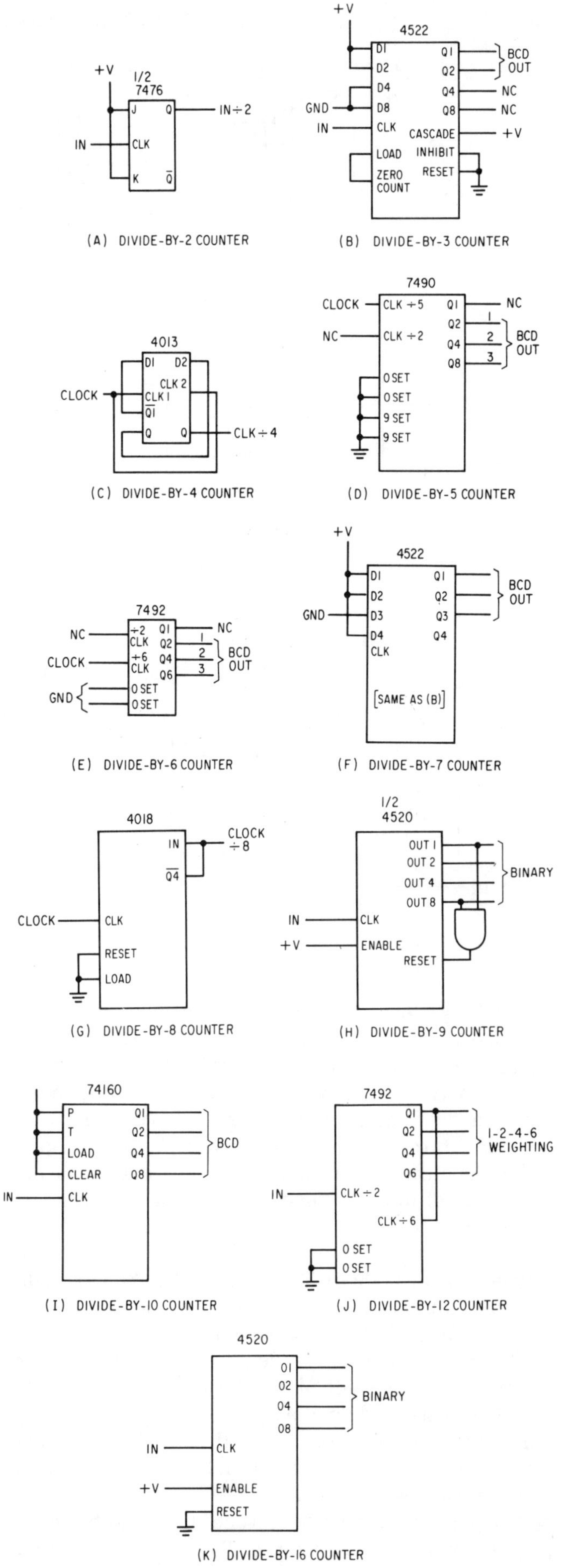

(A) DIVIDE-BY-2 COUNTER

(B) DIVIDE-BY-3 COUNTER

(C) DIVIDE-BY-4 COUNTER

(D) DIVIDE-BY-5 COUNTER

(E) DIVIDE-BY-6 COUNTER

(F) DIVIDE-BY-7 COUNTER

(G) DIVIDE-BY-8 COUNTER

(H) DIVIDE-BY-9 COUNTER

(I) DIVIDE-BY-10 COUNTER

(J) DIVIDE-BY-12 COUNTER

(K) DIVIDE-BY-16 COUNTER

Fig. 4-18 Low-modulo counter circuits.

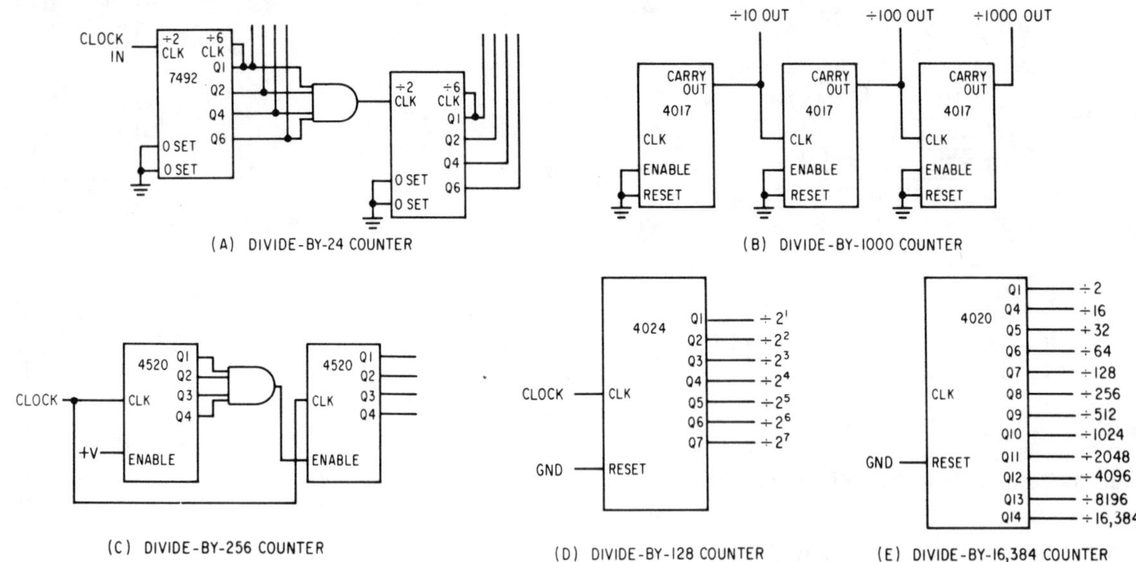

Fig. 4-19 High-modulo and cascaded counter circuits.

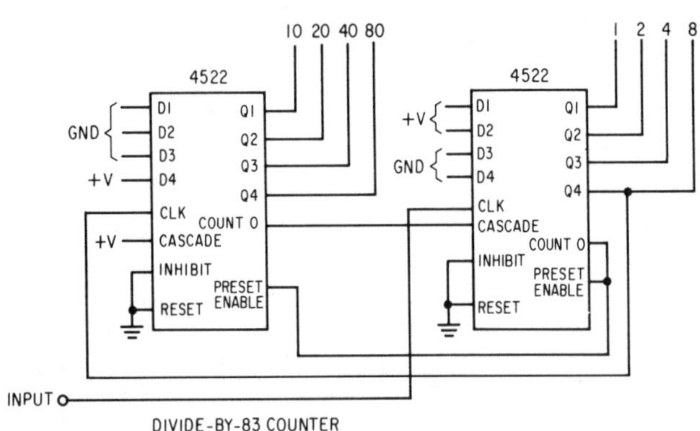

Fig. 4-20 Two 4522s cascaded by decades.

switching speed. While HTL is a speed improvement, it also is inadequate for many applications. Maximum clock rate for HTL is approximately 50 MHz.

A more significant improvement in TTL speed is obtained if Schottky diodes are used to prevent transistor saturation. The lower forward voltage drop of the Schottky diode speeds the current-bypassing function of the diode and increases speed considerably. Schottky TTL has a speed ranging up to 100 MHz. Schottky TTL is identified by an "S" in the type number (74S00). The type number 74LS00 indicates that the device is low-power Schottky TTL. Low-power Schottky is expected to replace standard TTL in popularity. The reason is that low-power Schottky retains all of the benefits of TTL while adding speed and decreasing power consumption by a factor of 5. Low-power operation is achieved at the expense of speed, but the LS devices will still operate at speeds of 60 MHz.

Operational speeds beyond the capability of Schottky are often required. For these, a different family of ICs must be used. This faster family is Emitter-Coupled Logic (ECL). The Motorola line of ECL is designated MECL and is available in four versions. MECL appeared earliest and provided speeds of 30 MHz with propagation delay of only 8 ns. MECL II improves considerably on this. It is capable of operation to 100 MHz and 4-ns propagation delay. MECL 10,000 is a compromise between MECL II and MECL III. It has a propagation delay of 2 ns and a clocking rate of 200 MHz. MECL III drops the propagation delay to 1 ns and ranges in speed up to 1.2 GHz. (GHz = 1,000 MHz). The benefit for speed of operation provided by these later versions of ECL is apparent. As indicated earlier, the major problem in using ECL is interface. ECL uses a negative logic swing. It is not saturated logic, and the voltage swing is not large. Interface adapters are available to match ECL with TTL or CMOS. Therefore, for high-speed circuits ECL becomes a logical "front end."

High-speed counting circuits often use ECL or Schottky TTL. These stages will divide the high-frequency input and the resulting lower speed output will be handled by slower, less expensive circuits. Figure 4-21 shows an ECL counter circuit designed as a decade counter capable of 100-MHz clocking. If the components are exchanged for MECL III units, the speed of the counter extends to 300 MHz (see Fig. 4-22).

High-speed devices are normally reserved for those applications that require their special attributes. Speed is usually accompanied with some design problems. In some applications, power consumption becomes significant. ECL uses negative supply voltage and a limited range logic swing (see Chap. 1). When extensive ECL design is contemplated, it is advisable to refer to manufacturer appli-

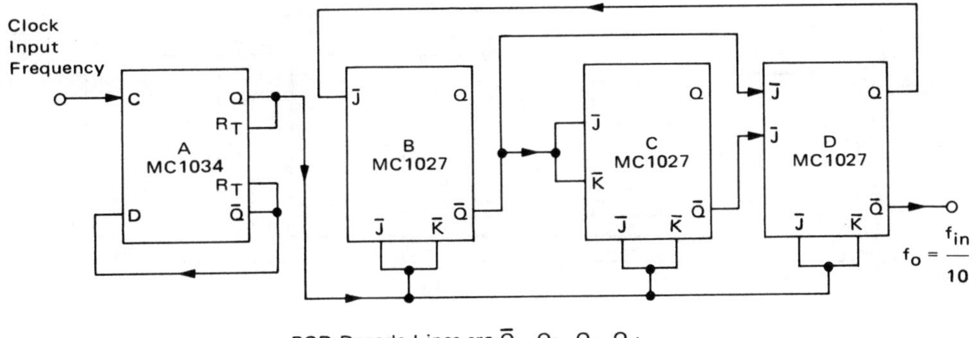

Fig. 4-21 MECL II counter circuit. (Courtesy of Motorola, Inc.)

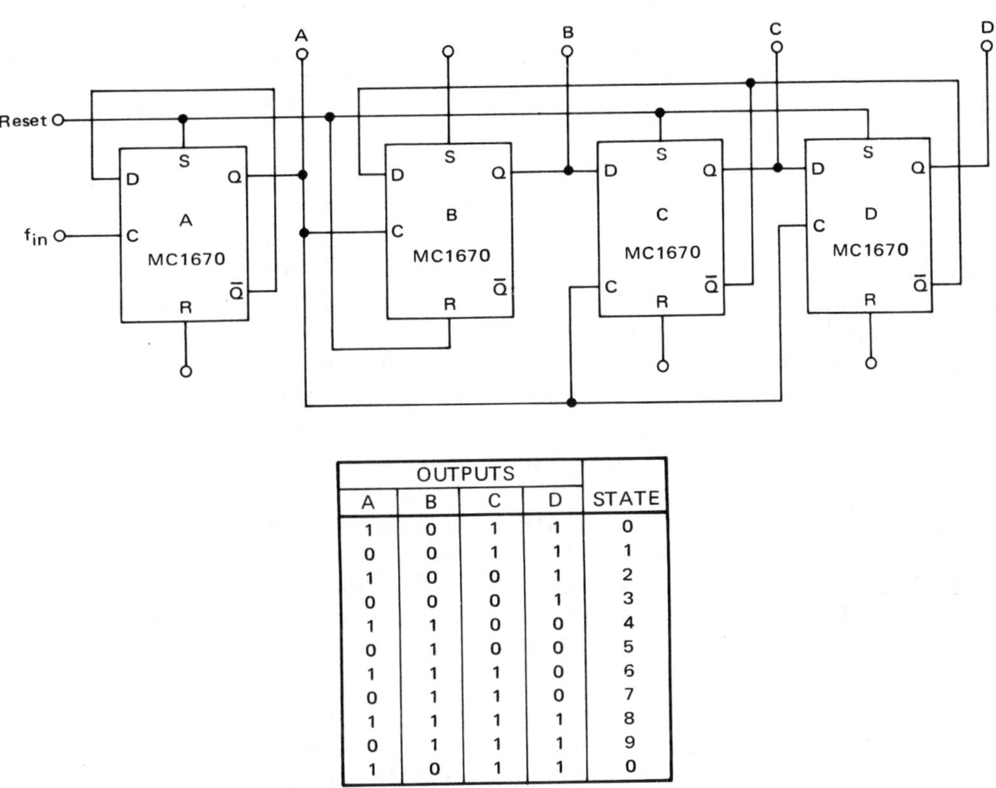

OUTPUTS				
A	B	C	D	STATE
1	0	1	1	0
0	0	1	1	1
1	0	0	1	2
0	0	0	1	3
1	1	0	0	4
0	1	0	0	5
1	1	1	0	6
0	1	1	0	7
1	1	1	1	8
0	1	1	1	9
1	0	1	1	0

Fig. 4-22 MECL III high-speed counter circuit. (Courtesy of Motorola, Inc.)

cation notes, data sheets, and similar information. Design data for ECL is currently available in a Motorola publication entitled *MECL System Design Handbook*, by William Blood, Jr. Information on MECL I, MECL II, MECL III, and MECL 10,000 is included.

Shift Registers

A *shift register* is composed of a group of flip-flops that are connected so that data on the input is shifted one stage at each clocking. A high logic signal that begins at the input of stage 1 will move one stage down the register each time a clock pulse is received until it is clocked out of the final register. The walking-ring counter discussed earlier is one form of shift register. It uses a self-loading technique to produce the ring-counting operation, but the basic circuit is a shift register.

Shift registers are classified according to their input/output characteristics. A simple form of shift register is the Serial In/Serial Out or SISO. This register simply clocks in data from the input and then clocks it out again. The output sequence is the same as the input sequence. A 6-bit SISO shift register is shown in Fig. 4-23. The unit is constructed from 7474 dual D-type flip-flops. The device is synchronous since the clock is connected to all flip-flops. The register sequence is begun by clearing all Q outputs to low logic states. At the first clock, the data at D of the first stage will be transferred to the

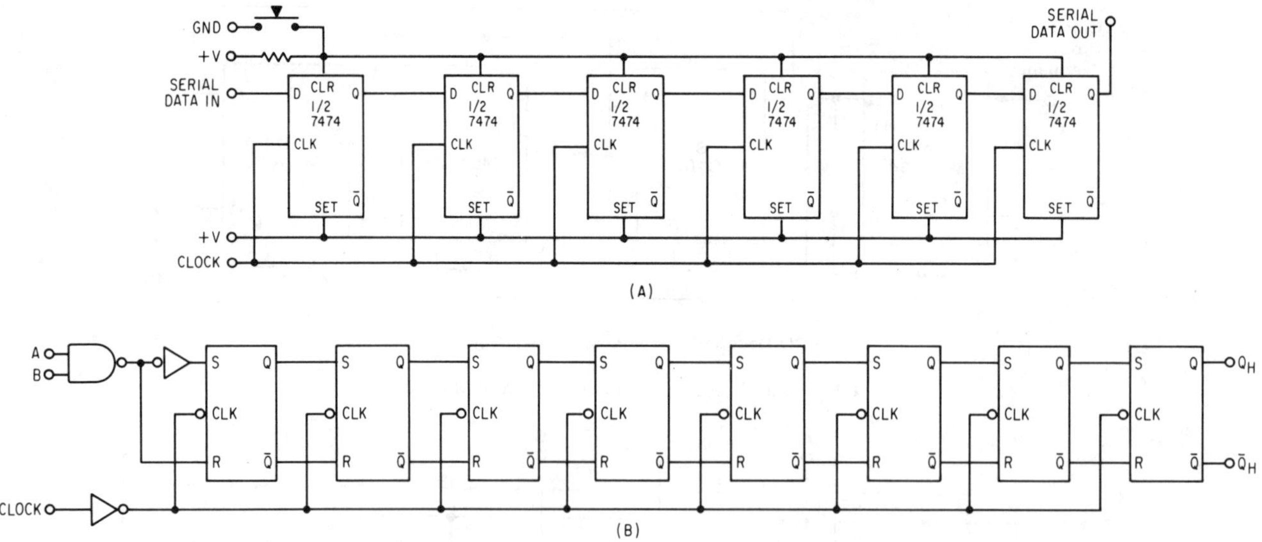

Fig. 4-23 (a) SISO shift register and (b) functional diagram of the 7491 SISO.

Q output of that stage. If the input at D is low, the Q output will remain unchanged. If the input at D is high, the Q output will go high. Remember that the D input of a D-type flip-flop will be transferred to the Q output only at clocking. The input can change between clocks and the Q output will remain at its existing level. At the second clock, the second stage will load the Q output of stage 1 into its Q output since the Q of stage 1 and the D input of stage 2 are interconnected. At the same time, the data at the D input of stage 1 will be clocked into the Q output of stage 1. This process will be repeated at each clock for each stage until the data clocked into the first stage at clock 1 will be loaded into the Q output of stage 6. At this time, the six stages will contain 6 bits of data. This data can be held for later use or passed on with successive clocks.

The value of an SISO shift register may not be readily apparent. Why clock data in and then clock it out unchanged? What is the purpose? There are several uses for the SISO that make its function beneficial. As already mentioned, storage is one use. The data can be clocked in and kept until needed; when it is needed, it can be clocked out. A long shift register could be used to store several words of data. A *word* is a group of high and low bits that are of a length equal to the word length used in the system. A microprocessor system might use words 4 bits long. A 64-stage SISO shift register could store 16 4-bit words. These words could then be clocked out one at a time or all at once as required.

An SISO shift register also can be used for delay purposes. An SISO shift register placed on a data stream will add a delay equal to its length. It also is possible to clock data into the shift register at one rate and then to clock it out at a second rate. Thus the data can be reduced or increased in speed.

An SISO shift register has limitations. The data, once inserted into the register, cannot be altered until it is clocked out. The order of words or bits in a long register cannot be changed. Input data must appear at the proper time and in the proper sequence or the output will be unusable.

Integrated SISO shift registers are available. Pin-outs and functional diagrams of several of these are included in Appendix C. The 7491A is an 8-bit shift register from the TTL family. As shown in the functional diagram (see Fig. 4-23b), R-S flip-flops are used to construct this register. D-type flip-flops are used in some registers. The CMOS 4031 is a 64-stage SISO shift register. This unit provides complementary outputs and a recirculate input. These registers can be cascaded to provide for longer shift register sequences.

The 4006 contains four separate shift register sections. Two of these can be used as 4-bit registers, and the other two can be used as either 4- or 5-bit registers. The registers can be interconnected for longer sequences if desired.

A second type of shift register is the Serial In/Parallel Out (SIPO) shift register. Figure 4-24 shows a functional diagram of a typical unit. Notice that the circuit is similar to the SISO register except that each Q output is made available. After the serial input has been shifted into the register, a word is available at the outputs. This kind of function is useful in computer systems in which the processor accepts words in parallel form (all bits at the same time). Many input devices operate in a serial mode. The SIPO shift register can be used to accept

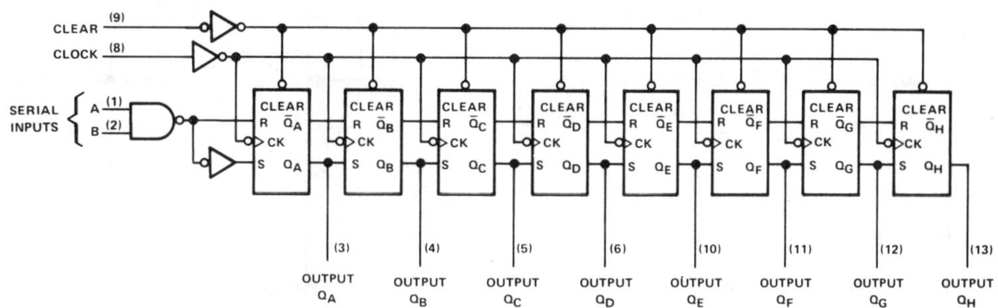

Fig. 4-24 The 74164 SIPO shift register. (Courtesy of Texas Instruments, Inc.)

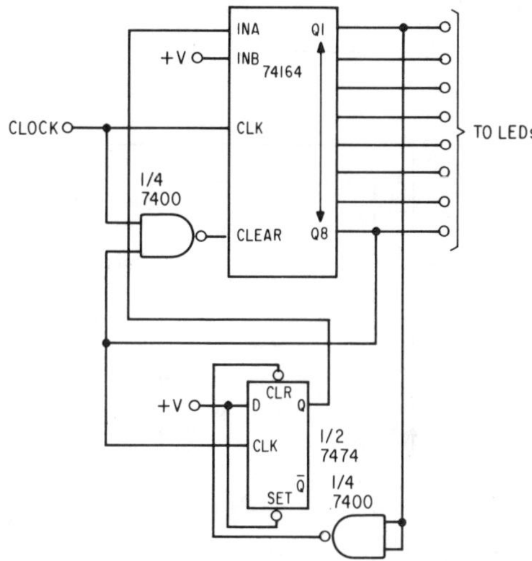

Fig. 4-25 Electronic stepper circuit.

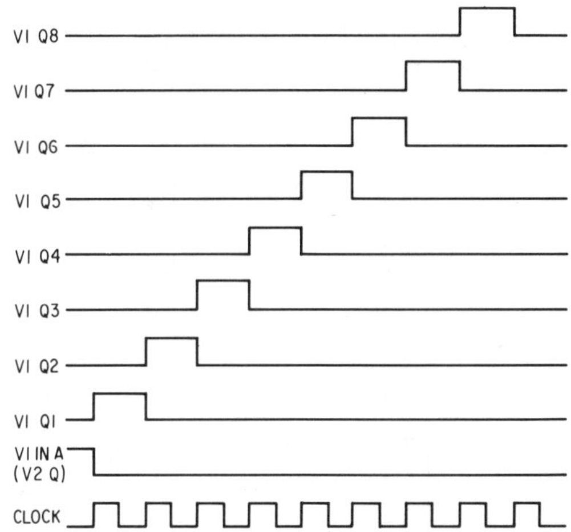

Fig. 4-26 Stepper circuit pulse train.

serial bits from the input device and then provide them to the processor as a parallel data word.

Integrated SIPO shift registers are included in Appendix C. The 74164 is typical. Data enters at one of the two inputs (pins 1 and 2). The second input is held high (see Fig. 4-24). Notice that the inputs are actually inputs to a two-input NAND gate. Input control can be achieved by selectively making the second input high or low. This may be useful for some applications. The data is shifted down the R-S flip-flops at clocking. The register is synchronous. Bringing the clear input low will load lows in all stages.

Experiment 4-4: Electronic Stepper

Data handling is not the only use for shift registers. An interesting application for the SIPO register is that it can be used as an *electronic stepper*. The specifications for the device include:

1. Only one high output at a time.
2. The high output will sequentially move down the register with each clock.
3. The stepper will continue to operate in a continuous pattern.
4. The stepper will clear at the beginning of each cycle.

Design a circuit that will accomplish all of these circuit goals. Compare your circuit with the one shown in Fig. 4-25. You may have designed your circuit differently, but the important question is: Does it meet all of the design requirements? Construct both circuits on the breadboard. If your circuit does not work properly, analyze it to determine why. A timing chart similar to the one shown in Fig. 4-26 may help.

The opposite of the SIPO shift register is the Parallel In/Serial Out (PISO) shift register. The use of such a device in processor systems is evident. The parallel data from the processor is routed to the parallel inputs and converted to serial output for use by serial input devices.

The TTL 74165 is a PISO shift register. Data is loaded from the parallel inputs by briefly bringing the load input low. The loaded 8 bits can then be shifted out of the register at the rate of 1 bit per clock. Complementary outputs are provided and a serial input can be used if SISO operation is desired. Shifting will not occur if the clock inhibit input is held high.

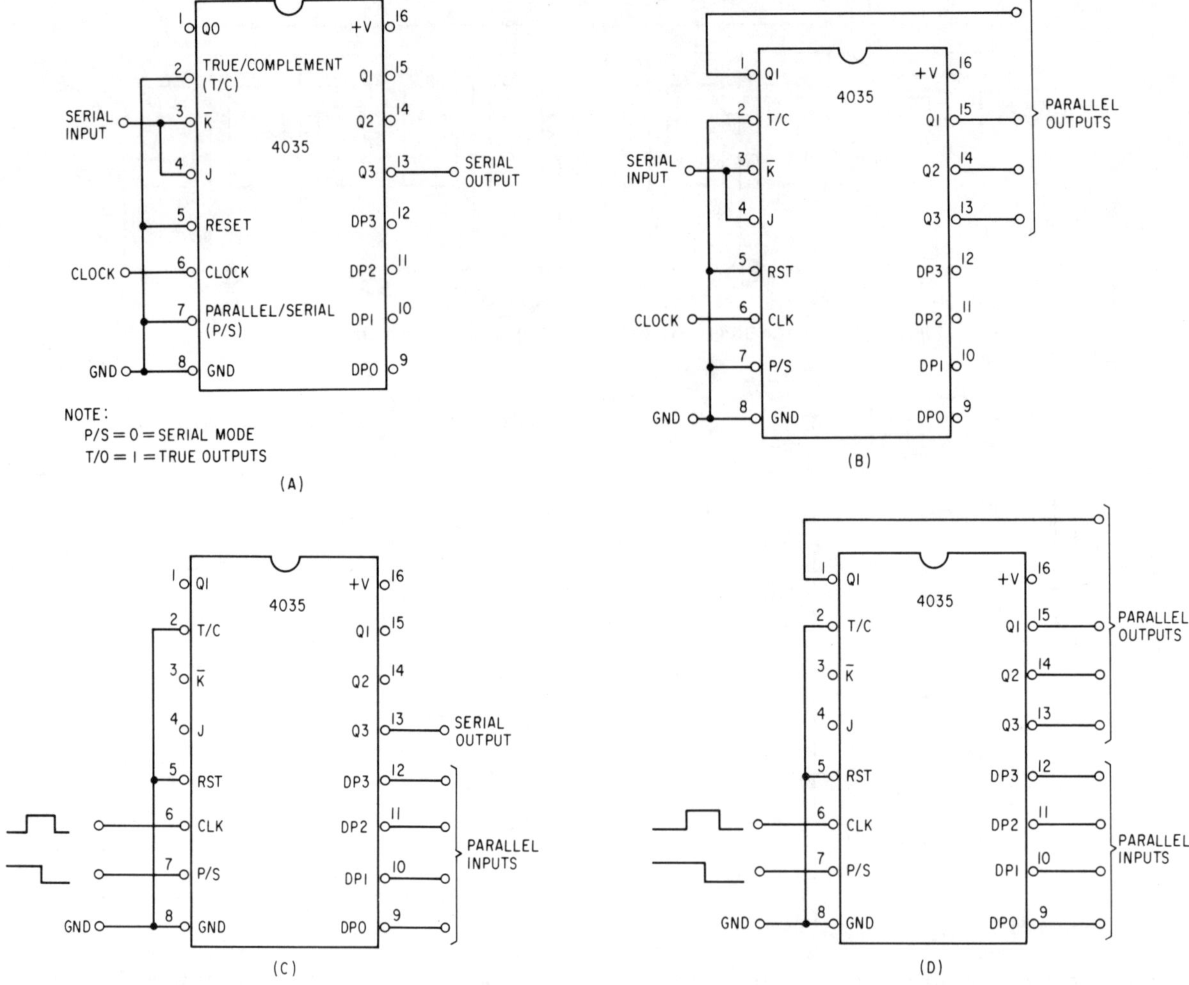

Fig. 4-27 The 4035: (a) used as an SISO shift register, (b) used as an SIPO shift register, (c) used as a PISO shift register, and (d) used as a PIPO shift register.

Some parallel load shift registers do not permit altering of stages that are high. To input a low, the register must be cleared and then the high inputs must be loaded. The inputs in this case function as presets rather than true data inputs.

A fourth type of shift register is the Parallel In/Parallel Out (PIPO) shift register. This register combines the actions of several of the other units. The versatile CMOS 4035 is an example of a 4-bit shift register. Actually, the 4035 can function as any of the four types of shift registers. Figure 4-27 shows the connections for each mode of operation. The 4035 provides for several interesting features. The RST input will force all outputs low if it is brought high. If the COMP input is made high, all outputs will be the complements of the register data. By making the J input high and the $\overline{K}$ input low, clocking will result in a change or complement of bit 1 of the register only. This feature is useful in some special applications. The flexibility of the 4035 is readily apparent.

The 74194 represents the ultimate shift register. It is called a 4-bit bidirectional universal shift register. It can operate in any mode—SISO, SIPO, PISO, and PIPO—and it is capable of shifting data in either direction. The inputs that determine the direction of the shift are the mode controls S1 and S0. If S0 is high and S1 low, data from the shift-right serial data input enters the register and is shifted with clocking from QA to QD. If S0 is low and S1 high, the shift is from QD to QA with data entering at the shift-left serial input.

FIVE

Data Manipulation and Display

Introduction to Data

Data has been mentioned several times in the previous chapters. In this chapter data will become the primary focus of interest. In subsequent chapters, data will be discussed as the raw material and the finished product of microprocessor operations. Data is a pervasive subject within logic circuits and for that reason deserves detailed consideration.

Data is information. It can be the output of sensors that detect changes or movements of some physical object or being. Data can be gained from active processes. Data may be the result of numerical operations. In short, data is information concerning some object, being, or event that can be used to identify, quantify, or analyze. Perhaps some specific examples of data will clarify these rather vague descriptions of it.

A recent news report discussed "boots" for cattle. The boots contained sensors that measured the weight of each of several test animals. The sensors transmitted their measurements via electronic data to a minicomputer. The minicomputer used the data from the sensors on the hooves of the cows to assist in decision making concerning feed quantities and composition for greatest weight gain. *Data is the information that makes decisions possible.*

A university could have a requirement to contact all students that had taken introductory English in the fall semester of 1970. To sort through thousands of records by hand would be time consuming and the risk of overlooking some students would be great. If all of this information had been reduced to computer data, the computer could locate the names and records of the former students in seconds without danger of overlooking any student whose data appeared in the file. Data is information that can be stored for later use.

Another example of the use of data is a radio-controlled airplane. The command from the transmitter on the ground is received by the receiver in the airplane. The receiver responds to the command by ordering a change in the position of the rudder. A closed loop is used that conveys data describing the required change to the rudder. At the same time, data is transmitted from the rudder back to the receiver indicating the rudder's actual position. As the change in rudder position takes place, the two sets of data are compared to determine when the required rudder position is achieved. The rudder mechanism is a servo system and is similar to many industrial control applications. *Data is the information that is used to control and monitor the progress of a process.*

From the previous examples, it can be seen that data is very versatile. Data may not be "all things to all people," but it does do a large number of things for a large number of people. The data by itself, however, is not capable of performing useful functions. For data to be used beneficially, it must be manipulated. Some device must either act upon the data or act in response to the data. This chapter discusses some devices that perform simple functions. The final two chapters will discuss the sophisticated data manipulation of the microprocessor.

Data Selectors

Data selectors are available in sizes ranging from 2 input lines to 16 input lines. The basic function of a data selector is to provide an appropriate output for any one of n inputs. The 74150 is a 16-input to 1-output data selector. The functional block diagram shown in Fig. 5-1 illustrates the data selection process. The circuit looks similar to the binary adder/decoder circuit in Chap. 2. Each input line feeds one input of a six-input AND gate. Four of the remaining inputs of the six-input gates are connected to selected address select lines or their complements. Input 6 of each AND gate connects to a strobe input. The *strobe input* performs an overriding function. If the strobe is high, a low will be placed on each AND gate and operation will be effectively inhibited. The output will be high regard-

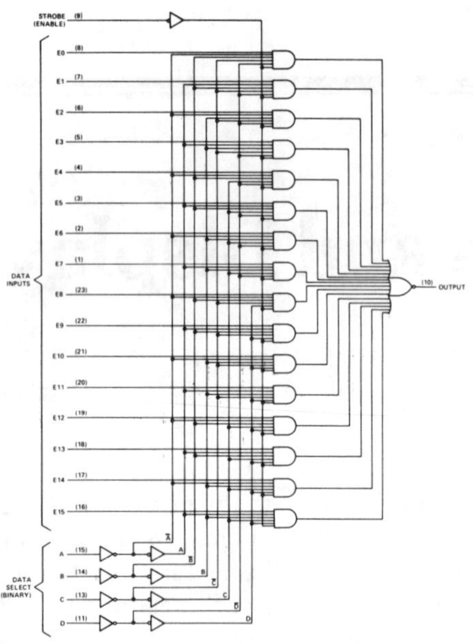

Fig. 5-1 Functional diagram of the 74150. (Courtesy of Texas Instruments, Inc.)

INPUTS					OUTPUT
SELECT				STROBE	W
D	C	B	A	S	
X	X	X	X	H	H
L	L	L	L	L	$\overline{E0}$
L	L	L	H	L	$\overline{E1}$
L	L	H	L	L	$\overline{E2}$
L	L	H	H	L	$\overline{E3}$
L	H	L	L	L	$\overline{E4}$
L	H	L	H	L	$\overline{E5}$
L	H	H	L	L	$\overline{E6}$
L	H	H	H	L	$\overline{E7}$
H	L	L	L	L	$\overline{E8}$
H	L	L	H	L	$\overline{E9}$
H	L	H	L	L	$\overline{E10}$
H	L	H	H	L	$\overline{E11}$
H	H	L	L	L	$\overline{E12}$
H	H	L	H	L	$\overline{E13}$
H	H	H	L	L	$\overline{E14}$
H	H	H	H	L	$\overline{E15}$

H = high level, L = low level, X = irrelevant
$\overline{E0}$, $\overline{E1}$ . . . $\overline{E15}$ = the complement of the level of the respective E input

Fig. 5-2 Function table for the 74150. (Courtesy of Texas Instruments, Inc.)

less of the address or data input states. If the strobe is brought low, a high will be placed on the AND gates and circuit action will be enabled. The strobe is sometimes called the *enable input*.

The address inputs and their complements connect to the 16 AND gates in varying patterns. This results in each AND gate having its four inputs high only when the binary address for that AND gate is applied to the four address inputs. When a gate is addressed and the strobe is low, the output of the gate will follow the selected input and the data selector output will be the complement of the selected input. Thus, a 4-bit binary code can be used to select one of 16 lines. The 74150 is a 24-pin DIP. Pin-outs are shown in Appendix C. The function table for the 74150 is shown in Fig. 5-2.

Selection of one of 16 inputs can be achieved with a data selector only half as large as the 74150. The 74151, for instance, is a 1-of-8 data selector. When used with its direct address inputs, the 74151 will select one of 8 inputs based on the three-digit binary number on the address lines. The circuit action is exactly the same as that of the 74150 except that the number of inputs has been reduced by half, the address inputs have been reduced by one, and a noncomplemented output has been added. The 74151 is constructed as a 16-pin DIP.

How can a 1-of-8 data selector be made to function as a 1-of-16 data selector? The answer is a technique called *folding*. A data selector provides an output based on the input data and the input address. Therefore, the effective number of inputs can be increased if the inputs are altered in an appropriate manner. Compare the function tables for the 74150 and the 74151, which are shown in Fig. 5-3. What is desired is that the output of the 8-to-1 selector agree with the output of the 16-to-1 selector. This means that the inputs to the 8-to-1 selector will require adjusting to obtain agreement on the extra eight inputs. An arbitrary input data pattern is used on the inputs. The 74150 provides the complement of this data pattern at its output. Since the 74151 has both a noncomplemented output and a complemented output, either can be chosen to use. The complemented output is the easier of the two for this application. The next step is to compare the octal pairs of address inputs. Notice that the A, B, C address pattern is repeated to obtain the 16 inputs. If the Q of the first address agrees with the Q of its matching address, both can be accommodated by applying the appropriate H or L input for both. Consider addresses 2 and 10. The three-digit addresses are the same, and the required outputs are the same. A high can be wired to these inputs, and no change will be required. Addresses 1 and 9 are different, however. The three-digit addresses are the same, but the required outputs are complementary. How can these different outputs be accommodated since only three address inputs are available? The answer is folding.

The D-address input can be used as the input to these two addresses. The result is that when D is low and the three-digit address is correct, $\overline{Q}$ will be high. This satisfies the condition of address 1. When the same three-digit address is called and D is high, $\overline{Q}$ will be low. This satisfies address 9. Addresses 1 and 9 use the uncomplemented D. The state of $\overline{Q}$

74150

	INPUTS					OUT
	DATA	SELECTOR				
		D	C	B	A	$\overline{Q}$
0	1	L	L	L	L	0
1	0	L	L	L	H	1
2	1	L	L	H	L	0
3	1	L	L	H	H	0
4	0	L	H	L	L	1
5	0	L	H	L	H	1
6	1	L	H	H	L	0
7	1	L	H	H	H	0
8	0	H	L	L	L	1
9	1	H	L	L	H	0
10	1	H	L	H	L	0
11	0	H	L	H	H	1
12	0	H	H	L	L	1
13	0	H	H	L	H	1
14	1	H	H	H	L	0
15	1	H	H	H	H	0

(A)

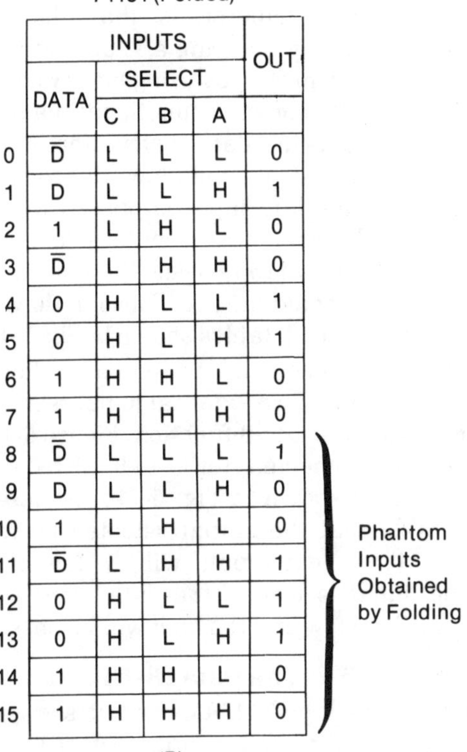

74151 (Folded)

	INPUTS				OUT
	DATA	SELECT			
		C	B	A	
0	$\overline{D}$	L	L	L	0
1	D	L	L	H	1
2	1	L	H	L	0
3	$\overline{D}$	L	H	H	0
4	0	H	L	L	1
5	0	H	L	H	1
6	1	H	H	L	0
7	1	H	H	H	0
8	$\overline{D}$	L	L	L	1
9	D	L	L	H	0
10	1	L	H	L	0
11	$\overline{D}$	L	H	H	1
12	0	H	L	L	1
13	0	H	L	H	1
14	1	H	H	L	0
15	1	H	H	H	0

Rows 8–15: Phantom Inputs Obtained by Folding

(B)

Fig. 5-3 (a) Truth table for the 74150 and (b) truth table for the 74151 with folding.

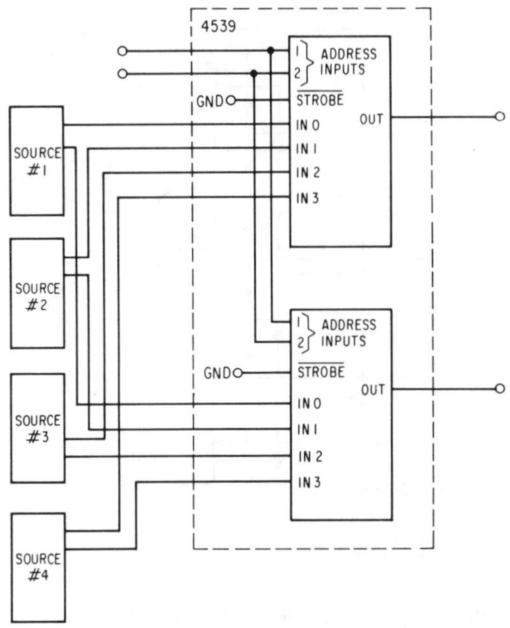

Fig. 5-4 Circuit diagram of the 4539 for selecting 2 bits from one of four 2-bit sources.

agrees with the truth table. Addresses 0 and 8 also are complementary. In this case, however, the uncomplemented D will not provide the correct output. Therefore, the complement of D is applied to these addresses. Only one other octal pair has complementary outputs. They are addresses 3 and 11. The D complement is required for their outputs to be correct. With these connections, the 74151 8-to-1 data selector has been converted to a 16-to-1 data selector.[1] By the same token a 16-to-1 data selector can be converted to a 32-to-1 selector and a 4-to-1 selector can accommodate eight inputs. The requirement is that there must be only one more address line than address inputs. When this requirement is met, input folding can be used.

Other data selectors are available. The CMOS 4539 is a dual four-input to one-output device. Since it is constructed as a 16-pin DIP, the select functions are shared. This may create difficulties for some applications that require two totally independent data selectors. The output of each selector assumes the state of the input that is addressed by binary code using pins 14 and 2. Both selectors will be addressed at the same time. The strobe inputs are individual with pin 1 controlling pin 7 and pin 15 controlling pin 9. The 4539 can be used to select 2 bits from one of four 2-bit sources. This application is demonstrated in Fig. 5-4. The 74157 functions similarly to the 4539, providing a 4-bit output selected from two 4-bit

[1]Folding can be used only when the input code is set and does not change, which is true of look-up tables.

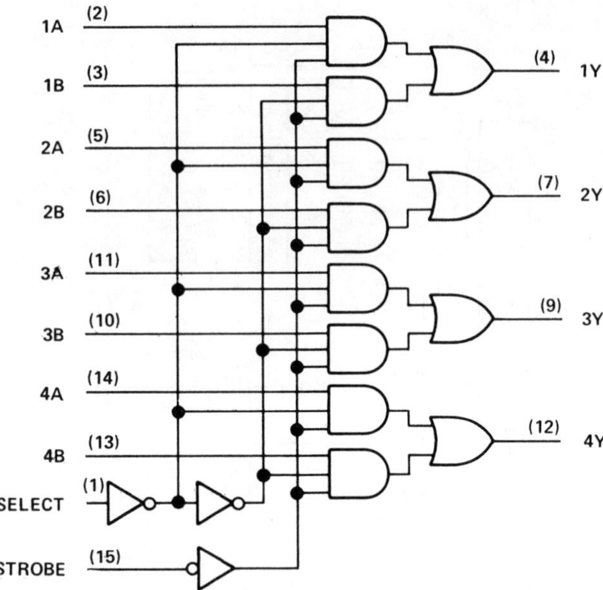

Fig. 5-5 Functional block diagram of the 74157 that provides a 4-bit output selected from two 4-bit sources. (Courtesy of Texas Instruments, Inc.)

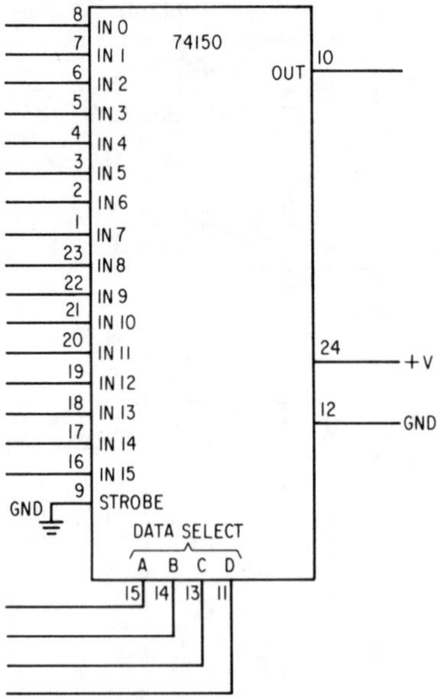

Fig. 5-6 A 74150 test circuit.

sources (see Fig. 5-5). The strobe is connected to all four gates and forces the outputs low when it is high. Addressing of the 74157 is accomplished using a single bit. If the select input is low, the outputs will follow the "A" input. If the select input is high, the outputs will follow the "B" inputs.

Experiment 5-1: The Data Selector

Use the breadboard to construct the circuit shown in Fig. 5-6 using a 74150. Apply data to the data inputs according to the truth table code shown in Fig. 5-7a. Vary the address lines in a binary fashion and record the output state in the truth table. The data selector action is evident from this experiment.

Add a fifth address line and develop a folded input scheme to accommodate the 32-bit code in the truth table shown in Fig. 5-7b. Vary the address inputs and verify that the outputs in the truth table are correct. If difficulty is encountered in the input-folding process, refer to Fig. 5-8.

Data Distributors

The data selector chooses one of "n" number of inputs and passes the binary value of that input to one output. The *data distributor* reverses this process and places the binary value of one input onto one of "n" number of outputs. The function of these devices is analogous to the function of a rotary switch (see Fig. 5-9). Just as a rotary switch is set to connect one of its inputs to its output, the data selector connects one of its inputs to its output (see Fig. 5-9b). Also, the rotary switch shown in Fig. 5-9c connects its input to one of its outputs, as does the data distributor shown in Fig. 5-9d. The major difference between the rotary switch and the data selector and distributor is that the switch is operated manually by using a knob and the ICs are operated electronically by using binary codes.

Figure 5-10a shows a functional diagram of the 74155 dual four-line data distributor. For half the device, with the strobe held low, the complement of the data on the data input is passed to the output selected by the binary code on the select inputs. The other half of the 74155 passes the uncomplemented data to the selected output. These actions are summarized in the truth tables shown in Fig. 5-10b.

The 74155 also can be used as a 1-line to 8-line data distributor. Two external jumpers are required to implement the 1-to-8 function. Figure 5-11a shows the 74155 connected for 1-to-8 distribution. The truth table is shown in Fig. 5-11b. Notice from the truth table that the output selected by the binary code at the select inputs will go low if the data (strobe) input is low. If the data input is high, no output will be low and the high will be passed on.

Experiment 5-2: The Data Distributor

Breadboard the 74154 circuit as shown in Fig. 5-12a. Apply binary codes to the select inputs of the 74154 as required by the truth table shown in Fig. 5-12b. Record the outputs with the data line held low.

What function is the 74154 fulfilling? The 74154 in this circuit is a 1-to-16 data distributor.

	DATA	SELECT D	C	B	A	$\overline{Q}$
0	1	L	L	L	L	
1	0	L	L	L	H	
2	1	L	L	H	L	
3	1	L	L	H	H	
4	0	L	H	L	L	
5	1	L	H	L	H	
6	0	L	H	H	L	
7	1	L	H	H	H	
8	1	H	L	L	L	
9	0	H	L	L	H	
10	1	H	L	H	L	
11	0	H	L	H	H	
12	0	H	H	L	L	
13	1	H	H	L	H	
14	1	H	H	H	L	
15	0	H	H	H	H	

(A)

	DATA	SELECT D	C	B	A	$\overline{Q}$
0	1	L	L	L	L	
1	0	L	L	L	H	
2	1	L	L	H	L	
3	1	L	L	H	H	
4	0	L	H	L	L	
5	1	L	H	L	H	
6	0	L	H	H	L	
7	1	L	H	H	H	
8	1	H	L	L	L	
9	0	H	L	L	H	
10	1	H	L	H	L	
11	0	H	L	H	H	
12	0	H	H	L	L	
13	1	H	H	L	H	
14	1	H	H	H	L	
15	0	H	H	H	H	
16	0	L	L	L	L	
17	1	L	L	L	H	
18	1	L	L	H	L	
19	0	L	L	H	H	
20	1	L	H	L	L	
21	1	L	H	L	H	
22	0	L	H	H	L	
23	0	L	H	H	H	
24	1	H	L	L	L	
25	1	H	L	L	H	
26	1	H	L	H	L	
27	0	H	L	H	H	
28	1	H	H	L	L	
29	0	H	H	L	H	
30	1	H	H	H	L	
31	0	H	H	H	H	

(B)

Fig. 5-7 (a) Truth table for the 74150 and (b) truth table with 32 data bits.

	DATA	$\overline{Q}$
0	$\overline{E}$	0
1	E	1
2	1	0
3	$\overline{E}$	0
4	E	1
5	1	0
6	0	1
7	$\overline{E}$	0
8	1	0
9	E	1
10	1	0
11	0	1
12	E	1
13	$\overline{E}$	0
14	1	0
15	0	1
16	$\overline{E}$	1
17	E	0
18	1	0
19	$\overline{E}$	1
20	E	0
21	1	0
22	0	1
23	$\overline{E}$	1
24	1	0
25	E	0
26	1	0
27	0	1
28	E	0
29	$\overline{E}$	1
30	1	0
31	0	1

Fig. 5-8 Data and output of folded 74150.

Multiplexing

A data selector can be used to reduce the number of bits required to represent "n" number of input lines. For example, the binary value of any one of 16 inputs can be determined by using a 4-bit address code. Using this process, the values of the 16 parallel lines can be placed on 1 line in serial fashion. The process of combining several lines onto a lesser number of lines is called *multiplexing*. In this case, the data on 16 lines is multiplexed onto 1 line. Used in this way, a data selector becomes a multiplexer.

Multiplexing can be reversed. The multiplexed serial stream of data is demultiplexed, and the appropriate values of the 16 output lines are recovered. The process is essentially an encoding and a decoding process. For this reason, data selectors used for multiplexing are often referred to as *encoders* or *multiplexers*. Data distributors are often called *decoders* or *demultiplexers*.

The operation of a decoder/demultiplexer will not necessarily differ from that of a data distributor. The difference is in the purpose of the device rather than the way it functions. The circuit of the 74154

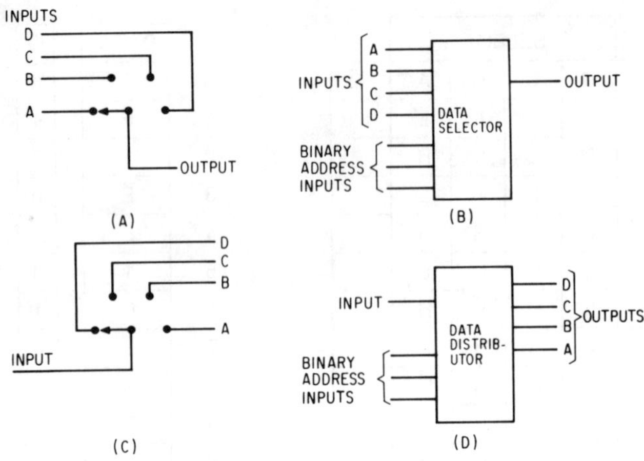

Fig. 5-9 Comparison of the data selector and distributor to rotary switches: (a) Four-input to single-output switch, (b) one-of-four data selector, (c) one-input to four-output switch, and (d) one to one-of-four data distributor.

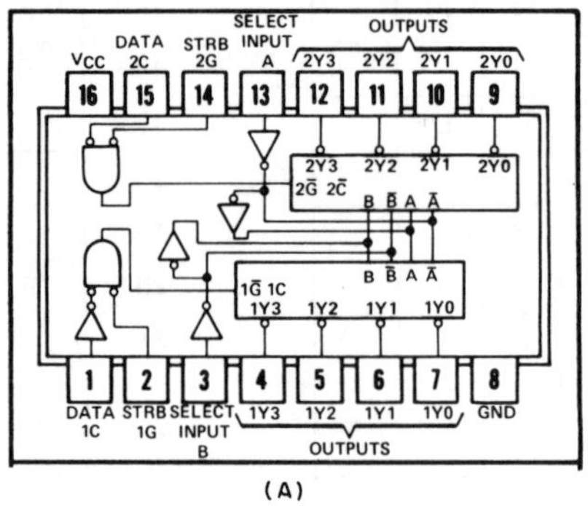

FUNCTION TABLES
2-LINE-TO-4-LINE DECODER
OR 1-LINE-TO-4-LINE DEMULTIPLEXER

INPUTS				OUTPUTS			
SELECT		STROBE	DATA	1Y0	1Y1	1Y2	1Y3
B	A	1G	1C				
X	X	H	X	H	H	H	H
L	L	L	H	L	H	H	H
L	H	L	H	H	L	H	H
H	L	L	H	H	H	L	H
H	H	L	H	H	H	H	L
X	X	X	L	H	H	H	H

INPUTS				OUTPUTS			
SELECT		STROBE	DATA	2Y0	2Y1	2Y2	2Y3
B	A	2G	2C				
X	X	H	X	H	H	H	H
L	L	L	L	L	H	H	H
L	H	L	L	H	L	H	H
H	L	L	L	H	H	L	H
H	H	L	L	H	H	H	L
X	X	X	H	H	H	H	H

(B)

Fig. 5-10 The 74155: (a) functional diagram and (b) truth tables summarizing its actions. (Courtesy of Texas Instruments, Inc.)

used in Experiment 5-2 requires only one change to convert it into a 4-line to 16-line decoder or demultiplexer (binary to hexadecimal). If the input marked "data input" in Fig. 5-12a is left low, the 74154 will

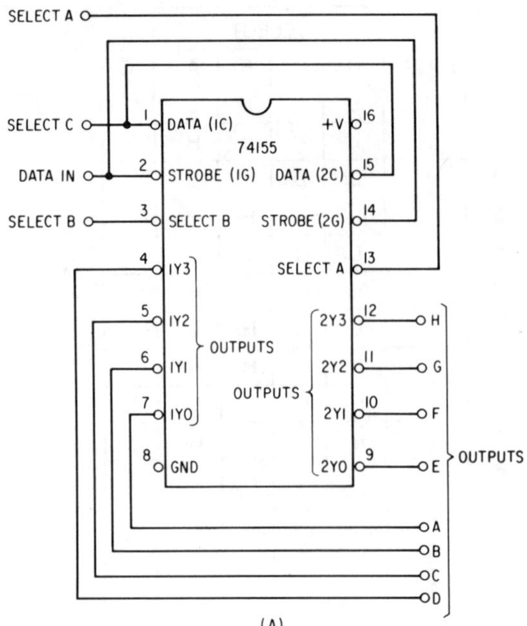

FUNCTION TABLE
3-LINE-TO-8-LINE DECODER
OR 1-LINE-TO-8-LINE DEMULTIPLEXER

INPUTS				OUTPUTS							
SELECT			STROBE OR DATA	(0)	(1)	(2)	(3)	(4)	(5)	(6)	(7)
C†	B	A	G‡	2Y0	2Y1	2Y2	2Y3	1Y0	1Y1	1Y2	1Y3
X	X	X	H	H	H	H	H	H	H	H	H
L	L	L	L	L	H	H	H	H	H	H	H
L	L	H	L	H	L	H	H	H	H	H	H
L	H	L	L	H	H	L	H	H	H	H	H
L	H	H	L	H	H	H	L	H	H	H	H
H	L	L	L	H	H	H	H	L	H	H	H
H	L	H	L	H	H	H	H	H	L	H	H
H	H	L	L	H	H	H	H	H	H	L	H
H	H	H	L	H	H	H	H	H	H	H	L

†C = inputs 1C and 2C connected together
‡G = inputs 1G and 2G connected together
H = high level, L = low level, X = irrelevant

(B)

Fig. 5-11 The 74155: (a) wiring diagram and (b) truth table for 1-to-8 distribution. (Courtesy of Texas Instruments, Inc.)

decode the four-digit binary input to force one of 16 outputs low. Thus, the 4-bit code is demultiplexed, restoring the original 16 values.

The 4539 dual 4-bit data selector discussed earlier can be used as an 8-to-1 multiplexer. The circuit shown in Fig. 5-13a will accomplish this function. Since the two halves of the 4539 share select inputs, all that is required is provision of a third select input. This is easily accomplished by using the strobe inputs with an inverter connected between them. In operation the circuit will follow the truth table shown in Fig. 5-13b. One of the eight input lines is selected by the binary code on the three select inputs. The half of the IC that has a high strobe will have a low output regardless of the select input. For the codes 000, 001, 010, and 011, the second half will have a high strobe and therefore a low output. For the codes 100, 101, 110, and 111, the first half of the 4539 will ignore the select inputs and provide a low output. The result is that the output will be the complement of the binary state of the one input line from among the eight that is selected by the three-digit code on the select and strobe inputs. If the three-digit code is incremented in a binary sequence, the device will convert the 8 parallel bits into an 8-bit serial data stream.

Multiplexing coding techniques are frequently used to convert from one numbering system to another. The functioning of the ICs is the same as those just discussed except that the nature of the inputs and the number of the outputs are different. The 4028, for instance, is a BCD-to-decimal decoder. A four-digit address that conforms to a binary coded decimal encoding is applied to the select inputs. Depending on the value of the BCD code, one of ten outputs goes high. As shown in the truth table (see Fig. 5-14), use of the binary codes above 1001 (decimal 9) will result in disallowed states. The truth table also shows that the 4028 can be used as an octal to one-of-eight decoder. If select input "8" (D input—pin 11) is grounded, the unit will force one of the eight outputs high as required by a three-digit binary code placed on select inputs "1" (A input), "2" (B input), and "4" (C input).

Display Decoders/Drivers

Decoders also are commonly used to drive displays. The 7447 is typical of decoders/drivers that are used with seven-segment displays. The use of decoders/drivers with LED displays was discussed briefly in an earlier chapter. A fuller discussion is in order at this point.

An LED is simply a diode that produces light upon the application of voltage. The LED shares several characteristics with other diodes. It permits the flow of electricity in the forward direction and opposes it in the reverse direction. For the LED to emit light, it must have voltage applied in the forward direction. In common with other diodes, the

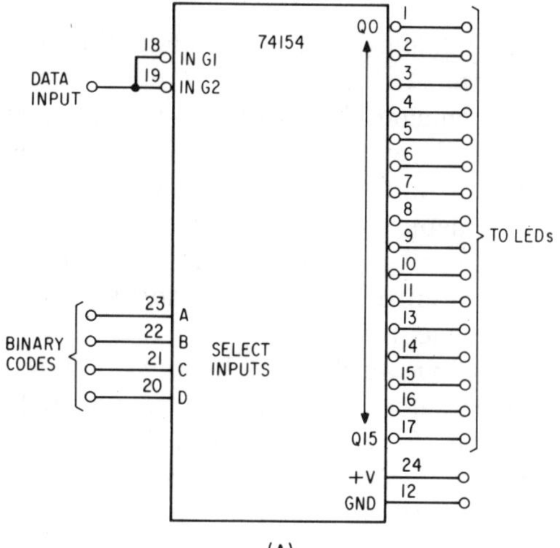

74154

INPUTS					OUTPUTS															
DATA	D	C	B	A	0	1	2	3	4	5	6	7	8	9	10	11	12	13	14	15
L	L	L	L	L																
L	L	L	L	H																
L	L	L	H	L																
L	L	L	H	H																
L	L	H	L	L																
L	L	H	L	H																
L	L	H	H	L																
L	L	H	H	H																
L	H	L	L	L																
L	H	L	L	H																
L	H	L	H	L																
L	H	L	H	H																
L	H	H	L	L																
L	H	H	L	H																
L	H	H	H	L																
L	H	H	H	H																

(B)

Fig. 5-12 (a) 74154 data distributor and (b) truth table.

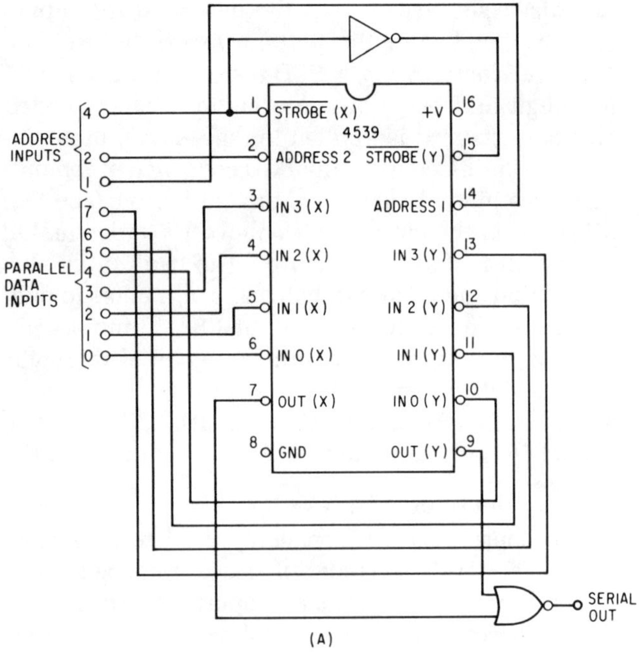

LINE NO.	ADDRESS			OUT
	STROKE	2	1	
1	0	0	0	$\overline{0}$(x)
2	0	0	1	$\overline{1}$(x)
3	0	1	0	$\overline{2}$(x)
4	0	1	1	$\overline{3}$(x)
5	1	0	0	$\overline{0}$(y)
6	1	0	1	$\overline{1}$(y)
7	1	1	0	$\overline{2}$(y)
8	1	1	1	$\overline{3}$(y)

(B)

Fig. 5-13 (a) 4539 wiring diagram and (b) truth table.

INPUT				OUTPUT									
D	C	B	A	Q9	Q8	Q7	Q6	Q5	Q4	Q3	Q2	Q1	Q0
0	0	0	0	0	0	0	0	0	0	0	0	0	1
0	0	0	1	0	0	0	0	0	0	0	0	1	0
0	0	1	0	0	0	0	0	0	0	0	1	0	0
0	0	1	1	0	0	0	0	0	0	1	0	0	0
0	1	0	0	0	0	0	0	0	1	0	0	0	0
0	1	0	1	0	0	0	0	1	0	0	0	0	0
0	1	1	0	0	0	0	1	0	0	0	0	0	0
0	1	1	1	0	0	1	0	0	0	0	0	0	0
1	0	0	0	0	1	0	0	0	0	0	0	0	0
1	0	0	1	1	0	0	0	0	0	0	0	0	0
1	0	1	0	0	0	0	0	0	0	0	0	0	0
1	0	1	1	0	0	0	0	0	0	0	0	0	0
1	1	0	0	0	0	0	0	0	0	0	0	0	0
1	1	0	1	0	0	0	0	0	0	0	0	0	0
1	1	1	0	0	0	0	0	0	0	0	0	0	0
1	1	1	1	0	0	0	0	0	0	0	0	0	0

Fig. 5-14 Truth table for the 4028. (Courtesy of Motorola, Inc.)

LED will be destroyed if the forward voltage reaches a level that results in excessive current flow. Damage to the device also will result if excessive reverse voltage is applied. The data shown in Fig. 5-15 is typical for LEDs.

An LED display is simply a collection of individual LEDs that are packaged to simulate numbers. The most common LED display is the seven-segment display. As shown in Fig. 5-16, the seven LEDs are arranged so that turning them on in selected patterns will display different numerals. The numeral "8" requires all seven LED segments to be lit. Other numerals require less—the numeral "1" requires only two segments to be lit.

The 7447 drives the LED display through voltage-dropping resistors. An LED has a typical forward voltage drop of 1.7 to 2.2 V. The resistor is used to drop the remaining supply voltage. Resistors are available in a 14-pin DIP. These are useful in simplifying design interconnection.

The 7447 accepts a BCD input as required for the indicated numeral. In addition, the 7447 provides for a lamp test, pin 3, which may be used to light all seven segments simultaneously. Two blanking pinouts provide for blanking of leading and trailing 0s. The blanking output, pin 4, will blank the entire display if brought low. This feature is used on occasion to reduce the brightness of the display by pulsing the blanking output. As the length of time that the display is off increases, the display appears to dim.

The LED has gained great popularity in recent years due to its compatibility with ICs and its convenient voltage range. Even so, other displays are in use. The display that was the standard for several

Type	Color: Light	Color: Lens	Luminance Intensity	View Angle	Forward Voltage @20 mA	Reverse Voltage @100 mA	Max. Forward Current	Max. Power Dissipation
4-43B (T-3/4)	Red	Red diffused	1.0 mcd	90°	1.7 V	5.0 V	50 mA	100 mW
4-50A (T-1)	Red	Clear trans.	1.4 mcd	90°	1.65 V	5.0 V	40 mA	80 mW
4-73A (T-1 3/4	Red	Red diffused	2.0 mcd	70°	1.7 V	5.0 V	50 mA	100 mW
4-344B (T-1 3/4)	Green	Green diffused	4.0 mcd	65°	2.0 V	5.0 V	35 mA	105 mW
4-482B (T-1 3/4)	Orange	Orange trans.	40.0 mcd	28°	2.0 V	5.0 V	35 mA	105 mW
4-584B (T-1 3/4)	Yellow	Yellow trans.	6.0 mcd	65°	2.1 V	5.0 V	35 mA	105 mW

Fig. 5-15 Typical LED specifications.

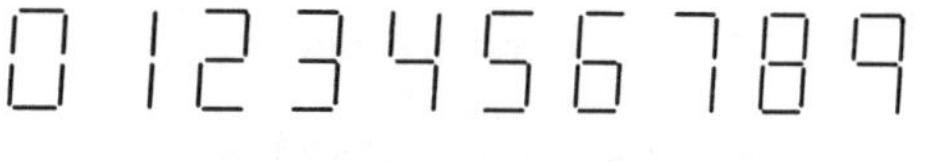

0 1 2 3 4 5 6 7 8 9

Fig. 5-16 Seven-segment LED numerals.

years is the neon display. The neon display is best known by the name the Burroughs Corporation made famous—Nixie.® The neon display uses the gas discharge principle. The tube contains electrodes shaped like the numerals 0–9 surrounded by neon gas. When a voltage of sufficient level is placed on the element, the neon gas around it is ionized and emits a glow. The electrodes are very thin outlines of the numbers. When the electrodes are placed one in front of the other, the glowing electrode is not significantly obscured by the others. Voltages for neon displays range from 140 to 170 V. A 1-of-10 decoder/driver is required for the neon display.

IC decoder/drivers also can be used with neon display tubes. The 7445 is a BCD-to-decimal decoder that is capable of handling increased current and higher voltages than the 7447. The decimal output requirement precludes inclusion of blanking and lamp test pin-outs. The 74142 includes a counter, a latch, and a decoder within a 16-pin DIP. Such complexity in a device of this size demonstrates the capabilities of MSI fabrication techniques.

The Liquid Crystal Display (LCD) is becoming popular because of its small current drain. The LCD does not emit light—ac voltage is used to alter the LCD so that it changes its light transmission characteristics. The polarization or reflectivity of the display will be altered, depending on the type of LCD, field effect, or dynamic scattering, and the digit will appear different from the surrounding crystal area. Since the current drain is very low, the LCD display can be left on continuously without concern about battery drain. Use of the LCD in the dark is not possible without supplementary lighting. LCD drivers are often included on watch ICs. Back lighting in the form of phosphorescence or low-power light provides for nighttime viewing.

Experiment 5-3: Multiplexing/Demultiplexing

Using a 74150, a 74154, a 74191, and the breadboard astable, design and breadboard a circuit that will fulfill the following requirements:

1. A 16-bit code will be multiplexed into a single serial stream of data.
2. The single serial stream of data will be demultiplexed and the 16 bits recovered.
3. The multiplexer and demultiplexer will be driven by a common binary counter.
4. The binary counter will be clocked at the rate of one pulse per second or slower.

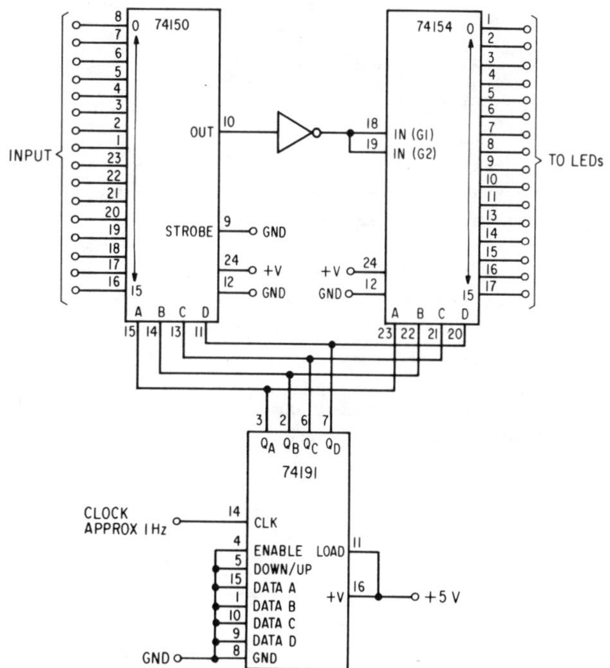

Fig. 5-17 Encoder/decoder test circuit.

5. Display the demultiplexed outputs and the clock to confirm that the output code agrees with the input code.

After the circuit is breadboarded, check it to be sure all of the design criteria were met. If problems are encountered, refer to Fig. 5-17 and compare that design with yours.

Transmission Gates

The CMOS family includes an interesting and useful set of gates called *transmission gates*. Transmission gates are capable of bidirectional operation. This results in increased versatility for devices such as the 4051, 4052, and 4053. These devices are multiplexers/demultiplexers. Figure 5-18 shows a functional diagram for the 4051. Notice that the 1-of-8 decoder connects to the eight transmission gates. A 3-bit code placed on the A, B, and C inputs will enable one of these gates. When enabled, the gate functions as a low-value resistor (typically 120 Ω for the 4051 with a 10-V supply). Signals can pass in either direction across the transmission gate. If data passes from one of the eight lines to the single line, the unit functions as a multiplexer or data selector. If the data passes from the single line to one of the eight lines, the unit functions as a demultiplexer or data distributor. When not enabled, the transmission gate presents a high resistance to the flow of data.

Transmission gates are often used for analog switching. For analog switching, the supply voltage

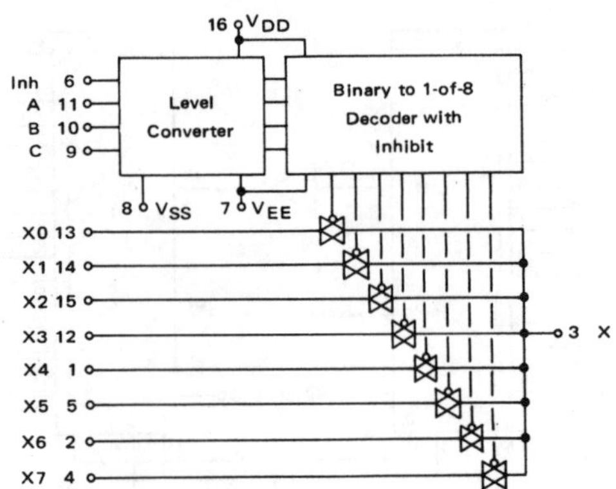

Fig. 5-18 Functional diagram of the 4051 multiplexer/demultiplexer. (Courtesy of Motorola, Inc.)

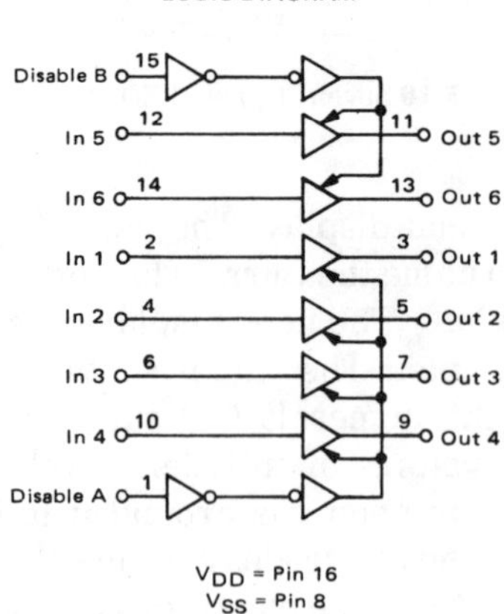

Fig. 5-20 The 4503B tri-state buffer. (Courtesy of Motorola, Inc.)

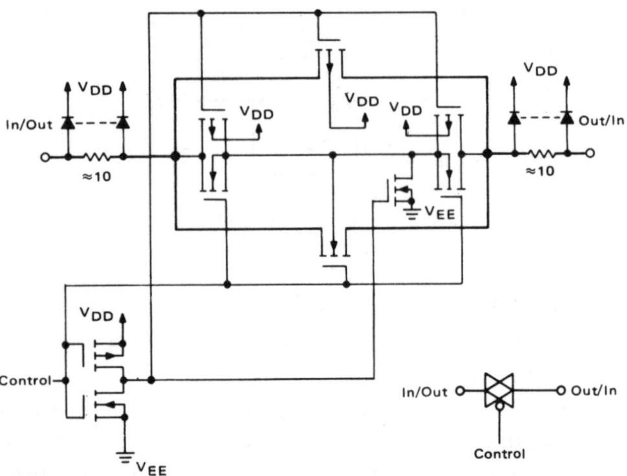

Fig. 5-19 CMOS transmission gate circuit. (Courtesy of Motorola, Inc.)

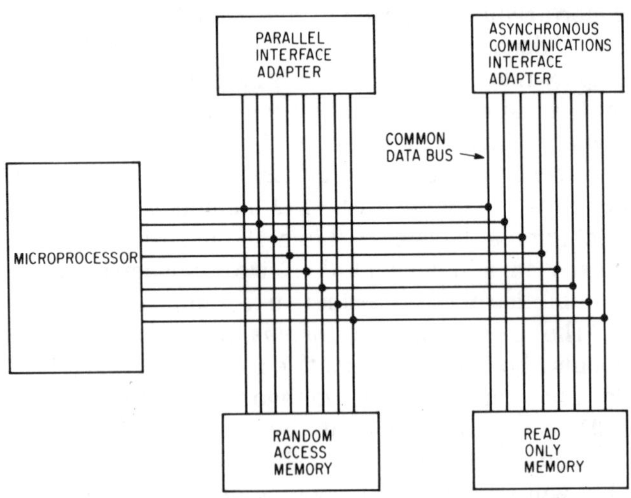

Fig. 5-21 Common bus MPU system.

must be altered slightly from the digital connections, and the analog signal must be held within the limits of the supply voltage to the device.

The circuit of the CMOS transmission gate is shown in Fig. 5-19. When an enable signal is present at the control input, the gates are biased into an on condition. When the enable signal is not present at the control input, the gates remain off, and the two data lines are effectively disconnected.

Tri-State Logic

Transmission gates have some application in common bus systems when peripheral devices both receive and supply data. However, there are limitations of the transmission gate that make an alternative logic beneficial. Tri-state logic permits the combining of several data sources onto one data bus. These data sources are selectively enabled so that only one source at a time is supplying data to the bus. This is necessary because the outputs of some logic families cannot be connected in the wired-OR configuration. The outputs of CMOS devices, for example, will conflict if interconnected. A high output and a low output will each attempt to achieve the normal state levels, and an undefined output will result.

Tri-state logic precludes this problem. Tri-state logic presents a high impedance to the line it is supplying to or receiving signals from, if the tri-state device is disabled. When enabled, the tri-state device operates as its traditional counterpart; i.e., a tri-state inverter operates as a standard inverter. Figure 5-20 shows the circuit diagram for one of the six tri-state buffers of a 4503B. When a high disable signal is present at the disable input, the complementary pair is placed into a high impedance (cutoff) state. When the disable input is low, the buffer is free to follow the input data.

Figure 5-21 shows a block diagram of a typical common bus system. Tri-state interfaces permit the several devices to use one set of buses. Each tri-state

ENABLE $\overline{G}$	DIRECTION CONTROL DIR	OPERATION
L	L	B data to A bus
L	H	A data to B bus
H	X	Isolation

H = high level, L = low level, X = irrelevant

Fig. 5-22 Function table for the 74LS245. (Courtesy of Texas Instruments, Inc.)

interface is enabled selectively so that only one device is connected to the buses at a time.

The 74LS245 is an improvement of the tri-state interface that incorporates the advantages of the transmission gate. The 74LS245 is an octal bus transceiver with tri-state outputs. It adds bidirectional operation to the tri-state device. This bidirectional operation is capable of control. Figure 5-22 shows the function table for the device. The tri-state function is controlled by the enable input. If the enable input is low, the device is enabled for normal transceiver operation. If the enable is high, the gates assume a high impedance condition. As shown in the function table, the direction control input, pin 1, determines whether data flows from A to B or from B to A. A low on both enable and direction control inputs will result in data flow from B to A. With enable low and direction control high, data will flow from A to B.

A similar device that has increased control functions is the 74S428. This device is designed for use with the 8080A microprocessor. It incorporates a controller with the bus transceiver. Figure 5-23 shows the 74S428 in an 8080A microprocessor system. Figure 5-24 defines the abbreviations used with the 74S428. A full discussion of the microprocessor system operation will not be given at this point. A brief explanation of the system operation is necessary to explain the data bus transceiver operation and will provide a foundation for later discussion.

The 8080A microprocessor operates according to processing cycles. The 8080A and all of the devices included in the microprocessor system are synchronized by the 8224 clock generator/driver. The microprocessor then performs functions based on commands and data that it receives. The commands are usually stored in a program memory that is addressed by the address bus. Each word of program instruction is addressed and then placed on the data buses. For the instruction to reach the 8080A, it must pass through the 74S428 bus transceiver. After the instruction is received, the 8080A will require data that also must be addressed and passed

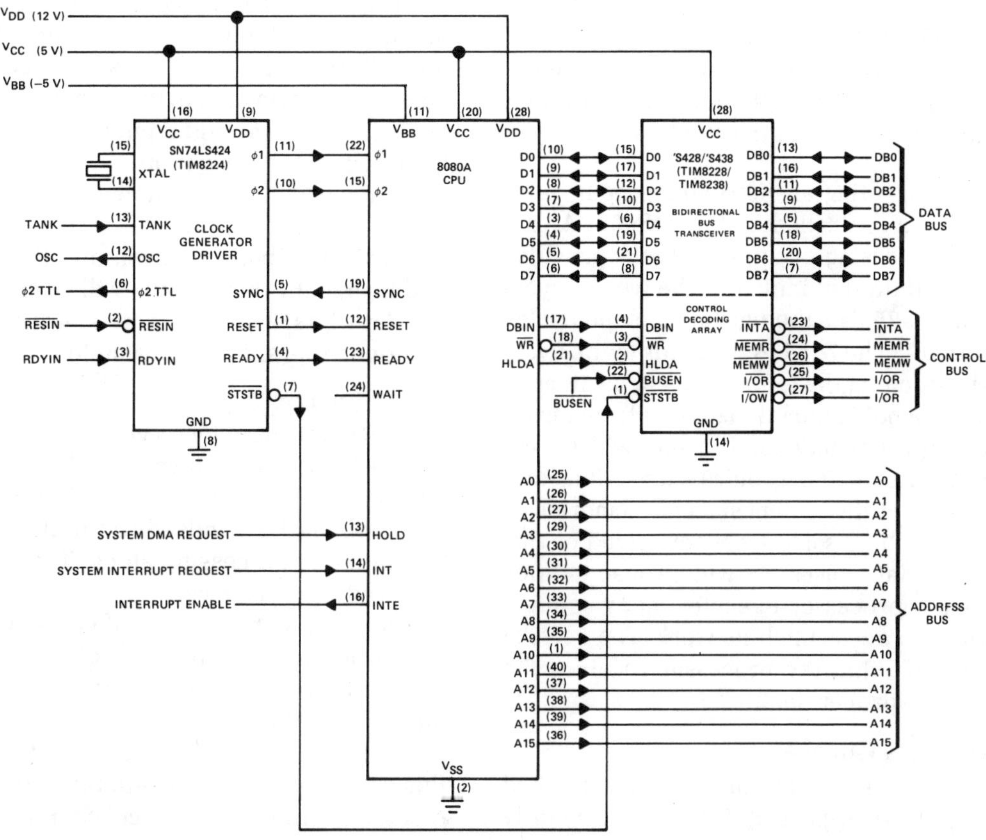

Fig. 5-23 The 74S428/8080A system. (Courtesy of Texas Instruments, Inc.)

PIN DESIGNATIONS

DESIGNATION	PIN NOS.	FUNCTION
D0 thru D7	15, 17, 12, 10, 6, 19, 21, 8	BIDIRECTIONAL DATA PORT (TO TMS 8080A)
DB0 thru DB7	13, 16, 11, 9, 5, 18, 20, 7	BIDIRECTIONAL DATA PORT (TO SYSTEM BUS)
$\overline{I/OR}$	25	READ OUTPUT TO I/O (ACTIVE LOW)
$\overline{IO/W}$	27	WRITE OUTPUT TO I/O (ACTIVE LOW)
$\overline{MEMR}$	24	READ OUTPUT TO MEMORY (ACTIVE LOW)
$\overline{MEMW}$	26	WRITE OUTPUT TO MEMORY (ACTIVE LOW)
DBIN	4	INPUT TO INDICATE TMS 8080A IS IN INPUT MODE (ACTIVE HIGH)
$\overline{INTA}$	23	INTERRUPT ACKNOWLEDGE OUTPUT (ACTIVE LOW)
HLDA	2	HOLD ACKNOWLEDGE INPUT (ACTIVE HIGH) FROM TMS 8080A
$\overline{WR}$	3	INPUT TO INDICATE TMS 8080A IS IN WRITE MODE (ACTIVE LOW)
$\overline{BUSEN}$	22	SYSTEM DATA PORT ENABLE INPUT (ACTIVE LOW)
$\overline{STSTB}$	1	SYNCHRONIZING STATUS STROBE INPUT FROM SN74LS424 (TIM8224)
V_{CC}	28	SUPPLY VOLTAGE (5 V)
GND	14	GROUND

Fig. 5-24 Pin designations used with the 74S428. (Courtesy of Texas Instruments, Inc.)

through the 74S428. After processing, the 8080A will output the result in the form of data that must pass through the 74S428 in the opposite direction and be routed to the destination addressed on the address bus.

The selective routing of instruction, input data, and output data is controlled by the 74S428. At the start of each processing cycle, the 8080A provides the 74S428 with data that indicates the type of cycle involved. Figure 5-25 lists the type of machine cycles used by the 8080A. The 74S428 responds to the cycle status data by latching the information in its storage latch and then decoding the latched data to generate appropriate control functions. The bus will be enabled to pass data in the required direction, and other components of the system such as memories or storage devices will be selectively connected or disconnected to the data buses. As will be seen in later chapters, a microprocessor executes many sequential instructions to accomplish its work. The 74S428 operates to ensure that the processor receives the proper data at the proper time.

Parity Generators/Checkers

The manipulation of data must be accomplished with accuracy. To transpose 1 bit is to significantly alter the data. When data is transmitted from one device to another, the possibility for errors is increased. Noise can be imposed on the data lines and be interpreted as a data bit. Complete accuracy is the goal in all data manipulation, but it is not always achieved. To safeguard against false data, a checking procedure is required. One method of checking accuracy is to send the data twice and compare the two sets of data to determine that they agree exactly. This is a very inefficient method of checking data accuracy.

Another method of checking data accuracy is to make all of the data words have an even number or all have an odd number of logic high bits. This method is called a *parity check*. If all words are to have even parity, the number of 1s in the word is first counted. If there is an even number of one bits, the word is transmitted as it is. If the word has an odd number of one bits, an extra bit (called a *parity bit*) is added to the data word. This results in the word having even parity. At the receive device, the parity is again checked. If the word has even parity, it is assumed to be accurate. If it has odd parity, the word is rejected and an error is reported to the transmitting station that will repeat the data. Odd parity also can be used for parity checking. The process is exactly the same as for even parity except that an odd number of one bits is used as the standard. The 74180 is typical of IC parity generator/checkers. Eight data bits are accommodated by the 74180. These are connected to the eight input lines. The use of either odd or even parity is possible. For even parity, the even input can be made high and the odd input low. This will result in a high at the even output if the data word has an even number of bits. Connecting the odd input high and the even input low will result in a high at the odd output. If both the even and the odd inputs are tied high, both sum outputs will go high regardless of the data bits. These actions are summarized in the function table shown in Fig. 5-26.

Data Storage

Electronic data by its very nature is transitory. It is generated. It is transmitted. It is acted upon (processed). At some point, however, data must be stored or it will be lost. Indeed, data storage is required within many processing steps. This storage is often referred to as *memory*.

In Chap. 3, the concept of memory was discussed in relation to R-S flip-flops. R-S flip-flops react to a data input by assuming some stable state. The data bit will set the flip-flop, and the flip-flop will "remember" that data bit until reset. The flip-flop in this case has stored the data bit. If the flip-flop is not reset, its output can be checked at a later time to obtain the value of the stored data bit.

STATUS WORDS

STATUS WORD	8080A STATUS OUTPUT								TYPE OF MACHINE CYCLE	'S428/'S438 COMMAND GENERATED
	D0	D1	D2	D3	D4	D5	D6	D7		
1	L	H	L	L	L	H	L	H	Instruction fetch	$\overline{MEMR}$
2	L	H	L	L	L	L	L	H	Memory read	$\overline{MEMR}$
3	L	L	L	L	L	L	L	L	Memory write	$\overline{MEMW}$
4	L	H	H	L	L	L	L	H	Stack read	$\overline{MEMR}$
5	L	L	H	L	L	L	L	L	Stack write	$\overline{MEMW}$
6	L	H	L	L	L	L	H	L	Input read	$\overline{I/OR}$
7	L	L	L	L	H	L	L	L	Output write	$\overline{I/OW}$
8	H	H	L	L	L	H	L	L	Interrupt acknowledge	$\overline{INTA}$
9	L	H	L	H	L	L	L	H	Halt acknowledge	NONE
10	H	H	L	H	L	H	L	L	Interrupt acknowledge at halt	$\overline{INTA}$
	INTA	$\overline{WO}$	STACK	HLTA	OUT	M1	INP	MEMR		
	STATUS INFORMATION									

Fig. 5-25 Status words for the 8080A. (Courtesy of Texas Instruments, Inc.)

FUNCTION TABLE

INPUTS			OUTPUTS	
Σ OF H's AT A THRU H	EVEN	ODD	Σ EVEN	Σ ODD
EVEN	H	L	H	L
ODD	H	L	L	H
EVEN	L	H	L	H
ODD	L	H	H	L
X	H	H	L	L
X	L	L	H	H

H = high level, L = low level, X = irrelevant

Fig. 5-26 Function table for the 74180. (Courtesy of Texas Instruments, Inc.)

INPUTS					OUTPUT
CLEAR	CLOCK	DATA ENABLE		DATA	Q
		G1	G2	D	
H	X	X	X	X	L
L	L	X	X	X	Q_0
L	↑	H	X	X	Q_0
L	↑	X	H	X	Q_0
L	↑	L	L	L	L
L	↑	L	L	H	H

When either M or N (or both) is (are) high the output is disabled to the high-impedance state; however sequential operation of the flip-flops is not affected.

Fig. 5-27 Function table for the 74173 4-bit D-type register. (Courtesy of Texas Instruments, Inc.)

For storage of small numbers of data bits, flip-flops or latches can be used. These devices have the benefit of being *static*, which means that they retain the stored bit without having to be periodically *refreshed*. Latches also are *volatile*. They can be changed (i.e., if power to the device is interrupted). A nonvolatile memory will retain the stored information permanently or in some cases until altered. This is useful for information that must be kept for long periods of time. For some nonvolatile storage devices, this is limiting in that the information once recorded cannot be altered. The CMOS 4043 and 4044 are quad R-S flip-flops with tri-state outputs. They can be used for storage in applications that do not require clocking. The tri-state enable control is common to all four flip-flops. The outputs present a high impedance to the load if the enable is low and follow the flip-flop outputs when enabled.

Registers

Shift registers were discussed in Chap. 4. These devices are useful for storage especially when serial-to-parallel conversion also is required. Some shift registers permit parallel loading as well as parallel output. The 74198 is typical of these. Eight bits of data are applied to the eight data inputs. Synchronous parallel loading is accomplished at the positive clock transition if both control lines (S0 and S1) are held high. The loaded data will then appear at the eight outputs. Parallel-to-serial data conversion also can be accomplished if desired.

Registers often combine functions. The 74298 is typical. The 74298, a quadruple two-input multiplexer with storage, is essentially an eight-line to four-line multiplexer that feeds R-S flip-flops. The data from source 1 will be chosen if the word select input is low when the clock input receives a negative-going transition. Data from source 2 will be selected if the word select input is high when the negative-going clock transition is received. Data can be changed on the inputs at any time. The data change will be passed to the outputs only at the negative clock transition.

Figure 5-27 shows the function table for the 74173 4-bit D-type register. The 74173 is a storage register only. Data select logic is not provided. However, it does provide tri-state outputs, making it suitable for common bus systems. Up to 128 74173s can be connected to a common bus and drive capability, for each device will remain adequate. Data is loaded

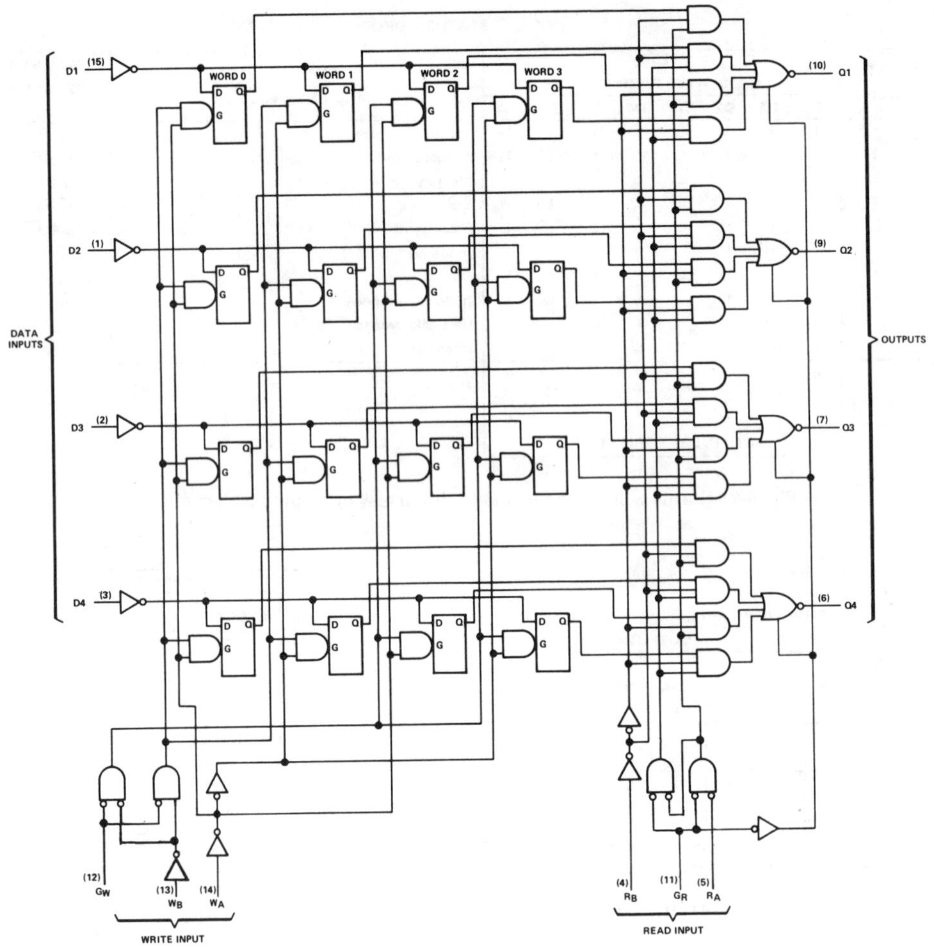

Fig. 5-28 Functional diagram of the 74LS670. (Courtesy of Texas Instruments, Inc.)

into the D-type flip-flops at the positive-going clock transition if both data enable inputs are low. If either or both of the output control inputs are high, the Q outputs will be disabled to a high-impedance state. If both output control inputs are low, the outputs will be the data last clocked into the flip-flop. Notice that the input and output control functions are independent of each other. The input can change even though the output is disabled. This type of register also is called a *bus buffer register*.

The *register file* is a device that incorporates several registers into one package. The 74LS670 is a register file that incorporates four registers of 4 bits each. The functional block diagram (Fig. 5-28) illustrates the complexity of the device. It incorporates the equivalent of 98 separate gates. The four registers, each composed of four D-type flip-flops, are apparent in the diagram. Notice that the read/write controls require three lines each, two address lines, and one enable line. The separate read/write controls permit simultaneous reading of one register and writing in a second. The output of the 74LS670 employs tri-state gating for use with common bus systems. The 74LS670 can be paralleled for increased word length and up to 120 outputs can be wired in an AND configuration to provide up to 512 words. If this amount of memory space is needed, however, use of other types of memory would be more economical and would simplify the design.

Experiment 5-4: The Register File

Breadboard the 74LS670 circuit shown in Fig. 5-29a. Clear registers zero and two by writing 0s in each location. This is accomplished by grounding the data inputs, addressing each register in turn, and enabling the write function (bring write enable, pin 12, low) after each address is applied. Place the code 1111 in registers one and three. The above process is followed with the data lines tied positive. Check for the written codes in the registers by addressing each register and bringing the read enable (pin 11) low. Remove and restore power to the 74LS670. Check each register again to see if the codes are as input. Write the data words shown in the table (see Fig. 5-29b) into the four registers. Read the four registers and confirm that the data stored is the same as the input data. Address register zero on both the read and write select lines. Tie the read input low, and note the data word located in the register. Encode 1010 on the data input and bring the write enable low briefly. Note that the change was read as it was accomplished. Address register one on the read select lines. Leave the write select at register zero. Encode 0101 on the input data lines. Connect the read and write enables to the same control line to enable them simultaneously. Return the read address to register zero and enable it. This will prove that the write function for register zero was accomplished while the read function for register one was in progress. From these experiments, the following characteristics of the 74LS670 register file can be identified:

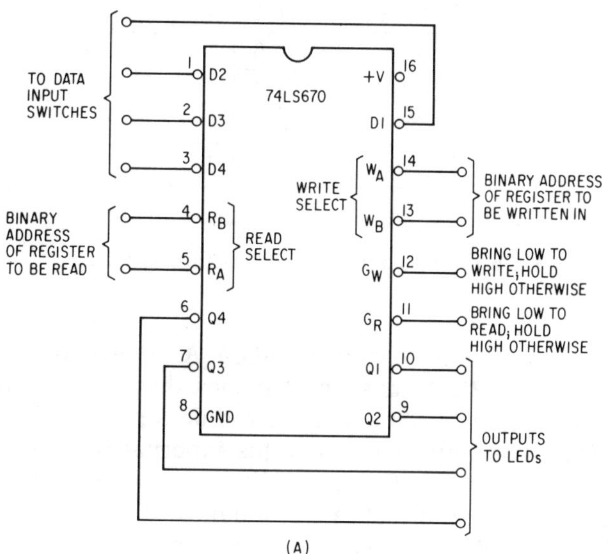

(A)

1. The read/write functions are independent.
2. The read function is nondestructive; i.e., the data is not disturbed by reading.
3. Random access to the registers is permitted.
4. Writing replaces the existing data with new data.
5. Storage is static; i.e., the data will be retained unless a new write command is received or the power is removed.

Registers are differentiated from memory primarily on the basis of size. A register is usually small. It is designed for temporary storage of data during processing. Registers are usually used as adjuncts to processors. An intermediate result may need to be stored while other processing is accomplished to obtain a result that will be used with the stored data. A mathematical calculation is an example of such an operation. Suppose the processor is to determine the result of the formula $(2 \times 3) + (2 \times 2)$. The processor will perform only one calculation at a time. It will multiply 2 and 3 and obtain the result 6. The 6 must be stored temporarily in a register so that the processor will be able to multiply 2 and 2 to obtain the result 4. The 6 can now be obtained from the register, and the addition can take place for a final result of 10. Registers would normally be used for this temporary storage due to speed considerations. It requires less time to write and read registers, especially if they are built into the processor chip, than it does to access memory.

Registers do not, however, provide for mass storage of data. Mass storage requires memory, of which there are several types. Memory can be volatile or nonvolatile. Nonvolatile memory is programmed permanently. It will retain the stored data

Test Data for the 74LS670

Input Data				Output Data			
REGISTER 00	REGISTER 01	REGISTER 10	REGISTER 11	REGISTER 00	REGISTER 01	REGISTER 10	REGISTER 11
0000	0000	0000	0000				
1111	1111	1111	1111				
0101	0110	1001	1010				
1010	0101	1010	0101				
1101	1011	0100	0010				

(B)

Fig. 5-29 The 74LS670: (a) wiring diagram and (b) table of test data.

when power is turned off. Read-Only Memory (ROM) is nonvolatile. ROM is used for program control in microprocessor and industrial control systems. Nonvolatile memory permits the processor to operate after power interruptions without reprogramming.

Volatile memory will not retain its stored data when power is removed. Loss of power is tantamount to erasing or at least disorganizing the memory. There is a benefit to this volatility. Volatile memory can be reprogrammed. The data within the memory can be altered and updated. For microcomputer applications that involve processing of data and storing the results of processing, volatile memory has distinct advantages. The microcomputer can alter the data as the program requires and store the altered data into memory for future use. The register file was described as a form of volatile memory. Another type of volatile memory is Random Access Memory (RAM).

Read-Only Memory

Read-Only Memory (ROM) permits reading of stored data but is not capable of alteration of the stored data by the electronic systems with which it is used. There are several types of memory that are classed as ROM. These types differ in the way that they are programmed. True read-only memory is programmed at the factory at the time of manufacture. The process involves the use of a mask that is designed to build in the data information required by the customer. Once this data is entered in the memory, it cannot be altered. Maskable ROM, as this is called, is the least expensive type of mass memory if large quantities of the same memory chips are involved. A manufacturer of a microprocessor-controlled cash register would want nonvolatile memory for his program that directs the cash register to perform all of its functions. If he intended to build 10,000 of the cash registers, maskable ROM would be the appropriate choice for memory.

Maskable ROM uses diodes for programming. The basic integrated circuit is constructed as described in Chap. 1, and a special mask is then used to create a metalization pattern that will connect or not connect p-n junctions created in the substrate. The metalization pattern is designed to create a diode pattern that fulfills the data need. The result is that when a certain memory location is addressed, the diode connections will provide the mask-created data at the output. This data pattern will not be subject to subsequent alteration.

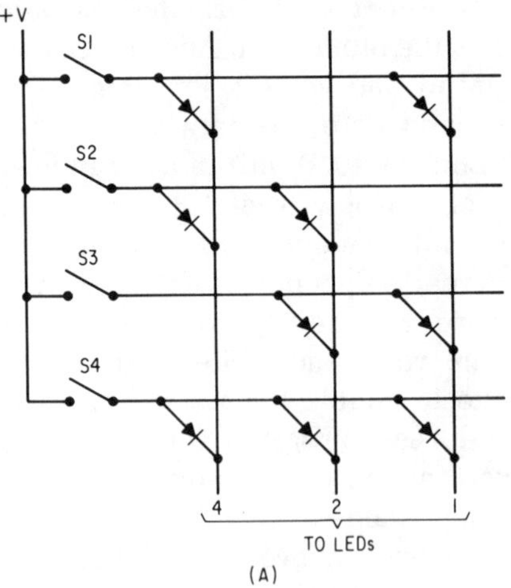

SWITCH CLOSED	OUTPUT CODE 1	2	4
S1	1	0	1
S2	0	1	1
S3	1	1	0
S4	1	1	1

(B)

Fig. 5-30 (a) Diode matrix ROM and (b) truth table.

Experiment 5-5: Read-Only Memory

Factory-produced masked ROMs are usually too sophisticated or expensive for experimentation. An alternative is to construct a ROM. As indicated previously, a ROM consists of diodes that have been connected according to a desired data pattern. The ROM to be built as a part of this experiment will use the same process. Construct the circuit shown in Fig. 5-30a. This is, of course, a standard diode matrix. When a switch is closed, a logical one will appear on appropriate output lines. Each switch will provide a different output code. Such a matrix could be used to program a bank of switches to provide a selected output code for each switch. Expand the matrix as shown in Fig. 5-31. The circuit is a binary-to-BCD converter that operates by storing the BCD words in the ROM. These BCD words are then accessed by one of 16 lines. The 4514 converts the binary word into one of 16 lines, and the process is complete.

This illustrates a frequent use of ROM. It is being used as a look-up table. Place the binary code 1111 (decimal 15) on the input of the 4514 and the BCD output code 0001 0101 will result. ROMs are used for many kinds of look-up tables. In microcomputer systems they decode keyboards. ROMs provide set sequences for games. ROMs control the operation of machines. For commercial applications, mask-programmed read-only memories are economical solutions to many problems. For prototype and hobby uses, the Programmable Read-Only Memory (PROM)

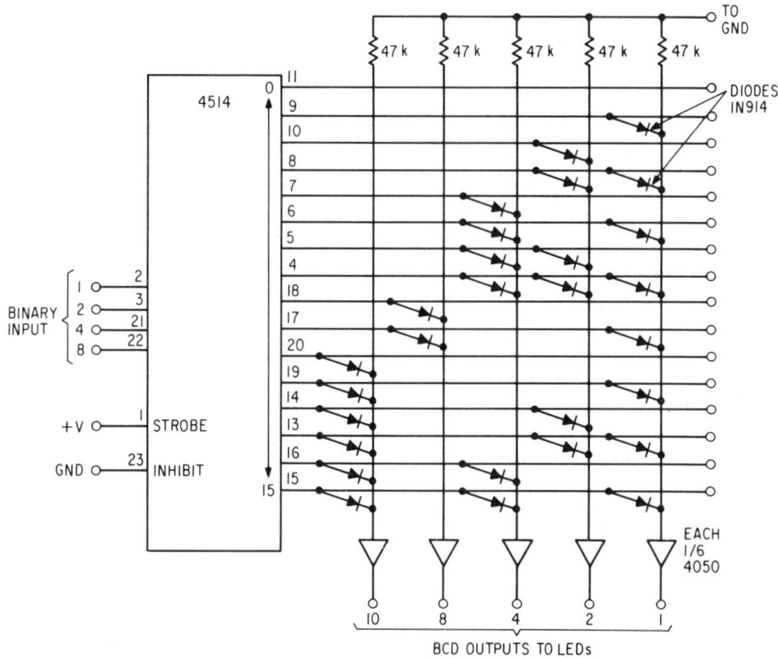

Fig. 5-31 Binary-to-BCD converter.

is the better choice. Applications requiring the features provided by these ROMs will find them simple solutions.

Programmable Read-Only Memory

A programmable read-only memory is not a contradiction of what was said earlier about ROMs. The PROM is programmed permanently, but it is programmed after manufacture. The PROM is constructed in a similar way to the ROM. In the case of PROM, however, all of the diodes are connected at the time of manufacture. When it is desired to program a set of data into the PROM, electric currents are used to burn out small fuses that are on each diode. These fuses are burned according to the truth table desired by the user. After programming, the PROM operates in exactly the same manner as the ROM. A memory location is addressed and the data word stored there is provided at the output.

The PROM is nonvolatile memory. Its programming is irreversible, even if the programmer makes a mistake. The PROM is said to be field programmable. A word of explanation is in order. PROMs are not all alike. They require different amounts of current for programming, and the current must be applied in different ways. The process is tedious since only 1 bit at a time can be programmed. The possibility of programming error is high. For this reason, many designers prefer to pay a nominal fee and have their PROMs programmed by a vendor who owns a PROM programmer. If the volume of PROM programming justifies the cost, the designer may choose to purchase a PROM programmer for his own use. Some hobby programmers have been the topic of articles in magazines that service the hobby computer and amateur radio fields. The do-it-yourselfer may find these circuits useful.

Erasable Programmable Read-Only Memory

A third type of read-only memory is the Erasable Programmable Read-Only Memory (EPROM). There is no contradiction here, either. The EPROM is nonvolatile. It is not programmable by the system. Even so, the EPROM offers unique opportunities for system prototype applications. The person who benefits most from the EPROM is the professional designer. He is able to purchase a microprocessor-controlled EPROM programmer. For others, it is necessary to send the EPROM to vendors for programming.

An EPROM charges a capacitive gate of a MOSFET cell. No discharge path is provided for the gate, and the charge once applied will remain. Programming the EPROM requires that enough charge be applied to ensure that it will not dissipate. At the same time, overcharging must be avoided. Microprocessor-based programmers operate by applying a small charge and then checking to determine the charge level. If the charge on the gate is inadequate, another charge is applied. This process is continued until the charge is at the proper level, and then programming moves to the next bit. EPROMs, even those with identical numbers, may program differently.

Some are faster and some slower. Some method of determining the proper charge level is needed.

EPROM is erasable. The process uses an ultraviolet light that dissipates the stored gate charge through a special quartz window installed above the EPROM chip. The erasing sequence is not instantaneous and requires an extended period of exposure to the ultraviolet light.

EPROM is very popular among prototypers. It is less expensive to program, test, erase, and reprogram these devices than to program and discard PROM elements. The EPROM, once programmed, has all of the attributes of ROM and PROM but adds the ability to erase and reprogram.

The 1702A is a popular EPROM. The unit is a 256-word by 8-bit memory. This means it has 256 memory locations, each of which will hold an 8-bit word. Tri-state output is provided.

Random Access Memory

Random Access Memory (RAM) differs from the various forms of ROM in its ability to be written into as well as read from by the system. RAM can be altered, updated, erased, or rewritten by system data and control signals. No ultraviolet light is required. RAM storage is not permanent. RAM is volatile memory. If the power supply is removed, the stored data will be lost.

RAM is available in two basic forms, *dynamic* and *static*. Dynamic RAM uses a charge on a storage cell similar in process to the EPROM just discussed. The major difference is that the dynamic RAM does not require ultraviolet erasure. The memory charge is limited so that it can be removed electronically. In fact, if the dynamic RAM is written into and then left alone, the stored data charge will leak away and the data will be lost. To prevent this, the dynamic RAM must be refreshed repeatedly. The Motorola MCM6604A is a dynamic RAM capable of storing 4,096 1-bit words. Each group of 1,024 bits is called 1K or one thousand due to its relation to the base two. Therefore, 4,096 is 4K bits, and in the MCM6604A there are 4K × 1 bit storage locations. The MCM6604A uses a one-transistor memory cell and tri-state outputs. The memory must be refreshed every 2 ms by sequentially cycling through the 64 row addresses. The block diagram of the MCM6604 is shown in Fig. 5-32. The address lines are common to both row and column. This multiplexing technique permits seven address inputs to address the 4,096

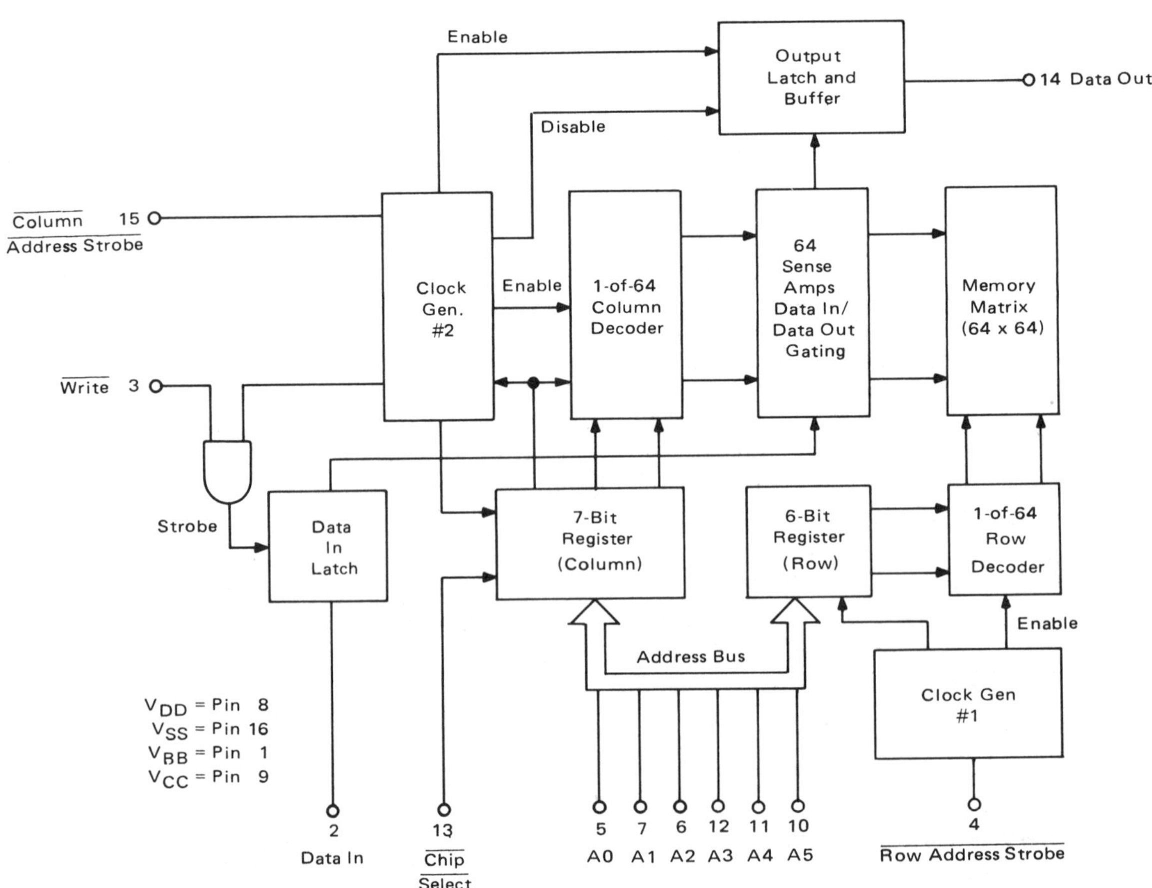

Fig. 5-32 MCM6604 functional block diagram. (Courtesy of Motorola, Inc.)

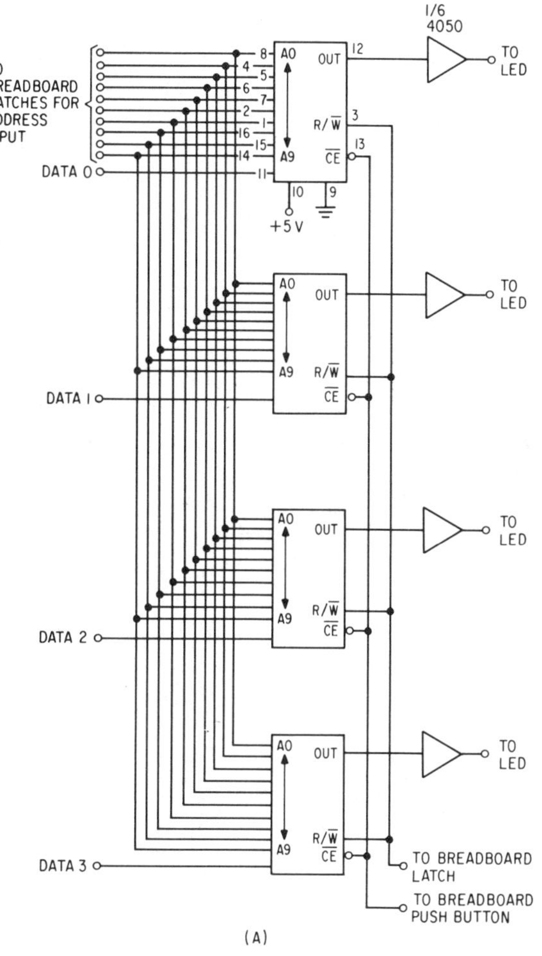

memory locations. Three control signals determine whether the memory chip is selected: the Chip Select ($\overline{CS}$) signal, the Row Address Strobe ($\overline{RAS}$), and the Column Address Strobe ($\overline{CAS}$). Refresh can be accomplished during any read, write, or read-modify-write cycle, but if refresh is to be accomplished during a write cycle, the chip must be unselected. Refresh can be accomplished without the $\overline{CAS}$ being applied to the chip.

Dynamic RAM has its obvious drawbacks. Refreshing must be accomplished or the stored data will be lost. In microprocessor systems, this is not hard to accomplish since the microprocessor can be programmed to carry out the refreshing. In other systems, dynamic RAM may not be usable. Power dissipation makes the refreshing acceptable. Power drain of dynamic RAM is extremely small when unselected. When active, power is a function of duty cycle. Low power dissipation makes a battery back-up system for the MCM6604 possible.

A second type of RAM is called static. Static RAM does not require refreshing. It uses static storage cells similar to those of the register files. Once the data is entered, it will be retained until a complementary data bit is written into the memory cell or until the power is removed from the chip.

A popular static RAM is the 2102. It provides 1K × 1 bit storage. Addressing is accomplished via ten address lines. The read/write function and tri-state

Data for Experiment 5-6

LOCATION			INPUT		OUTPUT		MODIFICATION		NEW
NO.	ADDRESS		WORDS		WORDS		WORD		OUTPUT
0	0000								
1	0001								
2	0010								
3	0011								
4	0100								
5	0101								
6	0110						1111		
7	0111								
8	1000								
9	1001						1111		
10	1010								
11	1011								
12	1100								
13	1101								
14	1110								
15	1111								

(B)

Fig. 5-33 (a) 2102 memory array wiring diagram and (b) data for Experiment 5-6.

output enable (chip enable) have separate control pin-outs. Data-in, data-out, + 5-V input and ground complete the 16 pin-outs of the DIP. Operation of the 2102 is simple. For writing, the desired memory location is addressed by applying high signals to the appropriate address inputs. The chip enable, pin 13, is brought low, and data is placed on the data input. When the R/$\overline{W}$ input is brought low, the data at the input will be written into the addressed memory cell. Returning the R/$\overline{W}$ input to its high state will permit the data in that cell to appear at data output, pin 12. Reading is nondestructive. The speed of the 2102 is determined by the final digit of its number as follows:

2102 = 1 μs
2102-1 = 500 ns
2102-2 = 650 ns

Many applications require data words longer than 1 bit in length. A 4-bit data word could be stored in RAM in a serial fashion. Each word would then occupy four adjacent locations in memory. The problem with this approach is speed. Each read from memory operation will require a length of time determined by the speed of the memory chip in use. To write or read a 4-bit data word into a 1-bit memory will require four memory cycles. In some applications, this delay may not be permissible. One solution to this problem is to use a memory chip that will permit parallel input of several data bits. The 2111 is a 256 × 4 bit static RAM. It has tri-state outputs and common data Input/Output (I/O). Operation of the 2111 is similar to the 2102 except for a few changes related to the common I/O. And output disable is held low for read cycles and high for write cycles. Dual chip enable inputs are held low to enable the chip. Dual controls permit easier selection of chips within multichip systems. The R/W input is held low to write into the chip and high for reading. Data is both input and retrieved using the same four data I/O lines. This requires that bidirectional bus components be used with the 2111 or that external data distribution be employed.

RAM can be obtained in several word lengths. Most popular, except in large-scale computers, are 1-bit, 4-bit, and 8-bit lengths. Memory chips also can be paralleled to obtain greater word lengths. The following experiment will demonstrate this.

Experiment 5-6: Random Access Memory

Breadboard the circuit shown in Fig. 5-33a using four 2102 RAMs. Input zeros at the four data inputs and tie the R/$\overline{W}$ line low. Step through the first 16 memory locations by establishing appropriate binary addresses on the ten address lines. After each new address is selected, briefly bring the $\overline{CE}$ low. Now repeat the process with the R/$\overline{W}$ input tied high. Zeros should appear at all 16 addresses. Next, write the 16 data words shown in Fig. 5-33b. After writing, read the 16 locations to confirm that the 16 words agree with those input. Modify locations 6, 9, and 11 by writing 1111 in these locations. Confirm that the rewriting process was accomplished.

This experiment demonstrates that RAM memory chips can be used in parallel to provide data words in multiple-bit lengths. It also demonstrates how the memory read, write, and rewrite functions are accomplished.

SIX

Data Processing

An Introduction to Processing

Data manipulation is an intermediate step in obtaining meaningful products for data. Selection, distribution, storage, and other manipulations of data may establish the conditions that permit data processing, but they do not accomplish that processing. Data processing is the function of systems that evaluate binary inputs and provide binary outputs that are the logical results of that evaluation. The evaluations that are most common with logical processors are related to binary addition. Binary addition was discussed in Chap. 2. The rules of binary addition were described as follows:

0 + 0 = 0
0 + 1 = 1
1 + 0 = 1
1 + 1 = 0 (with carry of 1, or = 10_2)

Digital logic can easily accommodate binary addition. An exclusive-OR gate, for example, follows the basic rules of binary addition. As indicated in Fig. 6-1a, a 0 at each input results in zero. A 0 at the first input and a 1 at the other results in an output of logic "one." Ones on both inputs result in a zero output. All that is lacking is the carry that can be provided by the AND gate used in the half adder shown in Fig. 6-1b. This adder circuit is a simple form of data processor. Data is input that is evaluated, and data is output that is the logical result of the input evaluation.

Addition is not the only process logic circuits must perform. Subtraction also must be performed. Subtraction is not handled by logic circuits as easily as addition. The rules for subtraction are

0 − 0 = 0
0 − 1 = 0
1 − 0 = 1
1 − 1 = 0

The exclusive-OR gate will reach the proper logical conclusions for all conditions except 0 − 1 = 0. Providing for this condition and borrowing would require more sophisticated gating circuits. To overcome this problem, a special form of binary addition is used. One type of this addition is called *1s complement addition*. The complement of the subtrahend (number to be subtracted) is obtained and added to the minuend (number from which subtracted). If the two numbers are of equal length and a 1 is in the overflow column of the sum of the two numbers, the 1 is added to the sum in an end-around carry to obtain the subtrahend difference. The overflow "1" indicates that the result is positive. If there is no overflow "1," it indicates that the result is negative. The procedure is as follows:

1. According to the rules of binary subtraction, the result of subtracting 1001 from 1011 is 1011 − 1001 = 0010.
2. Using 1s complement addition.
   a. The subtrahend complement is added to the minuend:

```
     minuend
  1011 ----->  1011
- 1001 ----->  0110
              -----
              10001
```

   b. Since there is an overflow "1," the result is positive and the overflow "1" must be carried around and added to the four-digit result:

```
  1011
  0110
 -----
 (1)0001  end around
   '-->1  carry
 -----
  0010
```

As can be seen, the 1s complement addition produces the same results as standard binary subtraction. This is advantageous when using IC adders.

A similar method of arriving at binary subtraction results from another addition process called *2s complement addition*. The 2s complement method requires that 1 be added to the complement

IN 1	IN 2	OUT
0	0	0
0	1	1
1	0	1
1	1	0

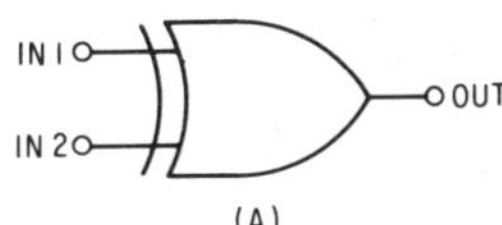

(A)

IN 1	IN 2	OUT 1	OUT 2
0	0	0	0
0	1	1	0
1	0	1	0
1	1	0	1

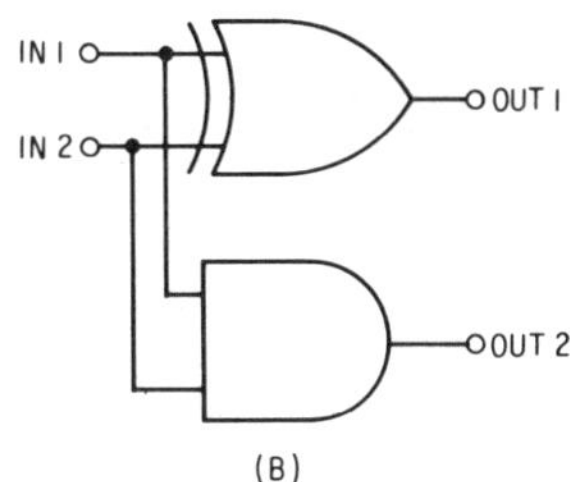

Fig. 6-1 (a) Exclusive-OR gate functions and (b) half adder circuit.

of the subtrahend prior to adding it and the minuend. The process would be as follows:

1. Obtain the complement of the subtrahend.

   1001 0110

2. Add 1 to the complement of the subtrahend.

   0110 + 1 = 0111

3. Add this number (2s complement) to the minuend:

   1011 + 0111 = 10010

4. Evaluate the overflow for sign of number and discard; i.e., an overflow of 1 = +, and overflow of 0 = –. The final result is 0010.

Arithmetic Logic Units

The Arithmetic Logic Unit (ALU) is a form of simple processor that is designed to process mathematical data. ALUs are capable of parallel bit addition and subtraction. Subtraction normally uses the 1s complement addition system. "Carry forward" for addition and "look backward" for subtraction are accomplished by companion logic blocks.

Figure 6-2 is a block diagram of a typical 16-bit ALU system. The 74181 ALU and the 74182 look-ahead carry generator form a numerical data processor system capable of 16 different arithmetic operations. Such a system also is referred to as a *function generator*. The function to be performed is determined by the binary code placed on the four function-select inputs. Table 6-1 indicates the select codes and their corresponding functions. These functions are performed for each bit of the data word. A single 74181 provides for arithmetic manipulation of 4-bit words. The devices can be paralleled to permit processing of data in longer word lengths. Internal carries are enabled when the mode-control input is held low. External carries can be accomplished using the external circuitry of the 74182, or ripple carries are possible that require no external circuitry. Ripple carry operation uses the ripple carry input (Cn, pin 7) and the ripple carry output (Cn + 4, pin 16).

Subtraction is accomplished using 1s complement addition. The complement of the subtrahend is generated internally with a forced end-around carry completing the process.

The ALU can be used for magnitude comparison. The subtract code is applied to the select inputs, and the ripple carry input is held high. When the two data words are of equal magnitude, the A = B pinout (pin 14) will go high. The Cn + 4 output (pin 16) will be high for $A \geq B$ and low for $A < B$.

The power of the 74181/74182 system is realized when logic functions are added to the arithmetic operations. If the mode-control input (pin 8) is tied high to disable the internal carry, the Boolean functions (logic functions) shown in Table 6-1 are performed by the system. In considering these functions, remember that the symbol "+" means "OR," *not* "addition." Also remember that the juxtaposition of the two variables is indicative of the "AND" function and does *not* mean "multiplication."

Experiment 6-1: The ALU

Breadboard the circuit shown in Fig. 6-3a using a 74181. Tie the mode-control input (pin 8) low, and set the ripple carry input (pin 7) high. This connection will enable the arithmetic operations shown in Table 6-1. Enter the code 1001 on the select inputs. Binary addition is now possible. Enter binary numbers at inputs A and B, as shown in Fig. 6-3b. Confirm that the output is the binary sum of the two numbers entered.

Enter the code 0110 at the select inputs. Binary subtraction has now been enabled. Enter binary numbers at inputs A and B, as shown in Fig. 6-3b. Confirm that the output in each case is the binary result of subtracting the two numbers. Notice in

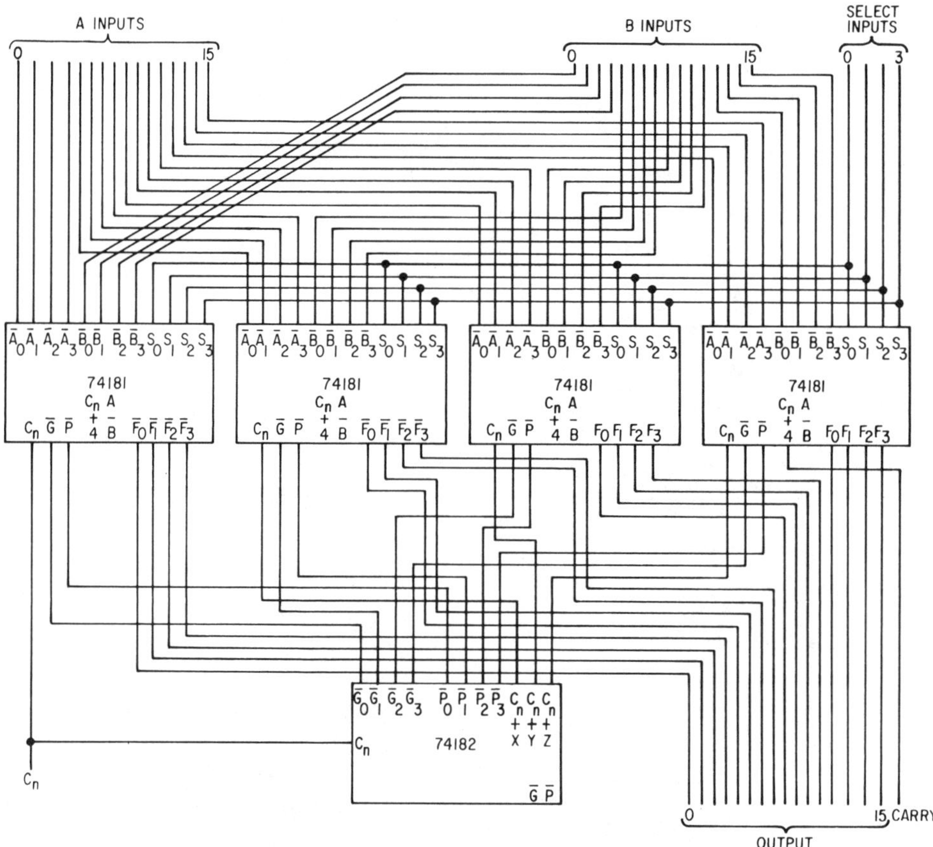

Fig. 6-2 Sixteen-bit ALU system.

Table 6.1 Arithmetic operations of the 74181/74182.

High Active Data						
Selection Code						
S3	*S2*	*S1*	*S0*	*Logic Functions*	$\overline{C}n$ = H	$\overline{C}n$ = L *(with carry)*
L	L	L	L	$F = \overline{A}$	$F = \overline{A}$	$F = A + 1$
L	L	L	H	$F = \overline{A + B}$	$F = A + B$	$F = (A + B) + 1$
L	L	H	L	$F = \overline{A}B$	$F = A + \overline{B}$	$F = (A + \overline{B}) + 1$
L	L	H	H	$F = 0$	$F = -1$	$F = 0$
L	H	L	L	$F = \overline{AB}$	$F = A + A\overline{B}$	$F = A + A\overline{B} + 1$
L	H	L	H	$F = \overline{B}$	$F = (A + B) + A\overline{B}$	$F = (A + B) + A\overline{B} + 1$
L	H	H	L	$F = A + B$	$F = A - B - 1$	$F = A - B$
L	H	H	H	$F = A\overline{B}$	$F = A\overline{B} - 1$	$F = \overline{AB}$
H	L	L	L	$F = \overline{A + B}$	$F = A + AB$	$F = A + AB + 1$
H	L	L	H	$F = A + B$	$F = A + B$	$F = A + B + 1$
H	L	H	L	$F = B$	$F = (A + \overline{B}) + AB$	$F = (A + \overline{B}) + AB + 1$
H	L	H	H	$F = AB$	$F = AB - 1$	$F = AB$
H	H	L	L	$F = 1$	$F = A + A$	$F = A + A + 1$
H	H	L	H	$F = A + \overline{B}$	$F = (A + B) + A$	$F = (A + B) + A + 1$
H	H	H	L	$F = A + B$	$F = (A + B) + A$	$F = (A + B) + A + 1$
H	H	H	H	$F = A$	$F = A - 1$	$F = A$

Table 6-1 that the operation for select code 0110 (LHHL) is referred to as "F = A − B − 1." The actual process is 1s complement addition.

Enter the code 0100 at the select inputs. This enables the operation "$F = A + A\overline{B}$." A truth table will help determine the expected value of F for values of A and B. Construct a truth table and confirm its validity by using the breadboard circuit. Many of these operations are specialized and may not be needed in a particular design. When attempt-

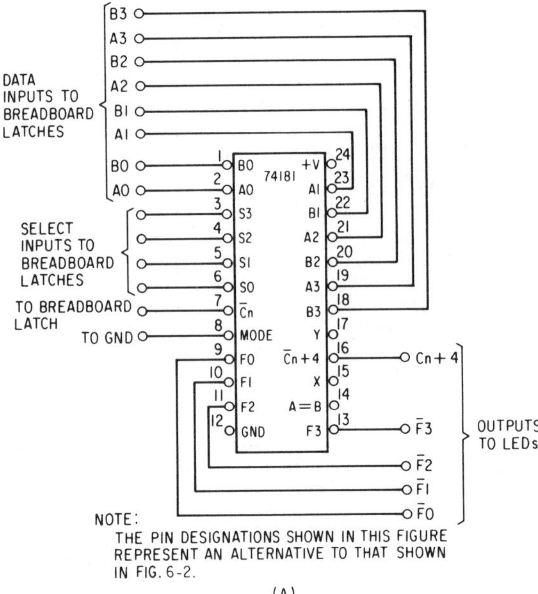

(A)

A L U Truth Table

INPUTS		OUTPUTS			
A	B	ADD	Cn+4	SUB Cn	− 4
0110	0110				
0111	0011				
1010	0101				
1100	1100				
1100	1000				
1000	0100				

(B)

INPUTS		OUTPUTS	
A	B	A = B	Cn+4
0100	0100		
1000	0100		
0010	0100		
1111	1111		
1010	1001		

(C)

Fig. 6-3 (a) 74181 ALU test circuit, (b) truth table for ALU, and (c) comparison results table.

ing to solve unusual logic problems, one of these operations may reduce the cost and complexity of the design significantly.

Enter the code 0110 at the select inputs. Set the ripple carry input (pin 7) high. Monitor the A = B output (pin 14) and the ripple carry output (pin 16). As shown in Fig. 6-3c, enter binary numbers at the A and B inputs. Confirm that the outputs reflect valid comparisons; i.e., A = B, A< B, or A> B (for A = B,

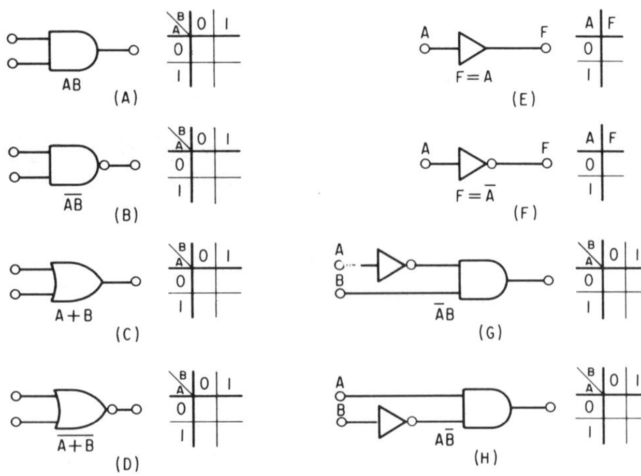

Fig. 6-4 Logic functions with truth tables.

A = B, Output = H, Cn + 4 = H; for A< B, A = B, Output = L, Cn + 4 = L; for A>B, A = B, Output = L, Cn + 4 = H).

Figure 6-4 shows several logic circuits. Complete the truth tables for each logic circuit. Tie the mode-control input high. From the logic functions shown in Table 6-1, obtain the appropriate select codes to enable each type of logic function and use the ALU to simulate these functions using inputs A_0 and B_0.

The ALU and the Microprocessor System

The power provided by the ALU's flexibility cannot be realized until an automatic control system is connected to it. When automatic control is added to the ALU, the operations accomplished by the ALU can be altered according to the requirements of each step of data processing. This permits the same unit to add, subtract, perform other arithmetic operations, or accomplish logic functions in response to the automatic control signal. External circuits must provide for the routing and coordination of the data to be processed by the ALU. When all of these functions are combined with a memory to form a processing system, the result is a computer. At the heart of every computer is an ALU.

This book is concerned with smaller processing systems that use a large-scale integrated circuit called a *microprocessor*. The microprocessor contains control and logic circuitry. A simplified block diagram of a typical microcomputer is shown in Fig. 6-5. The diagram shows that there are several component parts to a computer. The ALU is one of those parts, but to form a complete computer system other elements are required as well. Typical microprocessor chips combine the functions of the ALU with one or more of these other component parts on one LSI substrate.

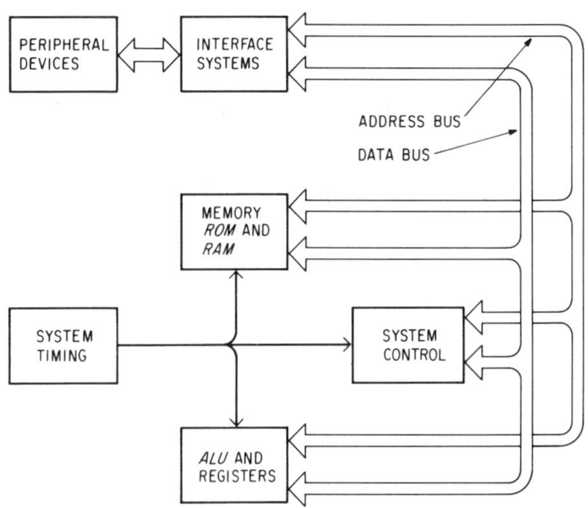

Fig. 6-5 Typical microcomputer system.

Some of the early microprocessors included the ALU and the system control sections of the microcomputer. More recent microprocessors have incorporated all of the system parts on the same chip. The result is a true computer on a chip. Of course, even these one-chip computers require external I/O devices to make them useful. After all, the most powerful microprocessor available is useless if it cannot be programmed and if it cannot receive and process data to accomplish meaningful results. Putting the products of processing in and getting the results of processing out of the microprocessor requires I/O devices.

Microprocessors are classified in part by the word length they can handle in a parallel form. They range from 1-bit processors to 2-, 4-, and 8-bit processors, with some very powerful units providing for 12 and 16 bits. (We can eventually look forward to processors with 32 bits or more.) Complexity of construction will increase with word length, but so also will speed and power. Complexity is directly translatable to cost. For this reason, the designer must match the processing task with the processor so that economy is preserved and complexity is no greater than necessary. It is inefficient to use a large, powerful, multibit processor to perform a simple task such as controlling traffic lights at a busy intersection or regulating air conditioning equipment in an office building. By the same token, a microcomputer system for hobby, business, industrial, or scientific applications would require greater capability than a 1-bit microprocessor could supply.

What does a microprocessor system do? Stated as simply as possible, the microprocessor system receives and analyzes data and carries out predetermined actions based on that analysis.

Consider a simple processing example. We must see if we can match two numbers. This might be true if we had to identify a person who desires to cash a check as being included (or not) on a list of known bad-check writers. To do this, the account number of the check casher is brought into the microprocessor. This input could come from a data entry terminal, a credit card reader, or some similar device. After the account number of the check casher is entered, it is compared with a list of account numbers of people who have written bad checks. This is done one account number at a time. If a match is found, the account number is printed out and the check is not cashed. If the numbers do not match, another number is drawn from the list and compared to the original account number. If no match is found in the list, the computer will indicate that the account number was not among the list of bad-check writers and the check will be cashed. The instructions that direct the processing are called the *program* of the microcomputer. The account numbers being processed are the *data*. Perhaps the best way to understand this processing action is to examine a simple microprocessor system.

The Programmable Logic Controller/Industrial Control Unit

The Motorola MC14500B is a 1-bit microprocessor that is referred to as an Industrial Control Unit (ICU). The primary reason for this designation is that a 1-bit microprocessor is poorly suited for complex calculations or parallel word data processing but ideally suited for decision- and command-oriented tasks. For these types of jobs, the 1-bit processor may outperform a multibit processor. Also called Programmable Logic Controllers (PLC), the 1-bit device has found application in industrial control systems that can benefit from reprogrammability. Environmental control systems for large buildings that alter air conditioning patterns based on weather, season, and day/night decisions can easily be accommodated with a PLC. Many tasks formerly accomplished by mechanical relays or hand-wired logic are now accomplished through software-directed PLC systems. The MC14500B is an example of a programmable logic controller. As will be seen, it has many characteristics in common with more sophisticated microprocessors, but retains a basic simplicity that makes it an excellent choice for an introduction to microprocessors. Here are some characteristics that make the MC14500B a good choice for a first microprocessor system:

1. Simple decision making: The 14500B operates on binary data 1 bit at a time. The decision is either yes or no, one or zero.

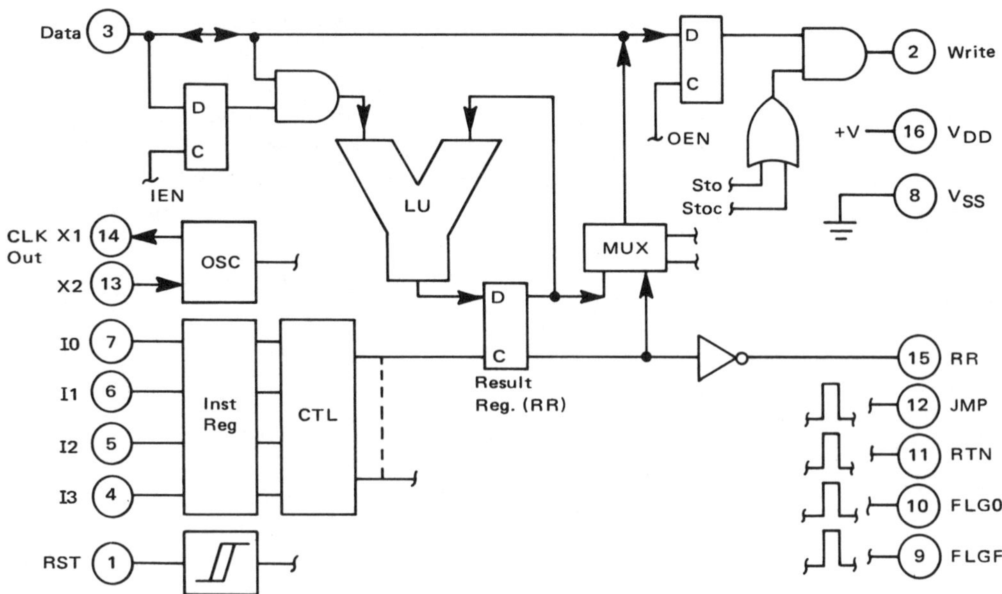

Fig. 6-6 MC14500B ICU simplified block diagram. (Courtesy of Motorola, Inc.)

2. Simple instruction set: The 14500B has only 16 instructions. All of the tasks to be performed by the controller will be defined by combinations of these 16 instructions.
3. Simple operation: The ICU system operates as a stored program processor. Instructions are stored in memory and are "fetched" and "executed" sequentially one instruction at a time. The process is looped so that the instruction sequence is repeated as required.
4. Simple timing: The 14500B is controlled by a built-in single-phased clock. Each instruction is accomplished in a single instruction cycle. At a clock frequency of 1 MHz, over 8,300 instructions can be executed in a 60-Hz power line half cycle.
5. Low cost.

The MC14500B ICU

Figure 6-6 shows a simplified block diagram of the MC14500B Industrial Control Unit (ICU). At the heart of the ICU is the Logic Unit (LU). Since the 14500B is a 1-bit device, the LU has the simple task of storing the results of Boolean manipulations in the 1-bit result register. The input data for the LU originates from two sources: the external data bus and the previous contents of the result register. The LU, therefore, upon command from the Control Logic (CTL) section, evaluates the contents of the Result Register (RR) based on the external data bus level. The result of this evaluation replaces the previous contents of the RR.

Timing of the system is via the built-in oscillator (OSC), although external timing can be accomplished using the "X2" input (pin 13). The frequency of the internal oscillator is set by connecting a resistor of appropriate size between the "X1" and "X2" inputs (pins 13 and 14). Processor speed varies from approximately 1 MHz for a 20-kΩresistor to 20 kHz for a 2-M resistor. The 14500B accomplishes one instruction during each machine cycle.

The operation of the ICU is controlled by instruction codes. Table 6-2 shows the 16 instruction codes used with the 14500B. These codes will be discussed in detail later. Instruction codes enter at pins 4, 5, 6, and 7. The instruction is latched into the instruction register (IR) on the negative-going edge of "X1." The instructions are decoded in the CTL section. The output of the CTL directs the LU to perform the appropriate actions. In addition, four flags are produced: FLGF (at pin 9), FLGO (at pin 10), RTN (at pin 11), and JMP (at pin 12).

The Input Enable register (IEN) and the Output Enable register (OEN) control the flow of data into and out of the ICU. The IEN signal routes the data from pin 3 to the LU when in the high state. A high OEN register enables the write signal at pin 2. The state of these registers is set via the data input. When the IEN register is loaded with a low, the input data will be interpreted as a low regardless of its actual state. The IEN register must be reloaded with a high to permit data to be transferred to the LU. In a similar way, a low in the OEN register will inhibit the write signal while a high will enable the write signal.

Pin 1 is a master reset. Bringing the pin high will cause several actions:

Table 6-2 MC14500B instruction set. (Courtesy of Motorola, Inc.)

	Instruction Code	*Mnemonic*	*Action*
#0	0000	NOPO	No change in registers. R → R, FLG0 ← ⎍
#1	0001	LD	Load Result Reg. Data → RR
#2	0010	LDC	Load Complement $\overline{\text{Data}}$ → RR
#3	0011	AND	Logical AND. RR · D → RR
#4	0100	ANDC	Logical AND Compl. RR · $\overline{\text{D}}$ → RR
#5	0101	OR	Logical OR. RR + D → RR
#6	0110	ORC	Logical OR Compl. RR + $\overline{\text{D}}$ → RR
#7	0111	XNOR	Exclusive NOR. If RR = D, RR ← 1
#8	1000	STO	Store. RR → Data Pin, Write ← 1
#9	1001	STOC	Store Compl. $\overline{\text{RR}}$ → Data Pin, Write ← 1
#A	1010	IEN	Input Enable. D → IEN Reg.
#B	1011	OEN	Output Enable. D → OEN Reg.
#C	1100	JMP	Jump. JMP Flag ← ⎍
#D	1101	RTN	Return. RTN Flag ← ⎍, Skip next inst.
#E	1110	SKZ	Skip next instruction if RR = 0
#F	1111	NOPF	No change in registers RR → RR, FLGF ← ⎍

1. All registers will be cleared.
2. All flag signals within the ICU will be held at zero.
3. Oscillator pin 14 (X1) will be held high, inhibiting the oscillator.

The ICU System

The 14500B requires the functions of several other ICs to permit operation as a microprocessing system. Figure 6-7 illustrates a typical 14500B ICU system. The associated ICs include a program counter, memory, input selectors, and output latches. The ICU accepts data from its 1-bit-wide data line, processes this data using the results stored in the internal result register, and returns the output of the data manipulation to the data bus. These actions are executed sequentially in a looping fashion. This is accomplished by using a MC14516B as the program counter. The 14516B is a divide-by-16 binary up-down counter. In this case, the count is in an upward direction. The two 14516Bs produce a 7-bit output. The 7-bit binary output represents the memory ad-

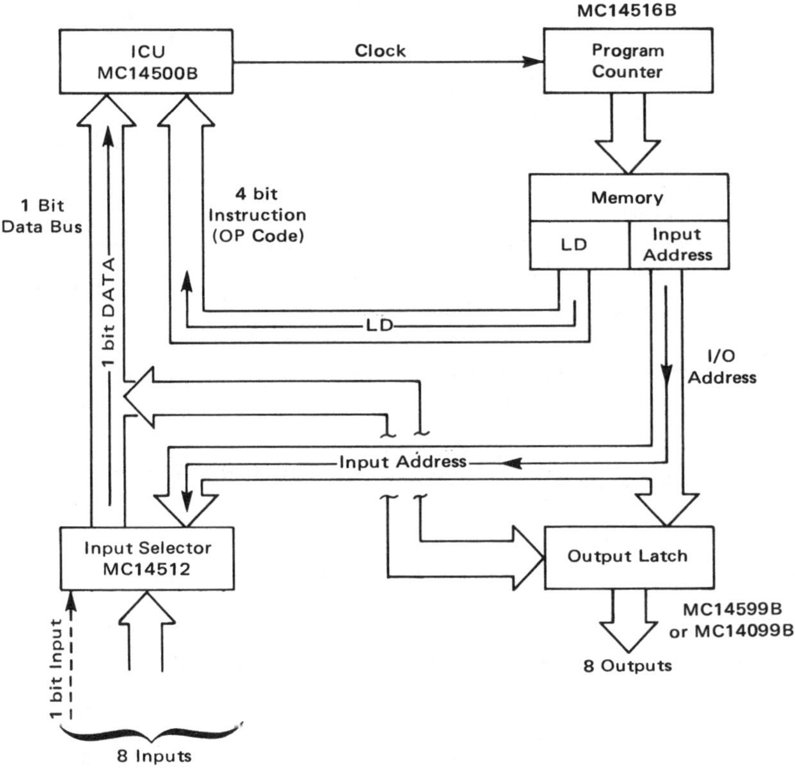

Fig. 6-7 Typical 14500B system block diagram. (Courtesy of Motorola, Inc.)

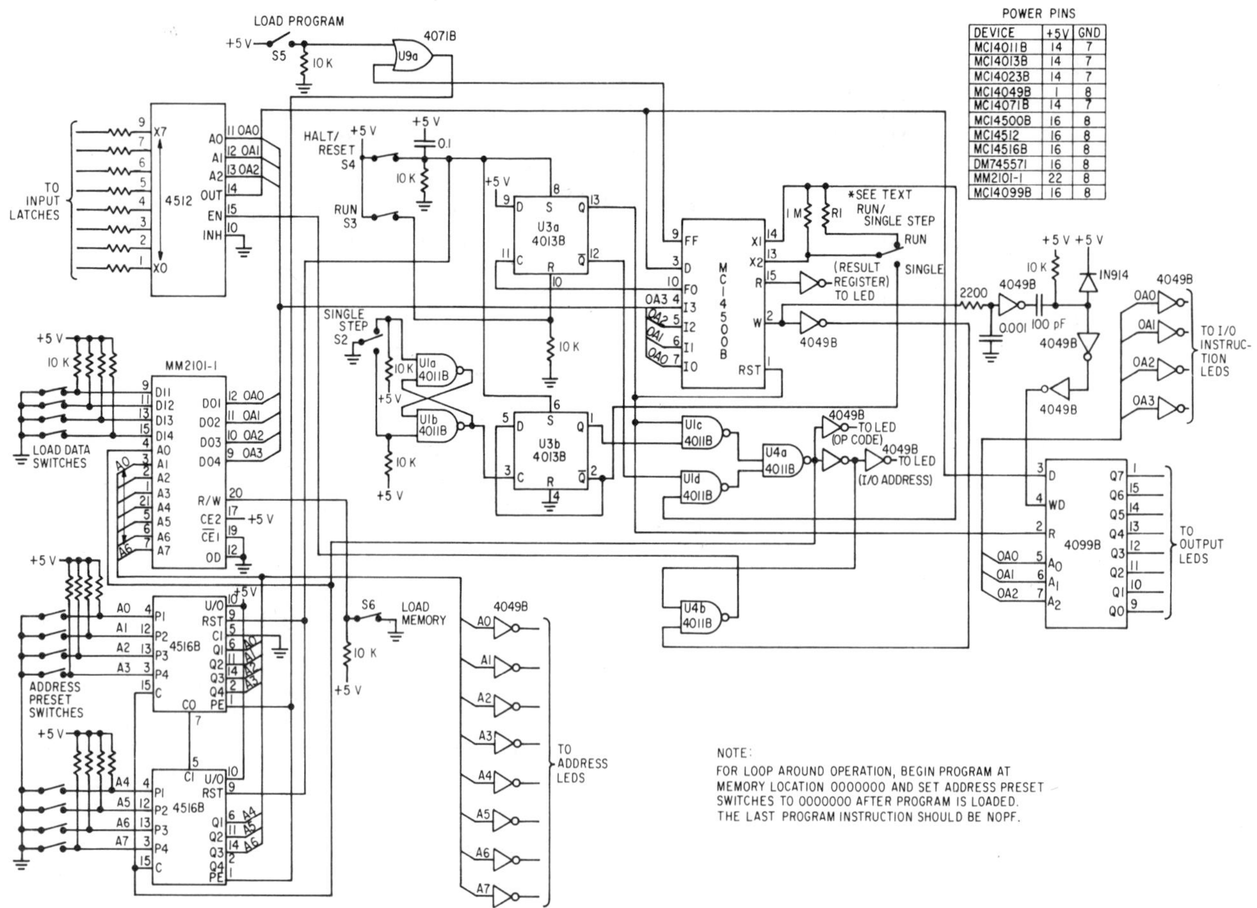

Fig. 6-8 ICU system schematic.

dress that contains the next instruction. The memory is used to hold instruction codes. These instructions are placed in memory in a sequential form and will therefore be executed in proper sequence as the binary count advances.

The memory provides signals to the ICU, the input selector, and the output latch. As dictated by the instruction code provided by the memory, data will be "fetched" from the appropriate input, processed by the ICU, and/or latched onto the appropriate output line. Data is transferred via the 1-bit data bus, and the addresses are selected using the multibit address bus.

The input selector decodes the address bus signal and transfers the selected input line state onto the data bus. Of course, this action is controlled by the ICU. The output register also is controlled by the ICU. When the output register is enabled by the write command from the ICU, the status of the data bus is latched at the output selected by the address bus code. A data manipulation sequence can be accomplished only once, or by automatically resetting the program counter at the end of the sequence, the program will be repeated in a looping fashion. Other program operations such as subroutines are within the capabilities of the 14500B. These more sophisticated routines require different system configurations, which will be described later.

Experiment 6-2: The MC14500B ICU Microprocessor System

Discussion of the 14500B instruction set will be enhanced if a system is available that will permit demonstration of its various operations. Figure 6-8 shows an ICU microprocessor system that uses the 14500B. Breadboard the circuit and retain it for the rest of the experiments in this chapter. If the Smitty breadboard is used, many of the monitoring LEDs will be available. The circuit provides for normal operation at the speed determined by the value of the resistor R1 or single step operation, which advances the processing one instruction at a time. Switch S1 selects the operating mode. Switch S2 actually causes the process to advance by individual steps. S2 has connections to U1a and U1b, which form a bistable multivibrator. This bistable debounces S2 prior to using it to clock U3b. The $\overline{Q}$ output of D-type flip-flop U3b advances the processor one step at a time if S1 is in the single step position.

Switch S4 activates halt/reset functions. Depressing S4 brings the set inputs of U3a, U3b, and the 4516Bs[1] high. High sets on U3a and U3b force a high at the Q pin-outs of each flip-flop. This condition will place a low on one input of U4a via U1c. This will inhibit processor functioning. The high at the reset of the 4516s will set their counters to 0000. Processor operation can be restarted by depressing S3. S3 places a high on the reset of U3a and forces a low at the Q output. A high will be present at the $\overline{Q}$ output, which drives one input of U1d. The clock of the 14500B can now pass through U1d and U4a to clock the system. A "software stop," a system halt controlled by the program, is accomplished by clocking U3a with the FLGO signal. This will transfer the high from the D input to the Q output. A high on the Q output will inhibit clocking by forcing U1c output low.

After the system has been halted and reset by depression of S4, if S1 is in "single step," single step operation can be accomplished by depressing S2. It was stated earlier that the $\overline{Q}$ output of U3b clocked the 14500B when in this mode. The Q output of U3b removes one high from U1c and forces a high at the U1c output. Since U1d is inhibited from passing the clock from the 14500B by the low on the $\overline{Q}$ output of U3a, the U3b output becomes a manual clock. The clock signal of the output of U4a feeds the clock inputs of the 14516B program counters and the MM2101-1 memory. The status of the clock is available at the OP code LED.

The MM2101-1 memory is loaded with the instruction codes that are at the D inputs when S6 is depressed. Since this circuit is designed for use with the Smitty breadboard, it is planned that the built-in push buttons and external slide switches will be used for data and address inputs. Eight addresses for the 4516Bs and eight data inputs for the 4512s are needed. In addition, four data lines are needed for the loading of the MM2101-1. The address of the desired memory location to be loaded is placed on the inputs of the 4516B and placed on the outputs of the 4516B by depressing S5. That location will now be indicated by the address LEDs and will appear at the MM2101-1 address inputs. The instruction code on the four data input lines will be loaded into the selected address by depressing S6. The data in the address will be indicated by the I/O-instruction LEDs with the existing data indicated before and new data indicated after loading. After loading, the address to the memory can be incremented by depressing S5, and the next instruction can be loaded. During processing, automatic resetting of the 4516Bs to the loaded address is accomplished by using the FLGF signal to drive U9a. Both I/O addresses in octal and instruction codes in hexadecimal are placed on the I/O-instruction lines. How the data is used will be based on the status of the clock. The clock serves as the LSB of the memory address. The 14500B will accept an instruction from memory only on the falling edge of the clock. With the clock low, the correct memory location is addressed to supply the I/O address to input and output circuits. Addresses are placed in memory immediately after their corresponding instructions.

The write operation loads data into the output latches (4099B). The address to be loaded is determined by the status of the address lines. The data on the data bus is transferred to the selected latch when the ICU enables the write function. The signal flow for the processor will become clearer when sample programs are run. First, however, the 14500B instruction set must be discussed.

The MC14500B ICU Instruction Set

A microprocessor instruction set is a list of instruction codes that tell the processor which operation to perform. Each microprocessor has its own unique instruction set. One measure of the power of a microprocessor is the size and sophistication of the instruction set. The 14500B has 16 instructions in its instruction set. More sophisticated microprocessors will have instruction sets that range from 75 to over 100 different instructions. The power and flexibility permitted by this large instruction set is significantly greater than that of a simple instruction set such as the one used with the 14500B. It also is true, however, that the larger instruction set increases the time required to become proficient at programming.

Table 6-2 shows the instruction set used by the 14500B. Programming is simplified if the mnemonics and their functions are memorized. Instructions are loaded into the MM2101-1 memory in the sequence needed by the program objective. Each instruction is a 4-bit code (see Table 6-2). The instructions can be divided into several different types. There are two NOP or "no operation" instructions. These instructions do not require any data transfer or register changes. In the demonstration system, the NOPO (0000) instruction forces FLGO high and subsequently halts microprocessor action via U3a. The NOPF (1111) instruction forces FLGF high. The high FLGF signal passes through U9a and forces the address lines to the number specified by the address switches. If this address is 0000 0000, the program

[1]The Motorola number for CMOS devices includes a "1" before the generic number; i.e., 4516B = Motorola MC14516B.

AND Function

INPUT	INITIAL RESULT REGISTER CONTENTS	NEW RESULT REGISTER CONTENTS
0	0	0
0	1	0
1	0	0
1	1	1

Fig. 6-9 Truth table for AND instruction.

ANDC Function

INPUT	INITIAL RESULT REGISTER CONTENTS	NEW RESULT REGISTER CONTENTS
0	0	0
0	1	1
1	0	0
1	1	0

Fig. 6-10 Truth table for ANDC instruction.

OR Function

INPUT	INITIAL RESULT REGISTER CONTENTS	NEW RESULT REGISTER CONTENTS
0	0	0
0	1	1
1	0	1
1	1	1

(A)

ORC Function

INPUT	INITIAL RESULT REGISTER CONTENTS	NEW RESULT REGISTER CONTENTS
0	0	1
0	1	1
1	0	0
1	1	1

(B)

Fig. 6-11 (a) Truth table for OR instruction and (b) truth table for ORC instruction.

counter will be reset to zero, and the sequence can begin again. If other codes are placed on the inputs of the 14516B, the address corresponding to this code will be loaded onto the outputs, and the data stored in that memory location will be passed on.

Four I/O instructions are included in the instruction set. The load control (LD; 0001) is used to place the value of the data bus into the result register. The IEN register must be loaded with a high logic level to permit a high to be loaded by the LD instruction. The load complement (LDC; 0010) places the complement of the value of the data bus into the result register. Again, the IEN register must be loaded with a high to permit a low to be loaded by the LDC instruction. These are the two instructions used to bring data into the processor. The store function (STO; 1000) places the value of the result register on the data line and places a high on the write pin (pin 2) if the OEN register has been previously loaded with a high. The store complement instruction (STOC; 1001) routes the data complement from the result register to the data line and places a high on the write pin (pin 2) if the OEN register has previously been loaded with a high. These two instructions are used to route data from the processor.

Five logic instructions are available in the 14500B instruction set. The AND instruction causes the value of the result register to be considered with the data input line in a logical AND operation. The truth table for an AND gate defines the result of the AND instruction (see Fig. 6-9). The result of the AND operation will be placed in the result register. The AND complement instruction (ANDC) instructs the

XNOR Function

INPUT	INITIAL RESULT REGISTER CONTENTS	NEW RESULT REGISTER CONTENTS
0	0	1
0	1	0
1	0	0
1	1	1

Fig. 6-12 Truth table for XNOR instruction.

processor to evaluate the value of the result register with the value of the complement of the data input according to an AND operation. The result of this operation is then placed in the result register. The truth table for this instruction is shown in Fig. 6-10.

The OR and the ORC instructions are similar to the AND and ANDC in that they perform logical functions. In this case the function is the logical OR manipulation of the result register and the data line for the OR instruction and the logical OR manipulation of the result register and the complement of the data input value for the ORC instruction. The truth tables for these instructions are shown in Fig. 6-11.

The XNOR instruction provides the exclusive-NOR function. Accordingly, if the result register and the value of the data line disagree, the result register is forced low. If the result register and the input data agree, a high is loaded into the result register. The truth table is shown in Fig. 6-12.

Two I/O enable instructions are used with the 14500B. These are the IEN and the OEN. These two

Input Enabling				Output Enabling	
INPUT	IEN REGISTER	DATA PASSED TO LOGIC UNIT	INSTRUCTION	OEN REGISTER	WRITE CONTROL SIGNAL
0	0	0	STO	0	INHIBITED
0	1	0	STO	1	WRITE
1	0	0	STOC	0	INHIBITED
1	1	1	STOC	1	WRITE

Fig. 6-13 Truth tables for IEN and OEN instructions.

instructions load the corresponding registers with the value of the data line. Depending on the value in the I/O registers, data will be passed or inhibited. If the IEN register is loaded with a high, data entry into the logic unit will be enabled. If a low is in the IEN register, data flow into the LU will be inhibited. By the same token, when a high is in the OEN register, data is permitted to flow to the output lines. If a low is in the OEN register, no data will flow out of the processor. Truth tables for the IEN and OEN instructions are shown in Fig. 6-13.

The three remaining instructions provide the 14500B system with sophisticated decision-making capabilities. The jump instruction (JMP; 1100) is used to signal the system to jump to a different location within the program. This instruction is normally used with the skip instruction (SKZ; 1110). Each time the JMP instruction is encountered in a program, transfer to a specified location will be accomplished. If a "jump" to the same location is made each time through the program sequence, the location to which the program jumped could be moved into the place of the JMP instruction. The program would then operate in the same way, only more efficiently. For the JMP instruction to be most effective, it must be part of a decision. "If some condition is met, then jump to location 0010. If the condition is not met, continue with the normal line of processing." The SKZ instruction permits this "conditional" jump. The SKZ will cause the 14500B to skip the following instruction if the result register is low. If the result register is high, the next instruction will be processed.

Suppose data input "one" is low and the following program is in the 14500B memory: (Note: IEN and OEN registers are high.)

Instruction No.	*Address**	*Instruction Code*	*Mnemonic*	*I/O and Destination Address*
	0000			
1	0001	0001	LD	0001
2	0010	1110	SKZ	
3	0011	1100	JMP	1000
4	0100	0011	AND	0010
5	0101	1000	STO	1100
6	0110	1111	NOPF	
7	1000	1000	STO	1110
8	1001	1111	NOPF	

*The address shown is the address set by the preset enable switches. The clock signal toggles A0, which selects instructions or addresses.

The program directs the ICU to:

1. Load the value of data input one into the result register.
2. If the RR is low, skip the next instruction. If it is high, continue with instruction 3. (It is high.)
3. Jump to location 1000 (go to step 7).
4. AND the data from location 0010 and the result register.
5. Place the RR output on output 1100.
6. Return to the top of program counter.
7. Place the RR output on output 1110.
8. Return to the top of program counter.

In this example, input 0001 is high. Instruction 1 loads this high into the result register. Instruction 2 directs the processor to skip the next instruction if the RR is low. Since the RR is high in this case, the SKZ instruction will be ignored and processing will continue with instruction 3. Instruction 3 directs the processor to jump over instruction 4 and go to instruction 7. If the RR had been low at instruction 2, the SKZ instruction would have bypassed instruction 3 and processed instructions 4, 5, and 6. The SKZ and JMP instructions have combined in this instruction to create a form of "computed go to" instruction. If the RR is low, one action is taken. If the RR is high, another action is taken. A flowchart for the operation is shown in Fig. 6-14. The NOPF instruction places a high on pin 9, which activates a return to the beginning counter location. The JMP, SKZ, and RTN instructions require additional circuitry that is not provided for in the demonstration system.

Algorithms

Prior to running programs on the 14500B system, some discussion of program design is in order. The computer profession often speaks of algorithms. An *algorithm* is simply a logical solution to a problem.

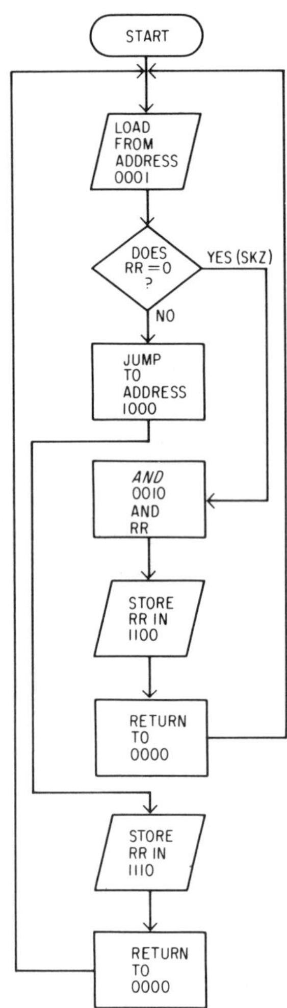

Fig. 6-14 SKZ/JMP flowchart.

Included in the concept of an algorithm is that the solution must be reduced to a series of single step instructions. An algorithm is a listing of program operations. The flowchart is the graphical representation of the algorithm. Each block on a flowchart is a finite part of the program. The symbols used in flowcharts are relatively standard. Some of these symbols are shown and discussed as follows:

1. The parallelogram is used to indicate that I/O operations are required.
2. The process symbol indicates that data manipulation such as calculation, logical operation, or data movement is to take place.
3. The hexagon indicates that data preparation is to be accomplished.
4. The diamond is the decision symbol. It asks the question, Is a condition met? If it is, one branch is accomplished. If it is not, a second branch is followed.

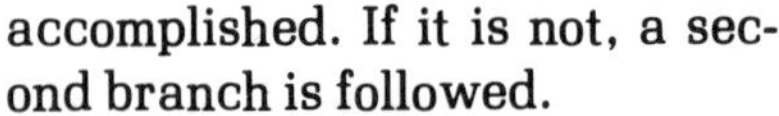

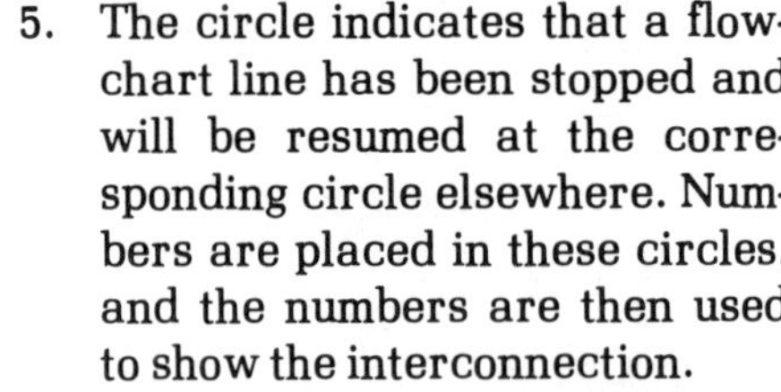

5. The circle indicates that a flowchart line has been stopped and will be resumed at the corresponding circle elsewhere. Numbers are placed in these circles, and the numbers are then used to show the interconnection.

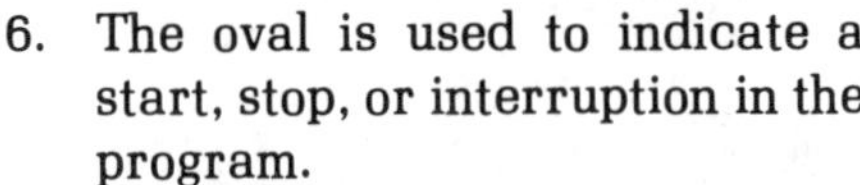

6. The oval is used to indicate a start, stop, or interruption in the program.

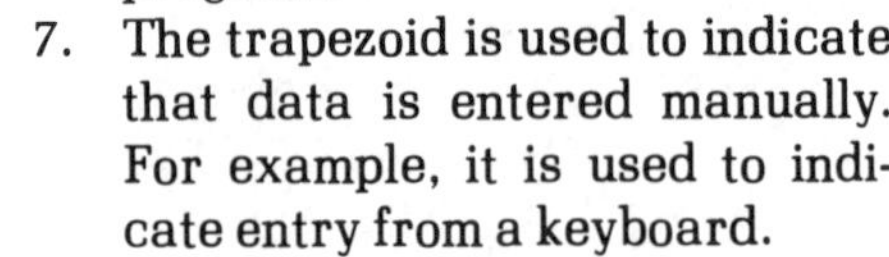

7. The trapezoid is used to indicate that data is entered manually. For example, it is used to indicate entry from a keyboard.

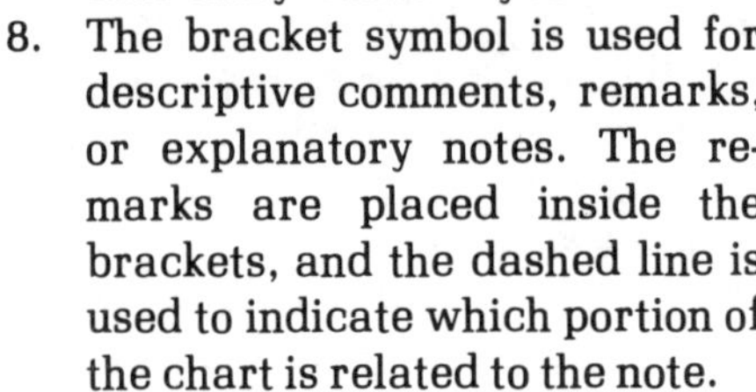

8. The bracket symbol is used for descriptive comments, remarks, or explanatory notes. The remarks are placed inside the brackets, and the dashed line is used to indicate which portion of the chart is related to the note.

There are numerous other symbols. Many of these relate to I/O devices such as tape drives, disks, paper tapes, and card readers. Since only a keyboard will be used in the demonstration circuit, these other symbols will not be considered. It is recommended that a standard flowchart template be used in drawing flowcharts. Neatness and clarity result.

Creating an algorithm for a problem is simply a matter of reducing the problem to a sequence of logical solution steps. Binary addition can be used as an example. From previous chapters, it is known that binary addition follows this form:

$$\begin{array}{r}1\\ +\ 1\\ \hline 10_2\end{array} \quad \text{or} \quad \begin{array}{r}0\\ +\ 1\\ \hline 1_2\end{array} \quad \text{or} \quad \begin{array}{r}1\\ +\ 0\\ \hline 1_2\end{array} \quad \text{or} \quad \begin{array}{r}0\\ 0\\ \hline 0\end{array}$$

To explain this logically, several steps would be included:

1. Obtain the first binary number.
2. Obtain the second binary number.
3. Add the two numbers.
4. Write the results.

To solve this problem with integrated circuit logic requires additional steps. Obtaining the binary sum is accomplished with the exclusive-OR gate, as indicated earlier. The problem is the carry. An AND gate fixes this. The half adder circuit shown in Fig. 6-1b is needed. Converting this circuit into software for the microprocessor takes several instructions. A flowchart will assist in designing the program. The flowchart in Fig. 6-15 shows the steps required. Two basic operations are required, an exclusive-OR

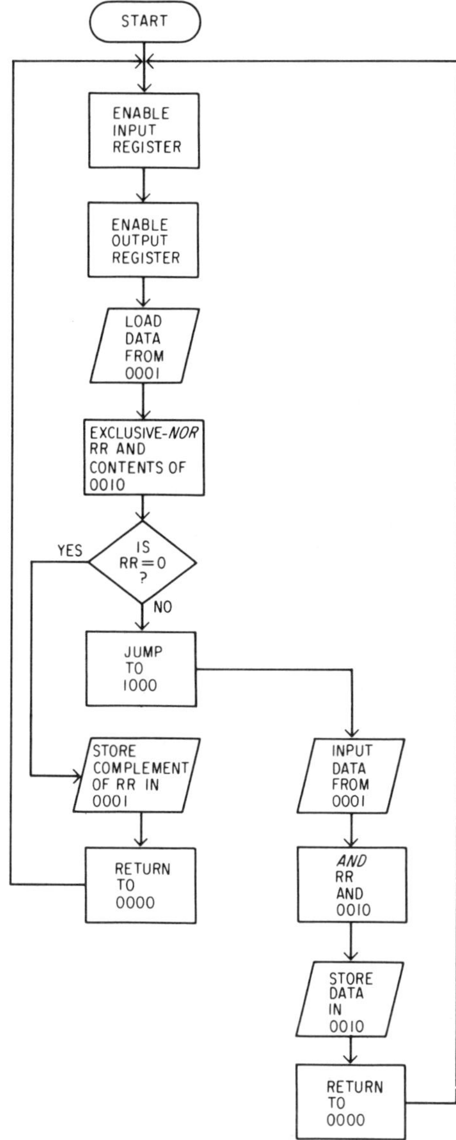

Fig. 6-15 Flowchart for half adder program using SKZ and JMP instructions.

operation to provide the basic binary addition and an AND operation to provide for the carry. From previous dealings with binary addition, it is known that a carry results only when both inputs are high. If both or either of the inputs are low, no carry will be required. In the flowchart, the decision to carry or not carry is accomplished by the SKZ/JMP combination. Notice that the 14500B provides an exclusive-NOR (XNOR) function. The result register will therefore contain the complement of the binary addition. This presents no problem as long as STOC is used to output the RR value and as long as the SKZ decision directs the processor to "AND" the inputs if the RR is low. Using this flowchart as a guide, the resulting program would be as follows:

Instruction		*Mnemonic*	*Address*
Code	*Address*		
1010	0000	IEN	0100
1011	0001	OEN	0100
0001	0010	LD	0001
0111	0011	XNOR	0010
1110	0100	SKZ	
1100	0101	JMP	1000
1001	0110	STOC	0001
1111	0111	NOPF	
0001	1000	LD	0001
0011	1001	AND	0010
1000	1010	STO	0010
1111	1010	NOPF	

Note: Set input 0100 high.

The same results can be obtained without the SKZ and JMP instructions. The efficiency gained by the SKZ/JMP operations will, of course, be lost (see Fig. 6-16). The program in this case would be as follows:

Instruction		*Mnemonic*	*Address*
Code	*Address*		
1010	0000	IEN	0100
1011	0001	OEN	0100
0001	0010	LD	0001
0111	0100	XNOR	0010
1001	0101	STOC	0001
0001	0110	LD	0001
0011	0111	AND	0010
1000	1000	STO	0010
1111	1001	NOPF	

Note: Set input 0100 high.

Using the algorithm, both the XNOR and the AND instruction sequence would have to be accomplished each time the addition program is run. With the earlier program, if the two inputs were different, the sequence would terminate after storing the product of the XNOR function. This represents programming economy since unlike inputs will not produce a carry.

The difference of economy between the programs for this simple example is not great. For other applications, the JMP and SKZ instructions may result in significant economy. This will be discussed in more detail when subroutines are covered in the next chapter. Notice that the economy (speed) is gained at the expense of a greater number of instructions.

The goal of flowcharting and the formulation of algorithms is to reduce programming problems. These devices assist in clarifying solutions to problems. As programs become more complex, they should be modularized. Each module will be consid-

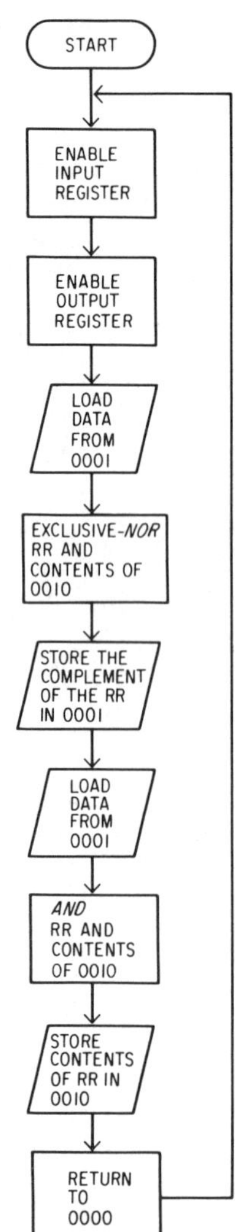

Fig. 6-16 Flowchart for half adder program not using SKZ and JMP instructions.

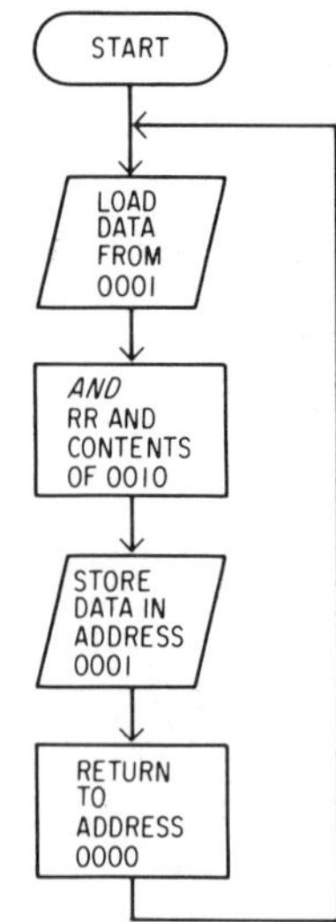

PROGRAM

0001*	Code for LD instruction
0001	Input address used by above instruction
0011	Code for AND instruction
0010	Input Address used by above instruction
1000	Code for STO instruction
0001	Output address used by above instruction
1111	Code for NOPF instruction (set flag F high to cause return to program START)

*The program must be preceded by these codes: 1010, 0110, 1011, and 0110. This will enable the inputs and outputs.

Fig. 6-17 Flowchart and program for AND gate function.

ered and flowcharted separately, and then all of the flowcharts will be incorporated into the final master program. There is a side benefit to this procedure. Modules can be lifted intact from one program and used in other programs. Just as a design file is used to decrease design time, a program file is used to speed programming.

Experiment 6-3: Programming the 14500B

Programming the 14500B is not difficult. An algorithm and flowchart for the problem to be solved is devised. This flowchart is converted to a set of program statements, and the program statements are entered into the program control memory. The AND operation of the previous example is a simple starting point. Figure 6-17 shows the flowchart and program that corresponds to the action of an AND gate. The program for the 14500B progresses in two-stage steps. The first stage of each clock cycle is used to accomplish the instruction. The second stage is used to indicate the address of the input or output that is associated with the instruction. In the example, the value at input 0001 is to be loaded into the result register by the first instruction. The first instruction of the program is the LD instruction, 0001. Following this instruction is the address of the input to be loaded, 0001 (input 1). The next code is the AND instruction, 0011. It is followed by the address 0010 (input 2). The result of these four codes is that the value at input 1 and the value at input 2 will "AND" together with the result becoming the RR value. The STO code 1000 is followed by the address code 0001. These two codes will route the value of the AND operation to output 1. The final code 1111 is the NOPF instruction and will loop to the top of the program.

Load this program into the MM2101-1 as follows:

1. Set the RUN/SINGLE STEP switch to SINGLE STEP.
2. Set all the PC switches to zero.
3. Press the HALT/RESET button. This resets the PC to zero, resets the ICU and the output latches, and sets the CLK signal high. The OP CODE light

will indicate that the CLK signal is high and an instruction should be loaded into memory.

4. Set the data switches to the code of the first instruction (binary 0001) and press the LOAD button. The 0001 pattern will be displayed by the data lights.
5. Press the SINGLE STEP button once. This toggles the CLK. The I/O address lights will indicate that the CLK is low and an I/O address should be loaded into memory.
6. Set the data switches to the ADDRESS of the input binary 0001 and press the LOAD button.
7. Press the SINGLE STEP button once. Note the PC has incremented and the CLK is high, indicating that the next complete statement should be entered.
8. Set the data switch to the bit pattern of the next piece of data to be entered, which in this case is 0011.
9. Press the LOAD button.
10. Press the SINGLE STEP button once.
11. Repeat steps 8–10 until the entire program has been entered.
12. Note: The NOPF instruction does not require that an I/O address be entered in memory. The I/O address location in memory for this instruction may be left unprogrammed.
13. Press the HALT/RESET button.
14. Stop.

The program entered can be verified as follows:

1. Press the HALT/RESET button. The PC will be reset to zero, the CLK will be high, and the first piece of data entered, 0001,[2] will be displayed by the data LEDs.
2. Press the SINGLE STEP button once. The second piece of data entered, 0001[2] will be displayed by the data LEDs. The entire program may be verified by sequencing through memory with the single step feature, while observing the data display LEDs. The address LEDs and the OP CODE and I/O address LEDs will aid in keeping track of particular ICU statements.
3. Press the HALT/RESET button.
4. Stop.

The program can be stepped through one step at a time. This is an instructive exercise. The action of each step will be indicated by the various LED monitors, and the processor action will become apparent. Set inputs 1, 2, and 6 high. Press the HALT/RESET button. With the RUN/SINGLE STEP switch in SINGLE STEP, press the RUN button and then sequence through the program by repeatedly depressing the SINGLE STEP button. Each time the SINGLE STEP button is depressed, the program will advance the clock ½ cycle. Instruction and addresses can be followed on the LEDs as the button is pushed. Output 1 will go high when the program reaches the next to the last code.

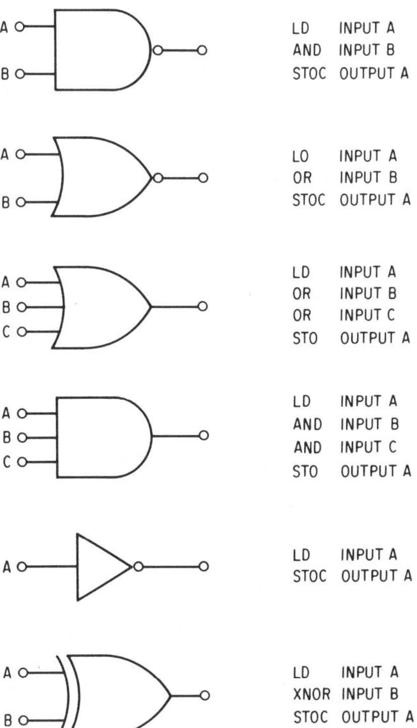

Fig. 6-18 Logic functions and program.

Place the RUN/SINGLE STEP switch to RUN. Press the HALT/RESET button. Develop a truth table for an AND gate, and set inputs 1 and 2 to the values indicated in the truth table. As each set of inputs is entered, depress the RUN button and record the resulting level of output 1. Check the resulting truth table to determine that the microprocessor is functioning as an AND gate.

Using a microprocessor as an AND gate is not very economical. However, a processor can be used to accomplish more complicated tasks. The 14500B can be used to perform the function of many different types of logic circuits. Figure 6-18 shows the instruction codes that result in some of these logic functions. Others can be programmed as well. The value of using a microprocessor in this way is that the functions can be altered by software (programs) rather than hardware. A microprocessor can perform the AND function at one point in the program and can be altered to perform a different function later in the program. Using more sophisticated tech-

[2]The IEN and OEN (input/output enable sequence) will be in the first four locations. Single step past them to reach data 0001.

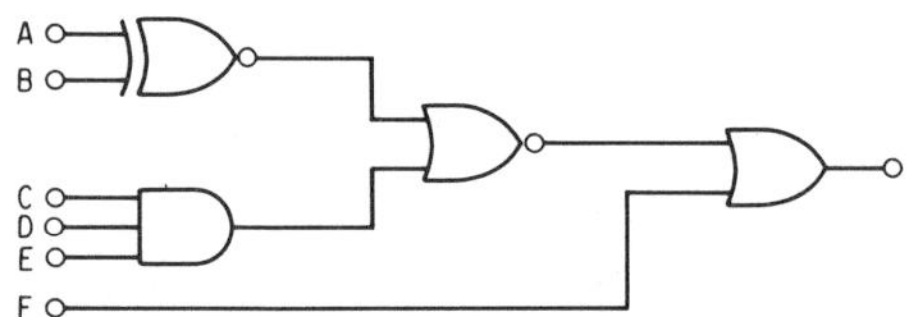

Fig. 6-19 Logic problem schematic.

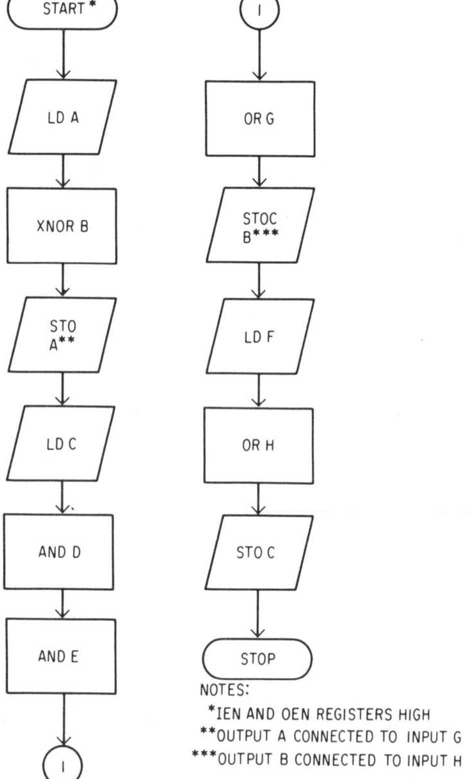

Fig. 6-20 Flowchart for logic problem.

niques, the processor can be directed to perform different functions based on the outcome of calculations. The power of the microprocessor is significantly increased as decision making is included in its processing.

Experiment 6-4: Logic Functions and the 14500B

Program the logic functions shown in Fig. 6-18. Test the programs on the 14500B system. Flowchart and program the logic circuit shown in Fig. 6-19. By way of suggestion, couple outputs back to inputs to permit storage and use of intermediate results. Vary the inputs and evaluate the processor action to confirm the flowchart. A truth table can be devised if desired. The flowchart shown in Fig. 6-20 may be helpful.

The 14500B is particularly useful for industrial control. An example suggested by Motorola is shown in Fig. 6-21. The program is shown in Fig. 6-22a. The

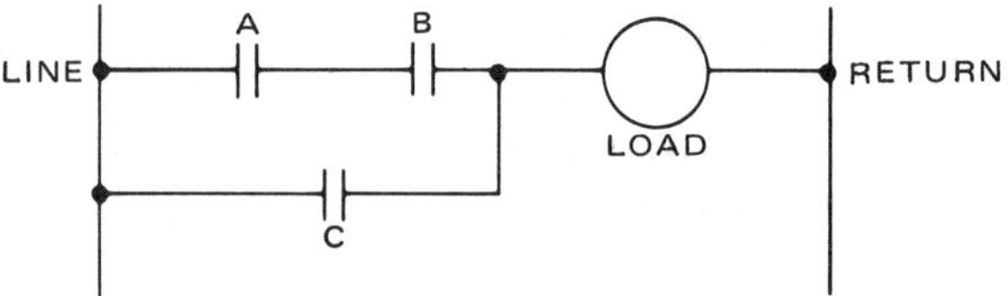

Fig. 6-21 Relay problem schematic. (Courtesy of Motorola, Inc.)

explanation for this program is shown in Fig. 6-22b. This is a typical motor control problem that might be found in industrial applications.

Relay control problems can easily be converted to logic circuits. The relay ladder shown in Fig. 6-23 is an example. The two relays in series, A and B, operate in the same way as an AND gate. Both relays must be energized to obtain a high output. The AND gate requires both inputs to be high to obtain a high output. If logic highs and relay closures are equated, the same decision will be made by each set of devices. In similar fashion, the parallel connection of relay C and the series circuit of relays A and B performs an OR function. The Boolean equivalent becomes AB + C = LOAD.

If the enable statements are ignored, the program to solve this control problem can be seen to be a combination of an AND function and an OR function as called for previously.

Make a flowchart and write a program to solve the relay control problem shown in Fig. 6-24. Some outputs may require connection to inputs to permit storage of intermediate values. Vary the inputs to verify that the processor operation conforms to the flowchart. If difficulty is encountered, refer to the logic diagram in Fig. 6-25 for assistance.

Three powerful programming sequences (or structures) provide unusual power for the ICU. These methods are (1) the if-then structure, (2) the if-then-else structure, and (3) the WHILE structure.

The if-then structure uses the output enable (OEN) instruction to permit or deny the execution of blocks of code based on inputs or calculations. As discussed earlier, the OEN register must contain a high to enable the output latches. This high can be loaded from any input or the result register. The result register pin-out (pin 15) can be connected to an input for addressable access. The if-then procedure loads the OEN register with the result of a calculation or with the value of an input. A greenhouse might be wired for automatic operation, for instance. The soil wetness could be monitored and used as an input. If the soil is wet enough, no watering is needed. If the soil is dry, watering is indicated. The if-then structure could handle the situation. Using mnemonics, the program would be as follows:

	Instruction	Operand	Notes
1 START	IEN	LOGIC 1	Enable the input register
2	OEN	LOGIC 1	Enable the output register
3	LD	A	Load the state of switch A into the Result Register
4	AND	B	Logically "AND" switches A and B
5	OR	C	Logically "OR" A • B with switch C
6	STO	LOAD	Transfer the result to the load to activate/deactivate it
7 END	NOPF		Causes the program to repeat this sequence

(A)

EXPLANATION OF PROGRAM

Statement #1 loads the IEN register with a logic 1. If the IEN register contained a logic 0, all future input data for the logical instructions would be interpreted as logic 0.

Statement #2 loads the OEN register with a logic 1 to enable the output instructions. If the OEN register contained a logic 0, the WRITE strobe from the ICU would be inhibited and the output latches could not be signalled to activate the load.

Statement #3 loads the Result Register with the state of switch A.

Statement #4 logically AND's the state of switch B with the contents of the Result Register; this result is then returned to the Result Register. The Result Register will now contain a logic 1 if and only if switches A and B were both high.

Statement #5 logically OR's the state of switch C with the content of the Result Register; this result is then returned to the Result Register. The Result Register will now contain a logic 1 if and only if switches A and B were high or switch C was high.

Statement #6 stores the content of the Result Register in the output latch. If the Result Register contained a logic 1, the output latch would receive a logic 1 to activate the load. The STO instruction does not alter the content of the Result Register.

Statement #7 creates a pulse on pin #9 of the ICU chip. This signal is used to preset the program counter to the beginning of the program. The entire sequence is then repeated.

(B)

Fig. 6-22 (a) Program for relay problem and (b) explanation of program. (Courtesy of Motorola, Inc.)

Step	*Mnemonic*	*Address*	*Action*
1	LD	SW	Load state of soil wetness switch into result register
2	OEN	RR	Enable outputs if RR = 1
3	STO	W/SW	Turn on water switch
4	STOC	HEAT	Turn off plant heaters
5	LD	CLK	Load value of timer
6	STO	W/SW	Turn off water after timer cycles to low
7	ORC	RR	Force RR to high
8	OEN	RR	Enable outputs

The if-then structure operates by disabling the outputs in step 2 if RR = 0. If the outputs are

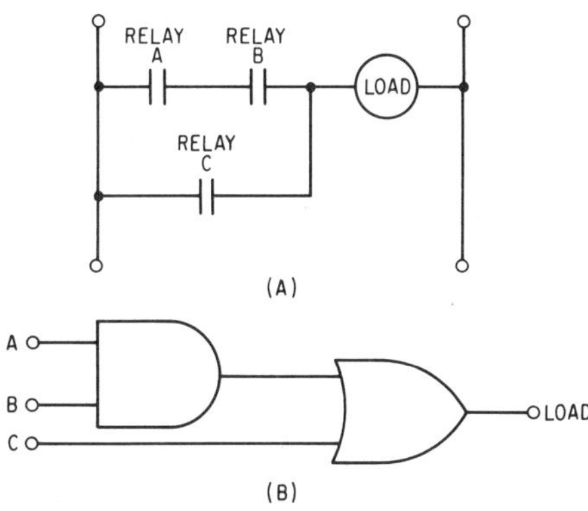

Fig. 6-23 (a) Relay ladder schematic and (b) logic equivalent.

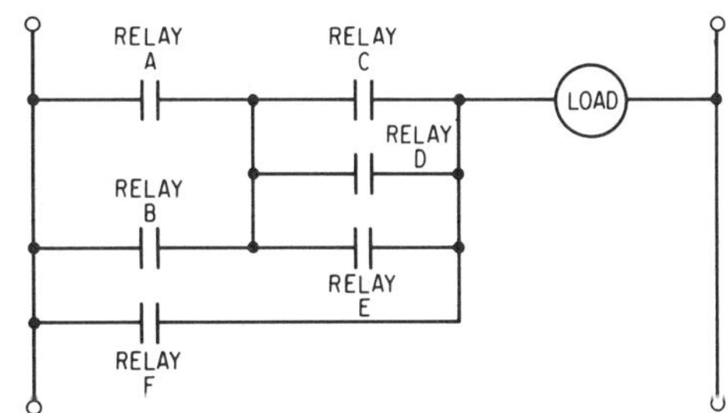

Fig. 6-24 Relay ladder schematic.

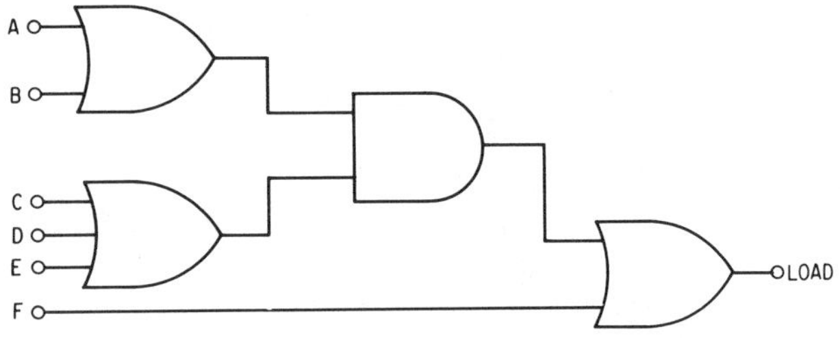

Fig. 6-25 Logic equivalent of relay ladder.

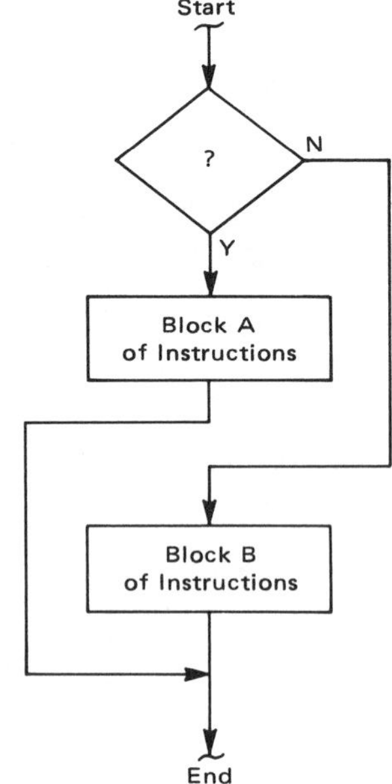

Fig. 6-26 Flowchart for the if-then-else function. (Courtesy of Motorola, Inc.)

disabled, the remaining statements will be cycled through, but no output changes will result. *If* RR = 1 in step 2, *then* the OEN will enable the outputs and the following statements will be effective in altering the outputs as required.

The if-then-else structure is similar to the if-then structure except that a second alternative (else) is provided (see Fig. 6-26). If the greenhouse were equipped with automatic temperature control, the if-then-else structure would be useful. If the temperature were too hot, the fan would be turned on, or if it were not hot, the louvers would be closed. The program could be:

Step	*Mnemonic*	*Address*	*Action*
1	LD	A	Load thermometer output
2	OEN	RR	Load OEN with RR
3	STO	FAN	If RR = 1, enable fan
4	LOC	A	Load complement of thermometer output
5	OEN	RR	Load OEN with RR
6	STO	louvers	Close louvers
7	ORC	RR	Force RR high
8	OEN	RR	Enable output

In this example, the thermometer status is used to determine whether to turn on the fan or close the shutter. A high thermometer status will enable the output at step 2 and will permit a high to be placed on the "fan" output. If the thermometer input is low, the output will not be enabled at step 2, and the fan output will remain low. Step 4 loads the complement of the thermometer input. A high input will load as a low and will disable the output at step 5. Step 6 will therefore result in no action. If the thermometer input is low, however, step 4 will load a high, the output will be enabled at step 5, and the "louver closing" output will go high at step 6. The if-then-else structure can be seen in the example: If the thermometer is high, the fan will be turned on; if the thermometer input is low, the louvers will be closed.

The last of these powerful programming possibilities is the WHILE structure. The WHILE structure operates to enable only one block of code during the time that a condition exists. There are two ways of accomplishing a WHILE structure: the loop-around-while and the jump-back-while.

The WHILE condition is activated by external conditions or internal instructions. The loop-around-while operates in a similar way to the if-then-else structure. The decision is "If the WHILE condition is satisfied, enable one block of code. If the condition is not satisfied, enable an alternate block of code." The choice can be made inside the WHILE structure. A temporary storage location can be assigned to each WHILE decision. The location would then receive the results of calculations or would be forced to some value by program instructions. Based on the value in the temporary storage location, the block of instructions would be executed or bypassed (see Fig. 6-27). An example of the loop-around-while could be a factory that places a certain number of pills in a bottle before putting on the cap. The WHILE test would be the count of the number of pills placed in the bottle. When the appropriate number is reached, the bottle will be capped. A flowchart for the control sequence is shown in Fig. 6-28. Write a program according to this flowchart. Run the program and determine that the objectives are accomplished. The output pulses for releasing the pill and putting on the cap will be very brief when the system is running at normal speed. Single

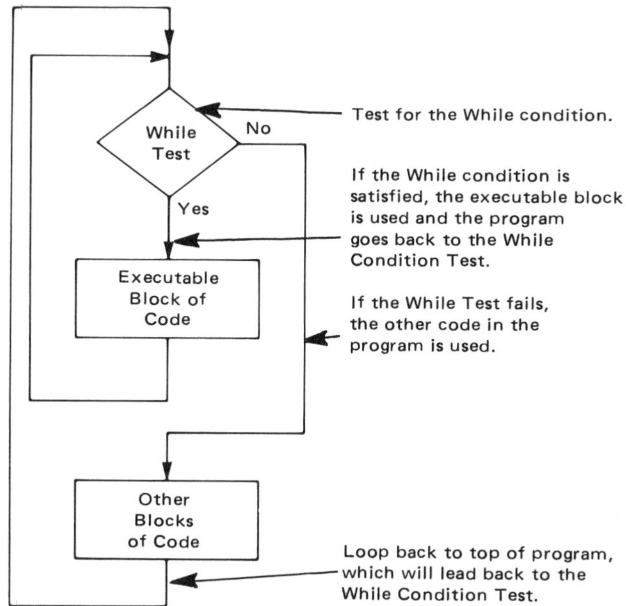

Fig. 6-27 Flowchart for the WHILE function. (Courtesy of Motorola, Inc.)

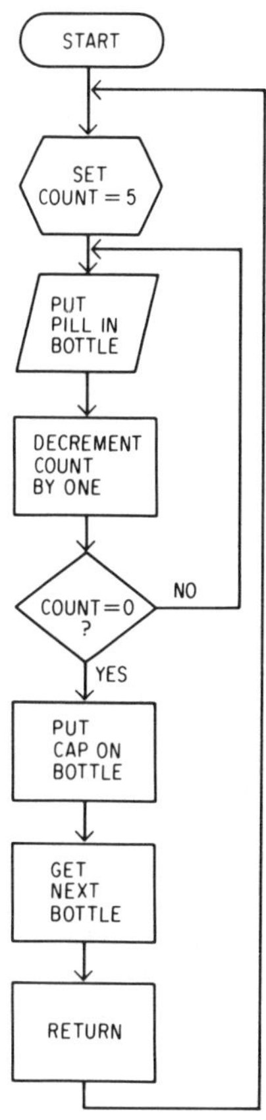

PROGRAMMING NOTES

1. Let one output be the pulse that releases a pill and one output install cap.
2. An external counter can be used to establish and monitor the pill count, or five outputs can be made high and then made low one at a time. The last would loop to an input, which would be checked to determine when "count = 0."
3. The OEN instruction can be used to implement the WHILE routine.

Fig. 6-28 Example flowchart.

stepping the program will display the program sequences more adequately.

The jump-back-while accomplishes the WHILE function by bypassing a JMP instruction as long as the WHILE condition is satisfied (see Fig. 6-29). This process was discussed earlier. The SKZ and JMP instructions are used together.

The following example will demonstrate the jump-back-while: A pump is to run until a tank is full. During the pumping time, we want no other action to take place. A flag switch (FS) will close, and FS = 1 when the tank is full.

```
1 START  LD    FS           LOAD FLOAT SW SIGNAL
2        SKZ                SKIP NEXT INST IF FS = 0
3        JMP   NEXT         JUMP TO NEXT
4        STO   PUMP         TURN ON PUMP
5        JMP   START        JUMP BACK TO START
6 NEXT   STOC  PUMP         TURN OFF PUMP
7              NEXT
               INSTRUCTION
```

Instruction 1 loads the FS signal into the ICU result register. If the switch is open, instruction 2 will cause instruction 3 to be ignored. Instructions 1–3 comprise the while question. The executable block is instructions 4 and 5, which drives the pump and jumps the program counter back to instruction 1. Instruction 6 turns off the pump when the WHILE condition fails. After instruction 6, the program counter steps through the balance of the program.

It should be noted that the JMP instruction requires external gating. The JMP instruction requires transfer to an alternate instruction location. To do this, the D inputs to the MM2101-1 memory must be determined by the JMP instructions and then loaded into memory. This capability has not been incorporated in the 14500B breadboard system in the interest of simplicity. The JMP instruction causes the JMP flag to go high for one clock period after receipt of the negative edge of the clock signal. This flag signal is used to enable the external gating circuits.

The two WHILE structures have their individual limitations and advantages. A few advantages of the loop-around-while are (1) external hardware is not required for jumping, and (2) two or more completely independent programs can run on the ICU at the same time. A few of its limitations are (1) its slower speed—all instructions are processed with some having no output enable, and (2) its more complex set of block-enabling conditions.

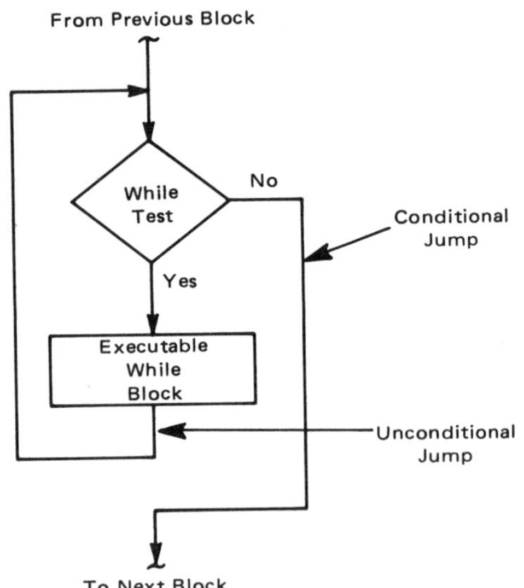

Fig. 6-29 Flowchart for the jump-back-while structure. (Courtesy of Motorola, Inc.)

Some advantages of the jump-back-while are (1) its simple structure, making it easy to write, (2) its simple operation, making it easy to follow, and (3) its increased speed. Some limitations are (1) external memory select hardware is needed, which can increase cost, and (2) only the WHILE block of instructions can be processed at any one time.

Perhaps the best answer is a combination of the two WHILE structures. The application will determine what is needed and what is cost effective.

Practice Exercises in Programming the 14500B

1. Make a flowchart and write a program that will turn a light on if the sun is up and will turn it off when the sun is down. Breadboard the circuit using a high on input 0 as an indication that the sun is up.

2. An automated drill press must drill four holes. The holes are to be at the corners of a square. The process must direct the drill press to drill a hole, move the material being drilled in one direction, drill a hole, move the material in another direction, drill, move, and drill. Different outputs can be used for each operation. The signal to move to the next operation will be a high on an input provided by an external timer. Flowchart, write, and run a program to accomplish this control process.

3. Write and run a program to accomplish binary subtraction of two three-digit binary numbers.

4. Sample a four-digit BCD input and display the digital value on a seven-segment LED readout. Drawing a logic circuit that will accomplish this will help. The displayed number will be fully lit only in the run mode. In the single step mode, only one segment at a time will be lit. Hint: Each number will be uniquely identified by a 4-bit code. Using the code and its complement, a high output can be obtained for any number from a four-input AND gate. By storing the high in the appropriate outputs, the LED readout segments for the number in question can be lit. The readout will have to be used in place of the output LEDs.

5. Write a program that will compare the magnitude of two 4-bit binary numbers and will display the larger numbers on the first four output LEDs and the smaller number on the last four output LEDs.

Going Further With the 14500B

The foregoing has not exhausted the power and capability of the 14500B. It has demonstrated the function of a basic microprocessor and the basic concepts of programming. The concepts of subroutines and similar program operations will be left for the multibit processor in the next chapter. For additional information about the 14500B, refer to the *MC14500B Industrial Control Handbook*, published by Motorola Semiconductor Products, Inc.

SEVEN

Multibit Microprocessors

Introduction

A microprocessor functions as a memory controller. The MC6802 microprocessor, for instance, regards all external devices as memory. For the processor to accept information (data) from an external source, the source must be made to look like memory. The microprocessor outputs its data as if the data were destined for memory. If input or output data is obtained from or destined for anything other than memory, conversion is required. The processor is not particular about the devices generating and receiving its data; it just insists that any data with which it deals appears to be memory data. The processor will use its data to control a motor as long as the motor circuit accepts data as if it were memory. The processor will accept data from an audio tape recorder or a magnetic disk file as long as that data appears to be coming from memory. In a microprocessor system, this requirement simply means that some interface circuitry must be provided to convert externally produced signals into processor compatible data and vice versa. Some of these interface devices will be discussed in this chapter.

Another result of this microprocessor idiosyncracy is that the operation of a microprocessor system can be understood without all of the expensive peripheral hardware that is normally associated with computers. A simple microcomputer, such as the one described in this chapter, will demonstrate the operational features of a multibit microprocessor at minimum expense with minimum complexity.

There are several differences between a system using a single-bit microprocessor, such as the 14500B, and a system built around a multibit microprocessor. Some of these differences are due to the nature of the processor construction and others are due to user-imposed limitations. A fundamental difference is word length. A single-bit system has only 1 bit per word, while multibit processors have from 2 to 16 or more bits per word. Longer word lengths usually mean increased efficiency for most applications. In a single-bit processor, each instruction cycle will process only 1 data bit. In a multibit system, the same cycle will process several bits. The efficiency is apparent, but efficiency is won at the expense of cost and complexity.

Multibit microprocessors are numerous and varied. Several word lengths are available. Among the more common are 4- and 8-bit units. These units function in similar ways except for the bit length capacity of the processors. The added power of these multibit machines make a nonlooping programming structure more desirable. The longer data words provide much greater flexibility in programming. The data can be used to alter the program memory sequence by jumping or transferring to new memory locations. The result is the ability to perform a variety of processing functions within the same program based on intermediate results within the data. This capacity is accomplished using a programming technique called *subroutines*. A subroutine is a smaller program that is contained within a larger program. The smaller program is usually capable of operating alone but is used within the larger program when its processing capabilities are needed along with others. The 14500B system can be designed to incorporate subroutine capability if external registers are added for temporary storage of data to form 4-bit words to be used in shifting the program memory to a new location. Such procedures are relatively time consuming when using 1-bit microprocessors. A minimum of five machine cycles would be needed for one memory shift. Four cycles would be used in filling the output registers with appropriate data. A fifth cycle would be needed to access the resulting memory location. The value of multibit processors is that memory locations can be identified with fewer data words and therefore fewer machine cycles.

Another advantage of multibit processors is the increased size of the instruction set. Most of these devices have more sophisticated data routines built into them. The result is less tedious programming.

Some multibit processors are part of a microprocessor family. IC chips within such families are fully compatible and provide for timing, control, interface, and memory functions. Using these family components, a sophisticated microcomputer system can be constructed with a minimum of design difficulties.

Overview of the Motorola MC6802 Microprocessor

The Motorola MC6802 microprocessor is representative of multibit microprocessors. The 6802 is a relatively recent addition to the MC6800 line of system conponents. It is promoted as a two-chip microcomputer. The MC6802 and the MC6846 can be combined to form a simple microcomputer system. The MC6802 contains all of the registers and accumulators of the earlier MC6800 and adds to them an internal clock oscillator and driver and 128 bytes of RAM. For the purpose of learning microprocessor design and operation, the 6802 is valuable since it is in large measure self-contained.

The functions of the MC6846 include ROM, I/O, and timer. These functions can be accomplished with other system chips. The I/O buffering function can be provided by the MC6820 Parallel Interface Adapter. This chip provides buffering to drive two 8-bit parallel, bidirectional data circuits. This is required since the 6802 does not possess the internal buffering required to drive peripheral devices and indicators.

The timing function can be accomplished by the MC6840 programmable timer chip. The 6840 provides three timers in a single package. Considering that the 6846 has only one timer, use of the 6840 will increase the system capability.

The ROM function can be accomplished with any of several memory chips depending on the type memory desired. Maskable ROM is primarily useful for large-quantity production runs and is available as the MCM6830A (1,024 × 8), the MCM6832 (2,048 × 8), the MCM68308 (1,024 × 8), and the MCM68317 (2,048 × 8). EPROM is available in the MCM68708 (1,024 × 8) alterable ROM. A variation of the MCM68708 can be programmed singly or in small quantities for system debugging, but cannot be subsequently erased.

The MCM6810A (128 × 8) static RAM can be included in the system and the MCM6604 (4,096 × 1) and the MCM6605A (4,096 × 1) provide dynamic RAM capability. Full information concerning these system chips is available in data sheets provided by Motorola.

The 6802 has several desirable features as follows:

- On-chip clock circuit
- 128 × 8 bit on-chip RAM
- 32 bytes of low-power retainable RAM
- Software compatibility with the MC6800
- Expandable to 65K words of memory
- Standard TTL-compatible inputs and outputs
- 8-bit word size
- 16-bit memory addressing
- Interrupt capability
- Single + 5 Vdc supply requirements

Among microprocessors, this list is impressive. All of the features necessary for a powerful microcomputer are included. Compatibility with the MC6800 provides access to much existing software. The 6802 is a good choice for a breadboard microcomputer as well as for a standard microcomputer.

Figure 7-1 shows a block diagram of the 6802. As can be seen in the figure, the 6802 is significantly more complex than the 14500B. The interconnecting bus network shown is actually three networks. These buses carry 8 parallel bits of data, 16 parallel bits of address data, and control and timing data. The interconnected elements of the 6802 include two 8-bit accumulators. These two accumulators hold operands and results from the Arithmetic Logic Unit (ALU). Also associated with the ALU is the condition code register. This is an 8-bit register in which the two most significant bits[1] are unused and are always high. Figure 7-2 shows the six registers of the 6802, including the condition code register. The meaning conveyed by each condition code register bit is identified. These bits are used to signify the status of conditions that are significant for the conditional branch and other instructions. Bit zero, for example, indicates whether a carry is produced during arithmetic operations.

The index register is a 16-bit (two bytes)[2] register that is used for memory addressing and temporary data storage. The use of this register in addressing will be discussed later. In Fig. 7-1, "H" and "L" in the register and counter blocks refer to "high" and "low" respectively. The "H" blocks are the 8 Most Significant Bits (MSB), and the "L" block is the eight Least Significant Bits (LSB).

The program counter also is 2 bytes in length The program counter points to the address of the next instruction to be executed. In a similar fashion, the 16-bit stack pointer contains the address of the next available location in a first-in/last-out register stack. This stack is located in external memory. It

[1]Most Significant Bit (MSB) and Least Significant Bit (LSB) are terms often used to indicate the two ends of a binary number with the MSB being the bits carrying the greatest binary value.

[2]The MC6802 uses an 8-bit word length. One byte equals 8 bits.

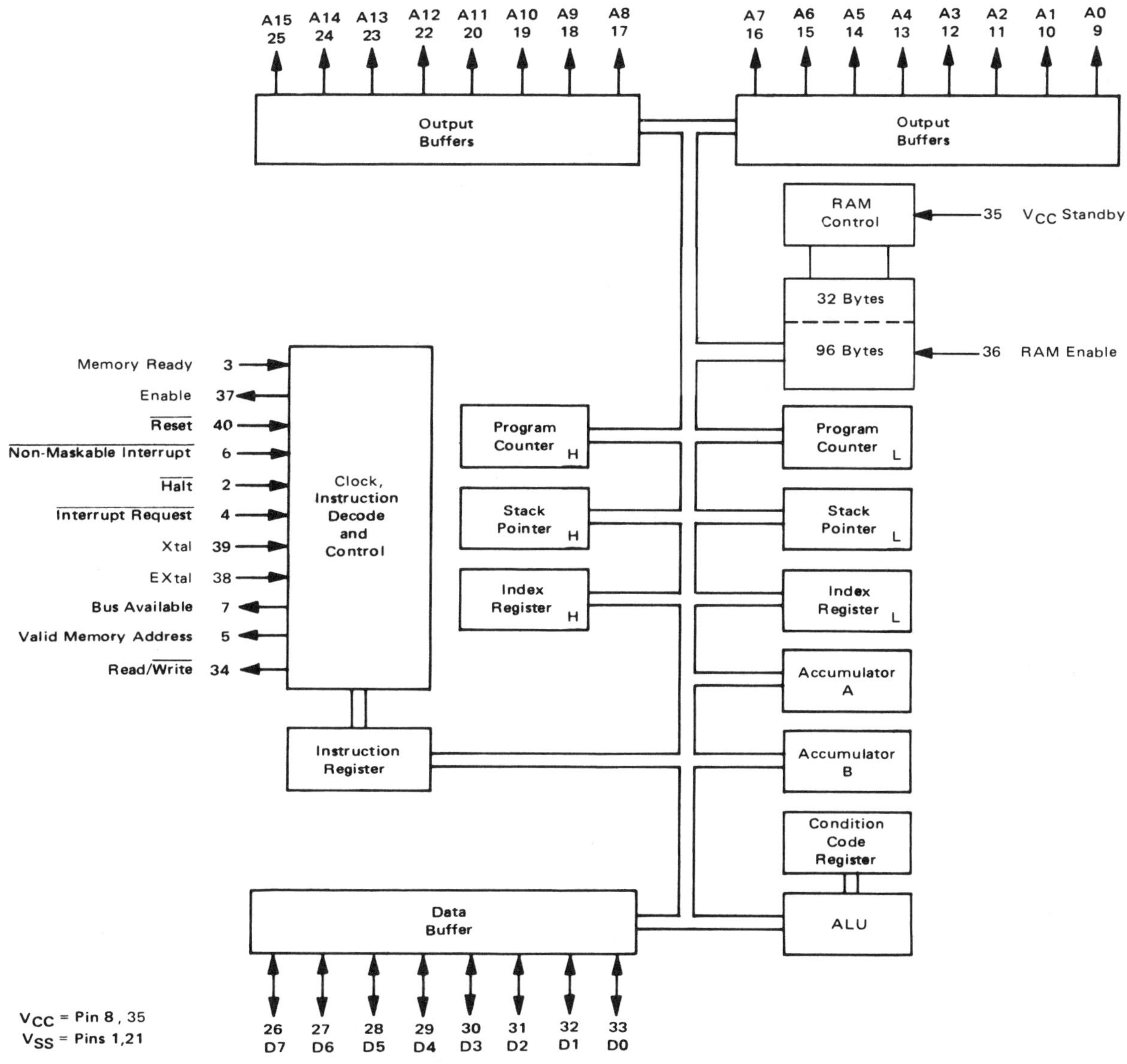

Fig. 7-1 Functional block diagram of the 6802. (Courtesy of Motorola, Inc.)

should be noted that many microprocessors incorporate the register stack into the chip itself. This has the advantage of speed but the disadvantage of space limitations. An "on-board" stack can contain only as many bytes as are included during manufacture. If data is entered after the stack has become full, the data word that was entered first will be pushed out of the stack and lost. With an external stack, the number of memory locations can be expanded as required.

The 6802 is unusual in that it incorporates 128 bytes of RAM on the chip itself. This is a step toward a one-chip microcomputer.[3] The 128 bytes of RAM are useful in small systems. It can be used as scratchpad memory during system development and can hold intermediate results of processing for small programs. The 32 bytes of retainable RAM are handy for storge of start-up routines. Since RAM is volatile, removal of power to the chip will result in loss of all stored data. If a battery supply is connected to the V_{CC} standby (pin 35), the lower 32 bytes of RAM (hex addresses 0000-001f) will be retained at very low current drain, 8 mA maximum at 5.25 V.

Access to external memory and peripheral devices is accomplished via the data and output buffers. Addressing is provided on 16 addressing lines. Data uses 8 lines. The 6802 is capable of driving capacitance loads up to 130 pF with current loads of up to 1.6-mA sink current and 100-μA source current. This drive capacity is enlarged by using system chips such as the MC6820 Peripheral Interface Adapter.

[3]One-chip computers already exist. For example, the Intel 8748 includes 1K × 8 EPROM, 64 × 8 RAM, 27 I/O lines, interval timer, and interrupt features.

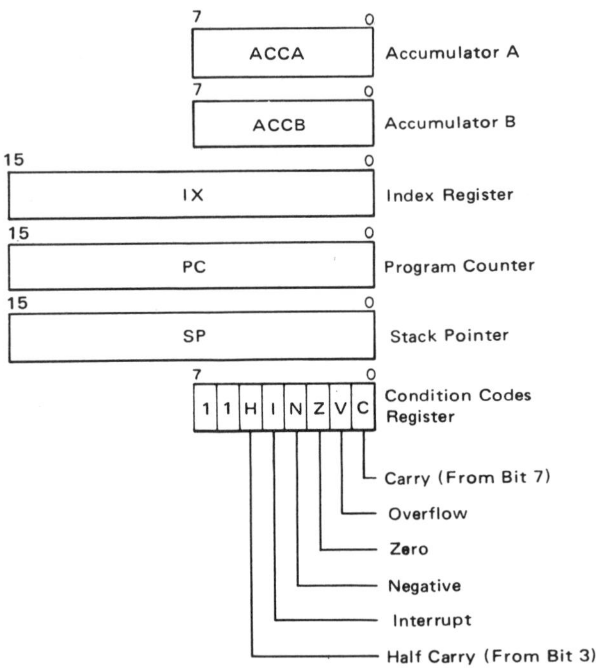

Fig. 7-2 Register and accumulator identification. (Courtesy of Motorola, Inc.)

The remaining blocks shown in Fig. 7-1 are associated with the on-chip clock and the instruction decoding and control functions. The frequency of the two-phase clock is determined by a crystal connected between pins 38 and 39. A 1-MHz crystal is typical for the 6802, but a divide-by-4 circuit has been added that permits use of a lower-cost 4-MHz crystal. Crystals used for time bases in color television systems are often used for this application. With a frequency of 3.58 MHz and greater availability, this type of crystal is preferable in some instances.

An external TTL clock signal can be used to drive the 6802 when connected to pin 38 with pin 39 floating.

The control-decoding function responds to external signals and is best understood in relation to them. As shown in Fig. 7-3, there are 11 control signals besides the two crystal connections. These control signals are identical to those of the MC6800 except that inclusion of the clock and RAM circuitry has resulted in deletion of the tri-state control, data bus enable, Φ1 and Φ2 pinouts, and resulted in the addition of RAM enable, memory ready, V_{CC} standby, and the two crystal connections. The 6800 had two unused pins and these plus the deleted and altered signals permitted addition of the required pinouts while still using a 40-pin DIP.

The $\overline{\text{Halt}}$ signal (pin 2) stops all activity in the processor when the pin is low. The function is level sensitive and must be brought low prior to the last 250 ns of the Φ2 output enable signal. When the halt mode is initiated, the processor will complete the instruction in process before halting. To restore operation, the $\overline{\text{Halt}}$ signal must go high for one clock cycle. When halted, the following conditions will be observed:

1. The Valid Memory Address (VMA) will be at a low state.
2. All three state lines other than the VMA will be in the three-state mode.
3. The address bus will display the address of the next instruction.

The Read/$\overline{\text{Write}}$ (R/$\overline{\text{W}}$) signal indicates which of the two modes the Microprocessing Unit (MPU) is in. If pin 34 (R/$\overline{\text{W}}$) is high, the MPU is in the read mode and will accept data. If R/$\overline{\text{W}}$ is low, the write mode is in progress. This is a TTL level signal capable of driving 1.6 mA and 90 pF. In the halt and standby modes, the R/$\overline{\text{W}}$ will be high.

The Valid Memory Address (VMA) is used to enable peripheral interfaces. When high, the VMA signal indicates that a valid memory address is on the address bus. The signal will drive a standard TTL load and 90 pF.

When the microprocessor is in the halt or wait mode, the Bus Available (BA) signal will be in the high state. The BA signal indicates that the MPU's

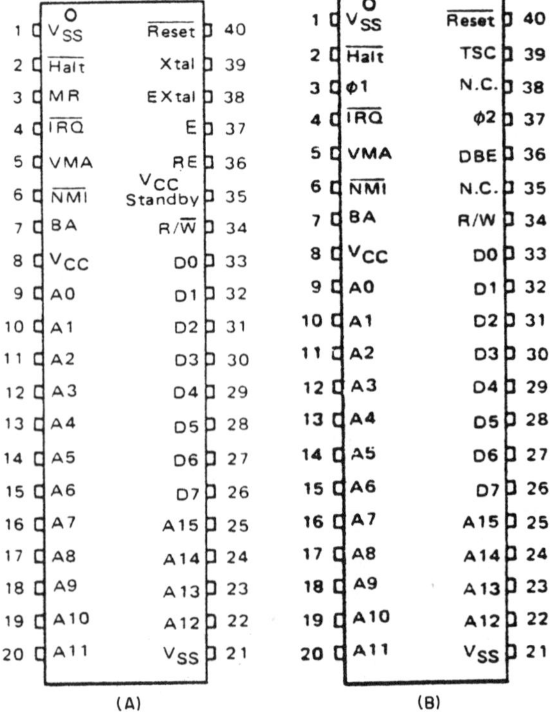

Fig. 7-3 (a) 6802 pin-outs and (b) 6800 pin-outs. (Courtesy of Motorola, Inc.)

tri-state output drivers are in the off state and other outputs will be off. The BA signal signifies that the bus is not in use by the MPU. If the MPU is in the wait state, a maskable or nonmaskable interrupt is required to terminate the wait mode. The BA signal will drive one standard TTL load and 30 pF.

The Interrupt Request ($\overline{IRQ}$) is used to halt the flow of data when required by an input. The $\overline{IRQ}$ signal directs the MPU to prepare for and execute an interrupt routine. The MPU will recognize the $\overline{IRQ}$ on a level sensitive basis. When the $\overline{IRQ}$ is received, the MPU will complete the current instruction prior to initiating the $\overline{IRQ}$ routine. At that time, if the interrupt mask bit in the condition code register is not set, the MPU will load the contents of index register, the program counter, the accumulators, and the condition code register on the memory stack. The interrupt mask bit will be set high to preclude further interrupts. A 16-bit address will then be loaded from memory locations FFF8 and FFF9, which will cause the MPU to break to an interrupt routine in memory. The interrupt will not be serviced unless the $\overline{\text{Halt}}$ line is high. If the $\overline{\text{Halt}}$ is low, the interrupt will be latched internally.

An external 3,000-Ω pull-up resistor is required between the $\overline{IRQ}$ pin (pin 4) and V_{CC} (pin 8). While a high-impedance pull-up device is included internal to the chip, the external resistor is needed for wired-OR connection and optimum control of interrupts.

The $\overline{\text{Reset}}$ signal is used to reset and start the MPU from a power-down condition. With the $\overline{\text{Reset}}$ line low, the MPU will be inactive and all register in-

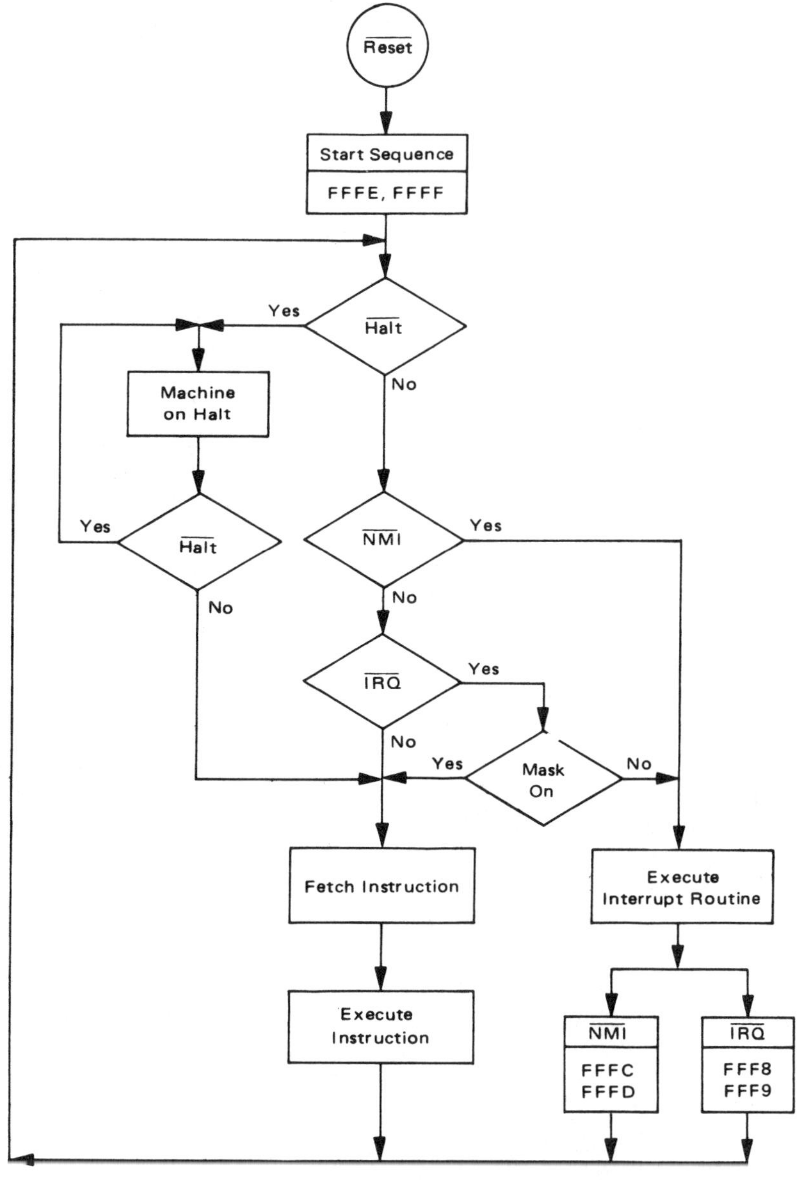

Fig. 7-4 Flowchart for the 6802 MPU. (Courtesy of Motorola, Inc.)

Table 7-1 Memory map for interrupt vectors. (Courtesy of Motorola, Inc.)

Vector		Description
MS	LS	
FFFE	FFFF	Restart
FFFC	FFFD	Non-maskable interrupt
FFFA	FFFB	Software Interrupt
FFF8	FFF9	Interrupt Request

formation will be lost. When a high state is detected on the $\overline{\text{Reset}}$ line, the restart sequence is initiated. As a part of this sequence, the interrupt mask bit is set, all of the higher order address lines will be forced high, and memory locations FFFE and FFFF will be used to load the program addressed by the program counter.

Another MPU routine is executed in response to a Nonmaskable Interrupt signal ($\overline{\text{NMI}}$). The $\overline{\text{NMI}}$ routine is similar to the $\overline{\text{IRQ}}$. Both the $\overline{\text{NMI}}$ and the $\overline{\text{IRQ}}$ are hardware interrupts that are sampled when the Enable (E) line is high and start the interrupt sequence when E goes low. During the $\overline{\text{NMI}}$ sequence, the index register, the program counter, the accumulators, and the condition code register are stored on the memory stack. At the end of the cycle, a 16-bit address in memory locations FFFC and FFFD will cause the MPU to branch to a nonmaskable interrupt routine in memory. The routine is referred to as being nonmaskable because the interrupt mask bit in the condition code register has no effect on the interrupt request. As with the $\overline{\text{IRQ}}$ signal, an external 3-kΩ resistor is required for wire-OR and optimum control of interrupts.

The flowchart in Fig. 7-4 shows the processing flow that results from the halt and interrupt routines that have been discussed. The address locations shown in the blocks beneath the $\overline{\text{Reset}}$, $\overline{\text{NMI}}$, and $\overline{\text{IRQ}}$ notations direct the MPU to go to that memory location to initiate the appropriate routine. These memory addresses are called *vector addresses* and are summarized in Table 7-1.

Two of the remaining control signals relate to the operation of the memory. The RAM Enable (RE) directs the on-chip RAM to respond to MPU controls when it is in the high state. When the RE is low, the on-chip RAM will be disabled. The RE also is used to prevent reading and writing the on-chip RAM during a power-down situation. To use this feature, the RE must be low at least 3 sec before Vcc goes below 4.75 V during power-down.

The Memory Ready (MR) is used to permit interface with slow memories. MR functions by altering the length of the Enable (E) control signal. With MR high, E is not affected. When MR is low, E may be stretched in integral multiples of half periods.

The Enable (E) signal is a TTL compatible output that is equivalent to the Φ2 clock signal of the MC6800. The E signal is used for system timing.

The final output is pin 35, Vcc standby. Vcc standby supplies power to the first 32 bytes of RAM and RAM Enable (RE) control logic. This will permit retention of these first bytes of RAM on a power-up, power-down, or standby condition. Maximum current drain in the standby mode is 8 mA at 5.25 V.

The use and integration of the registers, control lines, and signals will become clear as programming of the MPU system is discussed. Basic to the programming process is the MPU instruction set.

The MC6802 Instruction Set

Figure 7-5 shows an alphabetical listing of the 6802 instruction set. The increased capability and sophistication provided by this 8-bit processor above that of the 14500B is apparent when the sizes of the two instruction sets are compared. The 16 instructions of the 14500B have been expanded to 72 in the MC6802. In actuality, the 6802 will recognize and take action on 197 of the 256 possible instructions. This increased number is due to multiple addressing modes for some instructions. Discussing all of these instructions would be tedious and lengthy. Such detail is not required to accomplish the goals of this chapter. The intention of this chapter is to familiarize the student with the basic operation and design parameters of multibit processors. It is expected that the student will build on this foundation as he uses multibit processors in operating systems. The approach of the chapter will be, therefore, to discuss the elements of the instruction set in more generalized categories with discussion of specific instructions being reserved for those that are required in the experiments.

Program Statements

When writing programs to direct the operation of the MPU, use of binary numbers becomes tedious and time consuming. To reduce this problem, each of the instructions is identified as a three-letter mnemonic. When entered into the microcomputer, of course, the mnemonic must be converted to binary numbers. Mnemonics make initial construction of programs simpler. Program statements include an instruction that tells the microprocessor what to do. This instruction will be referred to as an *operator*.[4] The operator can be any of the instruction codes to which the MPU will respond. A second part of the

[4]The terms *operator* and *operand* are used in assembly language source program statements. An *assembler* is a program that converts mnemonics to binary code required by the MPU. This volume will not use an assembler but will code the programs manually.

ABA	Add Accumulators	INS	Increment Stack Pointer
ADC	Add with Carry	INX	Increment Index Register
ADD	Add	JMP	Jump
AND	Logical And	JSR	Jump to Subroutine
ASL	Arithmetic Shift Left	LDA	Load Accumulator
ASR	Arithmetic Shift Right	LDS	Load Stack Pointer
BCC	Branch if Carry Clear	LDX	Load Index Register
BCS	Branch if Carry Set	LSR	Logical Shift Right
BEQ	Branch if Equal to Zero	NEG	Negate
BGE	Branch if Greater or Equal Zero	NOP	No Operation
BGT	Branch if Greater than Zero	ORA	Inclusive OR Accumulator
BHI	Branch if Higher	PSH	Push Data
BIT	Bit Test	PUL	Pull Data
BLE	Branch if Less or Equal	ROL	Rotate Left
BLS	Branch if Lower or Same	ROR	Rotate Right
BLT	Branch if Less than Zero	RTI	Return from Interrupt
BMI	Branch if Minus	RTS	Return from Subroutine
BNE	Branch if Not Equal to Zero	SBA	Subtract Accumulators
BPL	Branch if Plus	SBC	Subtract with Carry
BRA	Branch Always	SEC	Set Carry
BSR	Branch to Subroutine	SEI	Set Interrupt Mask
BVC	Branch if Overflow Clear	SEV	Set Overflow
BVS	Branch if Overflow Set	STA	Store Accumulator
CBA	Compare Accumulators	STS	Store Stack Register
CLC	Clear Carry	STX	Store Index Register
CLI	Clear Interrupt Mask	SUB	Subtract
CLR	Clear	SWI	Software Interrupt
CLV	Clear Overflow	TAB	Transfer Accumulators
CMP	Compare	TAP	Transfer Accumulators to Condition Code Reg.
COM	Complement	TBA	Transfer Accumulators
CPX	Compare Index Register	TPA	Transfer Condition Code Reg. to Accumulator
DAA	Decimal Adjust	TST	Test
DEC	Decrement	TSX	Transfer Stack Pointer to Index Register
DES	Decrement Stack Pointer	TXS	Transfer Index Register to Stack Pointer
DEX	Decrement Index Register	WAI	Wait for Interrupt
EOR	Exclusive OR		
INC	Increment		

Fig. 7-5 MC6802 instruction set (alphabetic sequence.) (Courtesy of Motorola, Inc.)

statement is the data to be used with the operator or the location at which that data is to be obtained from or sent to. This second part of the statement is called an *operand.* In the following discussion, the operand will be written adjacent to the operator. In actual programs the operand will follow the operator. The nature and length of the operand will be determined by the operator and the addressing mode. Some operators require no separate operand. Some use one 8-bit operand while others use two 8-bit words as an operand. In some statements, the operand is data while in others, it is a memory location that contains data.

Addressing Modes

The instruction set of the 6802 is dependent in many cases on the addressing mode that is used with it. An addressing mode is the manner in which the program causes the MPU to obtain its data. There are several types of addressing modes available to the MC6802.

Inherent Addressing Mode This mode relates to addressing of the accumulators (accumulator A and accumulator B) and the index register, or to operators that have the address "built in."

Some operators have the address within the operator. The address is said to be *inherent.* ABA is such an operator. ABA contains all required information to permit its operation to be accomplished. A program statement using the ABA operator appears incomplete.

Operator	*Operand*	*Process Description*
ABA	—	Add accumulators A and B together and place the result in accumulator A.

All necessary addressing and instruction is included in the operator, however, and nothing else is required. A partial list of other inherent addressing operators include the following:

INX Increment the index register
CBA Compare accumulators
SBA Subtract accumulators
TBA}
TAB} Transfer accumulators
DEX Decrement the index register
DES Decrement the stack pointer
TXS Index register to stack pointer

Operators such as TSTA, ADDB, COMA, DECB, and SUBA appear to be four characters long. In reality, they each consist of an operator and an operand address label. TSTA is actually TST (operator) and A (operand). It means, test the contents of accumulator A. These operators also use the inherent addressing mode.

Immediate Addressing Mode This mode uses the operand to hold an actual data value. The following program statement illustrates this addressing mode:

Operator	*Operand*
ADDA	100_{10}

This statement directs the MPU to add the number "100" to accumulator A. This number is listed as an operand, not as a memory location. It is therefore immediately available. A decimal number is used for discussion. A binary number would be required for actual MPU operation. The operand value is limited to the range from 0 to 255. For the Compare Index Register (CPX), Load Index Register (LDX), and Load Stack Pointer (LSP), 2-byte (16-bit) operands are required. In the immediate addressing mode, the memory location of the operand immediately follows that of the instruction. Figure 7-6 illustrates the immediate addressing mode. Notice that the value 25 is loaded into accumulator A of the MPU from the memory location immediately following the instruction.

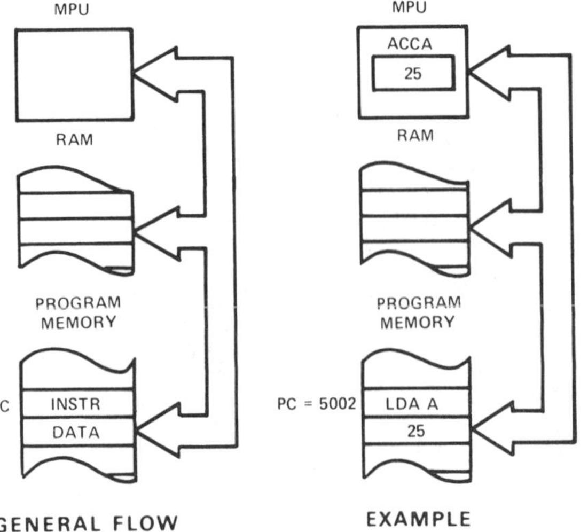

Fig. 7-6 Immediate addressing mode. (Courtesy of Motorola, Inc.)

Direct Addressing Mode This mode uses the operand field of the source statement as the address of the value that is to be operated on. In the direct mode, the 8-bit operand permits addressing of memory locations 0 through 255. A source program statement for direct addressing is as follows:

Operator	*Operand*
ADDA	2F*

*2F is a hexadecimal number that translates to binary 0010 1111.

This statement directs the MPU to add the contents of memory location 2F to accumulator A. Figure 7-7 shows an example of the direct addressing mode. In this case, the LDAA instruction directs the MPU to load accumulator A with the value of memory location 100. Memory location 100 contains the value 35, which is shown loaded into accumulator A of the MPU. If RAM is used for the first 255 memory locations, direct addressing can increase the speed of system operation.

Extended Addressing Mode This is another useful addressing mode. In this mode the number found in the memory location after the operator is used as the higher 8 bits of the address of the operand. The next memory location (second after the operator) contains a number that becomes the lower 8 bits of the address of the operand. Consider the following program segment:

Operator	*Operand*
ADDA	FFA0*
SUBB	E100

*$FFA0_{16}$ = $1111\ 1111\ 1010\ 0000_2$

The first statement calls for the MPU to add the contents of memory location FFA0 to accumulator A. Memory location (address) FFA0 is called the *extended memory address*. It will be found in two consecutive locations after the ADDA operator. The order will be FF followed by A0. Statement 2 will cause the contents of E100 to be subtracted from accumu-

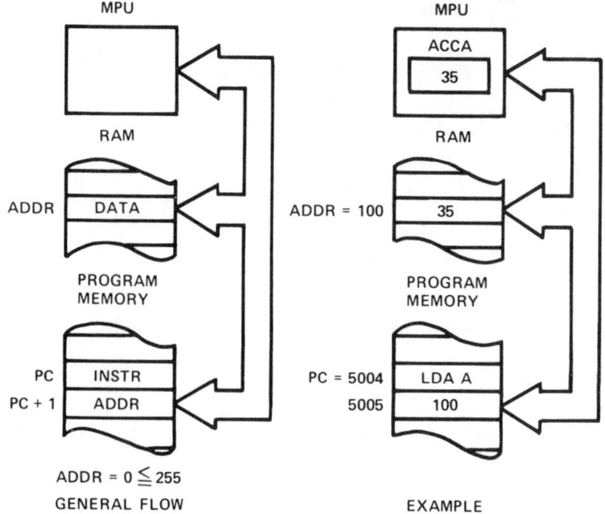

Fig. 7-7 Direct addressing mode. (Courtesy of Motorola, Inc.)

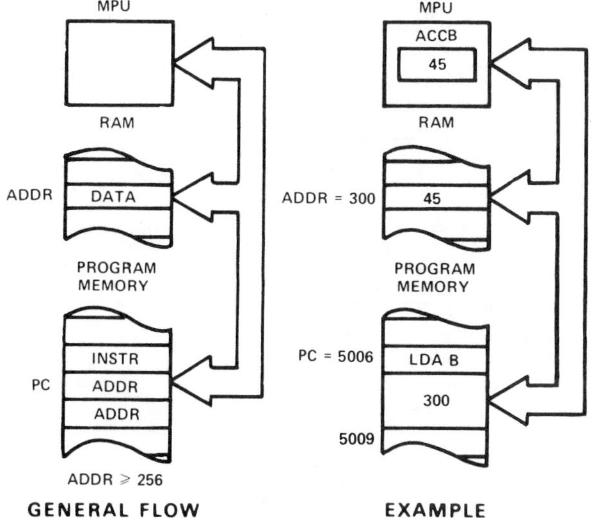

Fig. 7-8 Extended addressing mode. (Courtesy of Motorola, Inc.)

lator B. The location E100 in this case is the extended memory address.

Extended memory addressing permits the MPU to address up to 65,535 memory locations (16 bits). For systems that incorporate additional memory or that places ROM or EPROM in memory spaces beyond locations 0–255 (8 bits of address), extended addressing is necessary. Figure 7-8 shows an example of extended addressing. Memory location 300 is shown taking 2 bytes, since the binary code for 300 is 0000000100101100.

Relative Addressing Mode This mode operates in conjunction with the branch instruction (see Fig. 7-9). The branch instructions call for program branching to a new program location relative to the current location of the program counter. Program branches usually result from tests (see Fig. 7-5). If the test is met, a branch to a specified program counter location is executed. The branch can be to locations before or after the present program instruction. This bidirectional branching is accomplished by using a signed relative address. The eighth bit (MSB) of the address is used as a sign bit of either "0," which is interpreted as plus, or "1," which is interpreted as minus. The remaining 7 bits specify the address relative to the current program counter location + 2.[5] The branch can therefore be within the range from − 126 to + 129 memory locations of the branch instruction. Figure 7-9 will clarify the relative addressing mode. The BEQ (branch if the last instruction result was zero) is encountered in the program memory. The MPU is directed by the BEQ instruction to test the zero bit in the Condition Code Register (CCR) to see if the last result were zero (CCR zero bit = 1). If the bit is "0," indicating a nonzero result, the program advances to the second program counter location beyond the current location. (PC + 2) and continues with the next instruction. If the CCR zero bit is "1," the program branches to PC + 2 + 15 and continues. The branch always begins from the location of the next instruction past the branch instruction (PC + 2). This is the instruction location that would be selected if the test were negative. If program control must go beyond the branch instruction range, the jump and subroutine instructions will be used.

Indexed Addressing Mode This last addressing mode derives the operand or data address from the current contents of the index register. As an example, consider the following source program statement:

Operator	*Operand*
STAB	X

This statement directs the MPU to store the contents of accumulator B in the memory location indicated by the current value of the index register. The index register value can be altered by the program so that this addressing mode is "dynamic" in that program results can determine the program address.

It also is possible to use the operand to add a value to the contents of the index register. This is shown in Fig. 7-10. Notice that the value "5" is combined with the index register value "400" to give a

[5]The branch begins from the location where the next instruction would normally reside. The address of the branch follows the branch instruction, so the branch begins at program counter + 2.

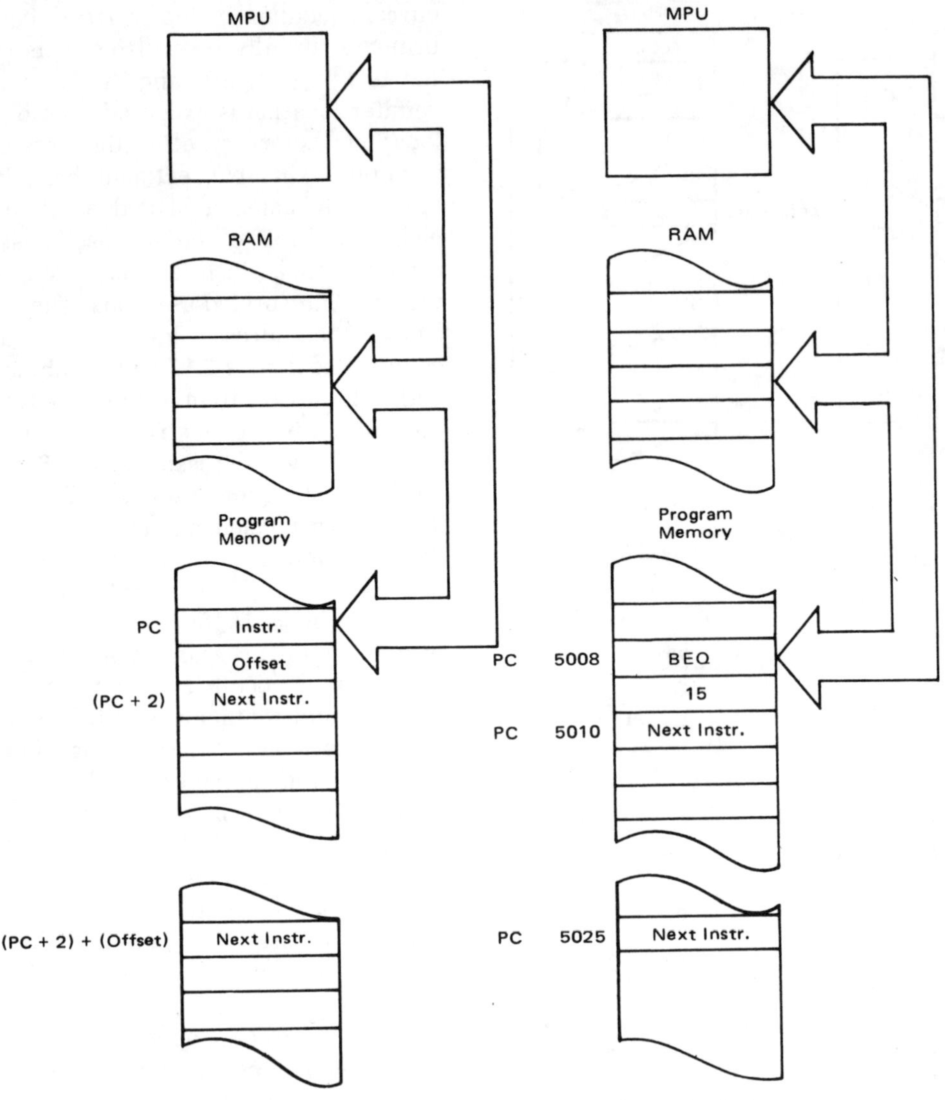

Fig. 7-9 Relative addressing mode. (Courtesy of Motorola, Inc.)

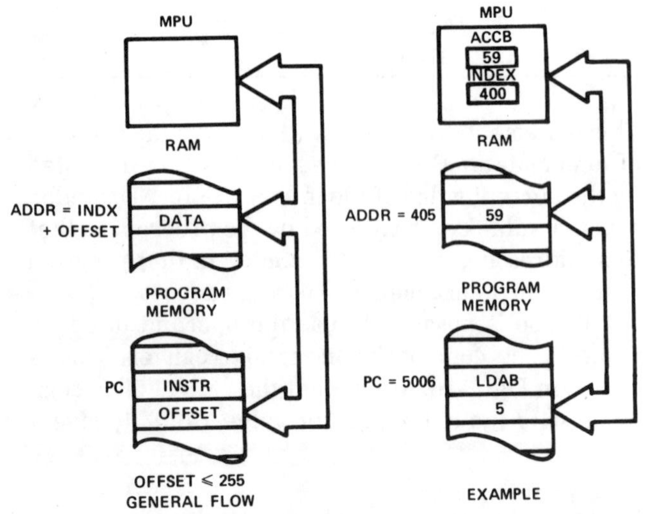

Fig. 7-10 Adding to the index register. (Courtesy of Motorola, Inc.)

memory address of 405. The contents of memory location 405 (59) is then loaded into accumulator B. The source program statement for the example would be:

Operator	*Operand*	*Process Description*
LDAB	5	Index addressing, load ACCB with contents of X + 5

The instruction set table shown in Fig. 7-11 will help clarify how these various addressing modes are selected. Notice that each operator has hexadecimal codes under the "OP" column of the various modes.

As shown in Fig. 7-12, the ADDA instruction can be used in all addressing modes except the inherent mode. The mode to be used will be determined by the op code. For this instruction, the immediate addressing mode results when the hexadecimal op code 8B

ACCUMULATOR AND MEMORY		ADDRESSING MODES															BOOLEAN/ARITHMETIC OPERATION	COND. CODE REG.					
		IMMED			DIRECT			INDEX			EXTND			INHER			(All register labels	5	4	3	2	1	0
OPERATIONS	MNEMONIC	OP	~	#	OP	~	#	OP	~	#	OP	~	#	OP	~	#	refer to contents)	H	I	N	Z	V	C
Add	ADDA	8B	2	2	9B	3	2	AB	5	2	BB	4	3				A + M → A	↕	•	↕	↕	↕	↕
	ADDB	CB	2	2	DB	3	2	EB	5	2	FB	4	3				B + M → B	↕	•	↕	↕	↕	↕
Add Acmltrs	ABA													1B	2	1	A + B → A	↕	•	↕	↕	↕	↕
Add with Carry	ADCA	89	2	2	99	3	2	A9	5	2	B9	4	3				A + M + C → A	↕	•	↕	↕	↕	↕
	ADCB	C9	2	2	D9	3	2	E9	5	2	F9	4	3				B + M + C → B	↕	•	↕	↕	↕	↕
And	ANDA	84	2	2	94	3	2	A4	5	2	B4	4	3				A • M → A	•	•	↕	↕	R	•
	ANDB	C4	2	2	D4	3	2	E4	5	2	F4	4	3				B • M → B	•	•	↕	↕	R	•
Bit Test	BITA	85	2	2	95	3	2	A5	5	2	B5	4	3				A • M	•	•	↕	↕	R	•
	BITB	C5	2	2	D5	3	2	E5	5	2	F5	4	3				B • M	•	•	↕	↕	R	•
Clear	CLR							6F	7	2	7F	6	3				00 → M	•	•	R	S	R	R
	CLRA													4F	2	1	00 → A	•	•	R	S	R	R
	CLRB													5F	2	1	00 → B	•	•	R	S	R	R
Compare	CMPA	81	2	2	91	3	2	A1	5	2	B1	4	3				A − M	•	•	↕	↕	↕	↕
	CMPB	C1	2	2	D1	3	2	E1	5	2	F1	4	3				B − M	•	•	↕	↕	↕	↕
Compare Acmltrs	CBA													11	2	1	A − B	•	•	↕	↕	↕	↕
Complement, 1's	COM							63	7	2	73	6	3				$\overline{M}$ → M	•	•	↕	↕	R	S
	COMA													43	2	1	$\overline{A}$ → A	•	•	↕	↕	R	S
	COMB													53	2	1	$\overline{B}$ → B	•	•	↕	↕	R	S
Complement, 2's	NEG							60	7	2	70	6	3				00 − M → M	•	•	↕	↕	①	②
(Negate)	NEGA													40	2	1	00 − A → A	•	•	↕	↕	①	②
	NEGB													50	2	1	00 − B → B	•	•	↕	↕	①	②
Decimal Adjust, A	DAA													19	2	1	Converts Binary Add. of BCD Characters into BCD Format	•	•	↕	↕	↕	③
Decrement	DEC							6A	7	2	7A	6	3				M − 1 → M	•	•	↕	↕	④	•
	DECA													4A	2	1	A − 1 → A	•	•	↕	↕	④	•
	DECB													5A	2	1	B − 1 → B	•	•	↕	↕	④	•
Exclusive OR	EORA	88	2	2	98	3	2	A8	5	2	B8	4	3				A ⊕ M → A	•	•	↕	↕	R	•
	EORB	C8	2	2	D8	3	2	E8	5	2	F8	4	3				B ⊕ M → B	•	•	↕	↕	R	•
Increment	INC							6C	7	2	7C	6	3				M + 1 → M	•	•	↕	↕	⑤	•
	INCA													4C	2	1	A + 1 → A	•	•	↕	↕	⑤	•
	INCB													5C	2	1	B + 1 → B	•	•	↕	↕	⑤	•
Load Acmltr	LDAA	86	2	2	96	3	2	A6	5	2	B6	4	3				M → A	•	•	↕	↕	R	•
	LDAB	C6	2	2	D6	3	2	E6	5	2	F6	4	3				M → B	•	•	↕	↕	R	•
Or, Inclusive	ORAA	8A	2	2	9A	3	2	AA	5	2	BA	4	3				A + M → A	•	•	↕	↕	R	•
	ORAB	CA	2	2	DA	3	2	EA	5	2	FA	4	3				B + M → B	•	•	↕	↕	R	•
Push Data	PSHA													36	4	1	A → M_{SP}, SP − 1 → SP	•	•	•	•	•	•
	PSHB													37	4	1	B → M_{SP}, SP − 1 → SP	•	•	•	•	•	•
Pull Data	PULA													32	4	1	SP + 1 → SP, M_{SP} → A	•	•	•	•	•	•
	PULB													33	4	1	SP + 1 → SP, M_{SP} → B	•	•	•	•	•	•
Rotate Left	ROL							69	7	2	79	6	3				M	•	•	↕	↕	⑥	↕
	ROLA													49	2	1	A C ← b_7 ← b_0	•	•	↕	↕	⑥	↕
	ROLB													59	2	1	B	•	•	↕	↕	⑥	↕
Rotate Right	ROR							66	7	2	76	6	3				M	•	•	↕	↕	⑥	↕
	RORA													46	2	1	A C → b_7 → b_0	•	•	↕	↕	⑥	↕
	RORB													56	2	1	B	•	•	↕	↕	⑥	↕
Shift Left, Arithmetic	ASL							68	7	2	78	6	3				M	•	•	↕	↕	⑥	↕
	ASLA													48	2	1	A C ← b_7 b_0 ← 0	•	•	↕	↕	⑥	↕
	ASLB													58	2	1	B	•	•	↕	↕	⑥	↕
Shift Right, Arithmetic	ASR							67	7	2	77	6	3				M	•	•	↕	↕	⑥	↕
	ASRA													47	2	1	A b_7 b_0 → C	•	•	↕	↕	⑥	↕
	ASRB													57	2	1	B	•	•	↕	↕	⑥	↕
Shift Right, Logic.	LSR							64	7	2	74	6	3				M	•	•	R	↕	⑥	↕
	LSRA													44	2	1	A 0 → b_7 b_0 → C	•	•	R	↕	⑥	↕
	LSRB													54	2	1	B	•	•	R	↕	⑥	↕
Store Acmltr.	STAA				97	4	2	A7	6	2	B7	5	3				A → M	•	•	↕	↕	R	•
	STAB				D7	4	2	E7	6	2	F7	5	3				B → M	•	•	↕	↕	R	•
Subtract	SUBA	80	2	2	90	3	2	A0	5	2	B0	4	3				A − M → A	•	•	↕	↕	↕	↕
	SUBB	C0	2	2	D0	3	2	E0	5	2	F0	4	3				B − M → B	•	•	↕	↕	↕	↕
Subract Acmltrs.	SBA													10	2	1	A − B → A	•	•	↕	↕	↕	↕
Subtr. with Carry	SBCA	82	2	2	92	3	2	A2	5	2	B2	4	3				A − M − C → A	•	•	↕	↕	↕	↕
	SBCB	C2	2	2	D2	3	2	E2	5	2	F2	4	3				B − M − C → B	•	•	↕	↕	↕	↕
Transfer Acmltrs	TAB													16	2	1	A → B	•	•	↕	↕	R	•
	TBA													17	2	1	B → A	•	•	↕	↕	R	•
Test, Zero or Minus	TST							6D	7	2	7D	6	3				M − 00	•	•	↕	↕	R	R
	TSTA													4D	2	1	A − 00	•	•	↕	↕	R	R
	TSTB													5D	2	1	B − 00	•	•	↕	↕	R	R

Fig. 7-11 Complete instruction set for the MC6800. (Courtesy of Motorola, Inc.)

INDEX REGISTER AND STACK POINTER OPERATIONS	MNEMONIC	IMMED OP	~	#	DIRECT OP	~	#	INDEX OP	~	#	EXTND OP	~	#	INHER OP	~	#	BOOLEAN/ARITHMETIC OPERATION	5 H	4 I	3 N	2 Z	1 V	0 C
Compare Index Reg	CPX	8C	3	3	9C	4	2	AC	6	2	BC	5	3				$(X_H/X_L) - (M/M+1)$	•	•	⑦	↕	⑧	•
Decrement Index Reg	DEX													09	4	1	$X - 1 \rightarrow X$	•	•	•	↕	•	•
Decrement Stack Pntr	DES													34	4	1	$SP - 1 \rightarrow SP$	•	•	•	•	•	•
Increment Index Reg	INX													08	4	1	$X + 1 \rightarrow X$	•	•	•	↕	•	•
Increment Stack Pntr	INS													31	4	1	$SP + 1 \rightarrow SP$	•	•	•	•	•	•
Load Index Reg	LDX	CE	3	3	DE	4	2	EE	6	2	FE	5	3				$M \rightarrow X_H, (M+1) \rightarrow X_L$	•	•	⑨	↕	R	•
Load Stack Pntr	LDS	8E	3	3	9E	4	2	AE	6	2	BE	5	3				$M \rightarrow SP_H, (M+1) \rightarrow SP_L$	•	•	⑨	↕	R	•
Store Index Reg	STX				DF	5	2	EF	7	2	FF	6	3				$X_H \rightarrow M, X_L \rightarrow (M+1)$	•	•	⑨	↕	R	•
Store Stack Pntr	STS				9F	5	2	AF	7	2	BF	6	3				$SP_H \rightarrow M, SP_L \rightarrow (M+1)$	•	•	⑨	↕	R	•
Indx Reg → Stack Pntr	TXS													35	4	1	$X - 1 \rightarrow SP$	•	•	•	•	•	•
Stack Pntr → Indx Reg	TSX													30	4	1	$SP + 1 \rightarrow X$	•	•	•	•	•	•

JUMP AND BRANCH OPERATIONS	MNEMONIC	RELATIVE OP	~	#	INDEX OP	~	#	EXTND OP	~	#	INHER OP	~	#	BRANCH TEST	5 H	4 I	3 N	2 Z	1 V	0 C
Branch Always	BRA	20	4	2										None	•	•	•	•	•	•
Branch If Carry Clear	BCC	24	4	2										C = 0	•	•	•	•	•	•
Branch If Carry Set	BCS	25	4	2										C = 1	•	•	•	•	•	•
Branch If = Zero	BEQ	27	4	2										Z = 1	•	•	•	•	•	•
Branch If ⩾ Zero	BGE	2C	4	2										$N \oplus V = 0$	•	•	•	•	•	•
Branch If > Zero	BGT	2E	4	2										$Z + (N \oplus V) = 0$	•	•	•	•	•	•
Branch If Higher	BHI	22	4	2										$C + Z = 0$	•	•	•	•	•	•
Branch If ⩽ Zero	BLE	2F	4	2										$Z + (N \oplus V) = 1$	•	•	•	•	•	•
Branch If Lower Or Same	BLS	23	4	2										$C + Z = 1$	•	•	•	•	•	•
Branch If < Zero	BLT	2D	4	2										$N \oplus V = 1$	•	•	•	•	•	•
Branch If Minus	BMI	2B	4	2										N = 1	•	•	•	•	•	•
Branch If Not Equal Zero	BNE	26	4	2										Z = 0	•	•	•	•	•	•
Branch If Overflow Clear	BVC	28	4	2										V = 0	•	•	•	•	•	•
Branch If Overflow Set	BVS	29	4	2										V = 1	•	•	•	•	•	•
Branch If Plus	BPL	2A	4	2										N = 0	•	•	•	•	•	•
Branch To Subroutine	BSR	8D	8	2										See Special Operations	•	•	•	•	•	•
Jump	JMP				6E	4	2	7E	3	3					•	•	•	•	•	•
Jump To Subroutine	JSR				AD	8	2	BD	9	3					•	•	•	•	•	•
No Operation	NOP										01	2	1	Advances Prog. Cntr. Only	•	•	•	•	•	•
Return From Interrupt	RTI										3B	10	1		⑩					
Return From Subroutine	RTS										39	5	1	See special Operations	•	•	•	•	•	•
Software Interrupt	SWI										3F	12	1		•	S	•	•	•	•
Wait for Interrupt	WAI										3E	9	1		•	⑪	•	•	•	•

CONDITIONS CODE REGISTER OPERATIONS	MNEMONIC	INHER OP	~	#	BOOLEAN OPERATION	5 H	4 I	3 N	2 Z	1 V	0 C
Clear Carry	CLC	0C	2	1	$0 \rightarrow C$	•	•	•	•	•	R
Clear Interrupt Mask	CLI	0E	2	1	$0 \rightarrow I$	•	R	•	•	•	•
Clear Overflow	CLV	0A	2	1	$0 \rightarrow V$	•	•	•	•	R	•
Set Carry	SEC	0D	2	1	$1 \rightarrow C$	•	•	•	•	•	S
Set Interrupt Mask	SEI	0F	2	1	$1 \rightarrow I$	•	S	•	•	•	•
Set Overflow	SEV	0B	2	1	$1 \rightarrow V$	•	•	•	•	S	•
Acmltr A → CCR	TAP	06	2	1	$A \rightarrow CCR$	⑫					
CCR → Acmltr A	TPA	07	2	1	$CCR \rightarrow A$	•	•	•	•	•	•

CONDITION CODE REGISTER NOTES:

(Bit set if test is true and cleared otherwise)

① (Bit V) Test: Result = 10000000?
② (Bit C) Test: Result = 00000000?
③ (Bit C) Test: Decimal value of most significant BCD Character greater than nine? (Not cleared if previously set.)
④ (Bit V) Test: Operand = 10000000 prior to execution?
⑤ (Bit V) Test: Operand = 01111111 prior to execution?
⑥ (Bit V) Test: Set equal to result of N ⊕ C after shift has occurred.
⑦ (Bit N) Test: Sign bit of most significant (MS) byte of result = 1?
⑧ (Bit V) Test: 2's complement overflow from subtraction of LS bytes?
⑨ (Bit N) Test: Result less than zero? (Bit 15 = 1)
⑩ (All) Load Condition Code Register from Stack. (See Special Operations)
⑪ (Bit I) Set when interrupt occurs. If previously set, a Non-Maskable Interrupt is required to exit the wait state.
⑫ (ALL) Set according to the contents of Accumulator A.

LEGEND:

OP	Operation Code (Hexadecimal);	00	Byte = Zero;
~	Number of MPU Cycles;	H	Half-carry from bit 3;
#	Number of Program Bytes;	I	Interrupt mask
+	Arithmetic Plus;	N	Negative (sign bit)
−	Arithmetic Minus;	Z	Zero (byte)
•	Boolean AND;	V	Overflow, 2's complement
M_{SP}	Contents of memory location pointed to be Stack Pointer;	C	Carry from bit 7
		R	Reset Always
		S	Set Always
+	Boolean Inclusive OR;	↕	Test and set if true, cleared otherwise
⊕	Boolean Exclusive OR;	●	Not Affected
$\overline{M}$	Complement of M;	CCR	Condition Code Register
→	Transfer Into;	LS	Least Significant
0	Bit = Zero;	MS	Most Significant

Fig. 7-11 cont'd Complete instruction set for the MC6800. (Courtesy of Motorola, Inc.)

ACCUMULATOR AND MEMORY OPERATIONS	MNEMONIC	IMMED OP	~	#	DIRECT OP	~	#	INDEX OP	~	#	EXTND OP	~	#	INHER OP	~	#
Add	ADDA	8B	2	2	9B	3	2	AB	5	2	BB	4	3			
	ADDB	CB	2	2	DB	3	2	EB	5	2	FB	4	3			

Fig. 7-12 ADDA instruction from the MC6800 instruction set. (Courtesy of Motorola, Inc.)

is used. In binary this is 10001011. Figure 7-12 shows that direct addressing is accomplished when an op code of 9B is used. Each instruction has its own unique op code for the various addressing modes.

A pattern within the accumulator and memory op codes can be identified. As shown in Fig. 7-11, the op codes for the instructions in the column for immediate mode will begin with hexadecimal 8 or C. If the codes are compared with the mnemonics, it will be seen that the "8" is associated with accumulator A and "C" with accumulator B. The same pattern holds true for the other modes with 9 and D being used for the direct mode, A and E for the index mode, and B and F for the extended mode. Instructions that are not associated with either accumulator utilize a "6" for the most significant 4 bits in the index mode and a "7" for the extended mode. The inherent mode does not conform to this pattern. Other patterns can be identified within the other instruction divisions. Recognizing these patterns will make programming easier since the amount of code memorization will be reduced.

Condition Code Register Operations

The 6802 instruction set can be divided into several categories. One category that affects and is affected by other categories of instructions is the Condition Code Register (CCR) instructions. The CCR, also called the *program status byte*, provides program flags or report bits that are used to indicate how the processing of programs should continue based on previous program operations. The bits of the CCR and their meaning for program control is shown as follows:

b_5	b_4	b_3	b_2	b_1	b_0
H	I	N	Z	V	C

H = Half-carry; set whenever a carry from b_3 to b_4 of the result is generated by ADD, ABA, ACD; cleared if no b_3-to-b_4 carry; not affected by other instruction.

I = Interrupt mask; set by hardware or software interrupt or SEI instruction; cleared by CLI instruction. (Normally not used in arithmetic operation.) Restored to a zero as a result of an RTI instruction, if I_m stored on the stack is low.

N = Negative; set if high-order bit (b_7) of result is set; cleared otherwise.

Z = Zero; set if result = 0; cleared otherwise.

V = Overflow; set if there was arithmetic overflow as a result of the operation; cleared otherwise.

C = Carry; set if there was a carry from the most significant bit(b_7) of the result; cleared otherwise.

The bits of the CCR are subject to alteration directly by using the instructions shown in Fig. 7-13 or

CONDITIONS CODE REGISTER OPERATIONS	MNEMONIC	BOOLEAN OPERATION	5 H	4 I	3 N	2 Z	1 V	0 C
Clear Carry	CLC	0→C	•	•	•	•	•	R
Clear Interrupt Mask	CLI	0→I	•	R	•	•	•	•
Clear Overflow	CLV	0→V	•	•	•	•	R	•
Set Carry	SEC	1→C	•	•	•	•	•	S
Set Interrupt Mask	SEI	1→I	•	S	•	•	•	•
Set Overflow	SEV	1→V	•	•	•	•	S	•
Acmltr A → CCR	TAP	A→CCR	①					
CCR → Acmltr A	TPA	CCR→A	•	•	•	•	•	•

R = Reset
S = Set
• = Not affected

① (ALL) Set according to the contents of Accumulator A.

Fig. 7-13 Condition code register instructions. (Courtesy of Motorola, Inc.)

ACCUMULATOR AND MEMORY OPERATIONS	MNEMONIC	BOOLEAN/ARITHMETIC OPERATION (All register labels refer to contents)	COND. CODE REG. 5 H	4 I	3 N	2 Z	1 V	0 C
Add	ADDA	A + M → A	↕	•	↕	↕	↕	↕
	ADDB	B + M → B	↕	•	↕	↕	↕	↕
Add Acmltrs	ABA	A + B → A	↕	•	↕	↕	↕	↕
Add with Carry	ADCA	A + M + C → A	↕	•	↕	↕	↕	↕
	ADCB	B + M + C → B	↕	•	↕	↕	↕	↕
Complement, 2's (Negate)	NEG	00 − M → M	•	•	↕	↕	①	②
	NEGA	00 − A → A	•	•	↕	↕	①	②
	NEGB	00 − B → B	•	•	↕	↕	①	②
Decimal Adjust, A	DAA	Converts Binary Add. of BCD Characters into BCD Format*	•	•	↕	↕	↕	③
Subtract	SUBA	A − M → A	•	•	↕	↕	↕	↕
	SUBB	B − M → B	•	•	↕	↕	↕	↕
Subract Acmltrs.	SBA	A − B → A	•	•	↕	↕	↕	↕
Subtr. with Carry	SBCA	A − M − C → A	•	•	↕	↕	↕	↕
	SBCB	B − M − C → B	•	•	↕	↕	↕	↕

*Used after ABA, ADC, and ADD in BCD arithmetic operation; each 8-bit byte regarded as containing two 4-bit BCD numbers. DAA adds 0110 to lower half-byte if least significant number >1001 or if preceding instruction caused a Half-carry. Adds 0110 to upper half-byte if most significant number >1001 or if preceding instruction caused a Carry. Also adds 0110 to upper half-byte if least significant number >1001 and most significant number = 9.

(Bit set if test is true and cleared otherwise)

① (Bit V) Test: Result = 10000000?

② (Bit C) Test: Result = 00000000?

③ (Bit C) Test: Decimal value of most significant BCD Character greater than nine? (Not cleared if previously set.)

Fig. 7-14 Arithmetic instructions. (Courtesy of Motorola, Inc.)

by operations of other instructions, as will be shown under the following instruction categories.

Accumulator and Memory Operations

The largest block of instructions fall under the category of Accumulator and Memory Operations. These operations can be further divided into four types.

Figure 7-14 shows the arithmetic instructions. The CCR bit locations with the arrows are those subject to change by the instruction. As an example, if the ADDA operation results in an arithmetic overflow, bit "V" will be set.

Figure 7-15 shows the logic instructions that can be used with the 6802. Again, the actions that take place within the CCR as a result of the operations are shown.

The instructions shown in Fig. 7-16 are data test instructions. These instructions compare or test the contents of one of the accumulators or a memory location to determine if the data at that location meets predetermined criteria; i.e., equal to zero or equal to another data word.

The instructions shown in Fig. 7-17 are used to manipulate data. As can be seen, these instructions will push and pull data, rotate it right or left, shift it, store it, or perform any of a number of manipulations. This data manipulation capability represents great programming flexibility and power.

Program Control Operation

The final category of instruction is listed as the Program Control Operations. Within this category are the index register and stack pointer instructions, as shown in Fig. 7-18.

It should be clear by this time that the 6802 is a sophisticated and powerful microprocessor. This is equally true of other popular multibit processors such as the 8080, the Z80, and others. To try to discuss such a device comprehensively would

ACCUMULATOR AND MEMORY OPERATIONS	MNEMONIC	BOOLEAN/ARITHMETIC OPERATION (All register labels refer to contents)	COND. CODE REG. 5	4	3	2	1	0
			H	I	N	Z	V	C
And	ANDA	A • M → A	•	•	↕	↕	R	•
	ANDB	B • M → B	•	•	↕	↕	R	•
Complement, 1's	COM	$\overline{M} \to M$	•	•	↕	↕	R	S
	COMA	$\overline{A} \to A$	•	•	↕	↕	R	S
	COMB	$\overline{B} \to B$	•	•	↕	↕	R	S
Exclusive OR	EORA	A ⊕ M → A	•	•	↕	↕	R	•
	EORB	B ⊕ M → B	•	•	↕	↕	R	•
Or, Inclusive	ORA	A + M → A	•	•	↕	↕	R	•
	ORB	B + M → B	•	•	↕	↕	R	•

Fig. 7-15 Logic instructions. (Courtesy of Motorola, Inc.)

ACCUMULATOR AND MEMORY OPERATIONS	MNEMONIC	BOOLEAN/ARITHMETIC OPERATION (All register labels refer to contents)	COND. CODE REG. 5	4	3	2	1	0
			H	I	N	Z	V	C
Bit Test	BITA	A • M	•	•	↕	↕	R	•
	BITB	B • M	•	•	↕	↕	R	•
Compare	CMPA	A − M	•	•	↕	↕	↕	↕
	CMPB	B − M	•	•	↕	↕	↕	↕
Compare Acmltrs	CBA	A − B	•	•	↕	↕	↕	↕
Test, Zero or Minus	TST	M − 00	•	•	↕	↕	R	R
	TSTA	A − 00	•	•	↕	↕	R	R
	TSTB	B − 00	•	•	↕	↕	R	R

Fig. 7-16 Data test instructions. (Courtesy of Motorola, Inc.)

require several lengthy volumes. Indeed, Motorola has a 714-page *Applications Handbook* and a 300-page *Programming Manual for the 6800 MPU System.* Treatment of the 6802 in this chapter must be understood to be cursory and introductory. With that in mind, the 6802 and its instruction set will be left at this point and will be returned to later in the chapter after some other system chips have been discussed. When the construction, programming, and operation of an actual 6802 system are undertaken, much of the foregoing material will become the foundation for understanding those processes.

The M6800 System Components

A microprocessor by itself is limited as far as functional capabilities are concerned. The microprocessor chip must be designed for flexibility to make it cost effective. If the chip is designed to interface with all possible types of peripheral equipment using on-board components, the cost and size of the chip would cause it to be unmarketable. To make the MPU capable of doing only one job or interfacing with only one type of peripheral device would limit the market size for the device and would again result in an uneconomical product. To prevent these problems, the M6800 system designers put everything in the 6802 chip that was needed for basic MPU and memory operation, but designed other chips to accommodate peripheral and expansion needs.

The MC6800 was the first Motorola MPU in the 6800 line. It differs from the MC6802 in that the oscillator and RAM are included in the 6802 and not in the 6800. The instruction set and functioning of the two MPUs are identical otherwise.

There are several system chips available for use in M6800/MC6802 systems. These were discussed earlier. A fuller treatment will assist in understanding the purposes and functions of each of these chips.

ACCUMULATOR AND MEMORY OPERATIONS	MNEMONIC	BOOLEAN/ARITHMETIC OPERATION (All register labels refer to contents)	COND. CODE REG. 5 H	4 I	3 N	2 Z	1 V	0 C
Clear	CLR	00 → M	•	•	R	S	R	R
	CLRA	00 → A	•	•	R	S	R	R
	CLRB	00 → B	•	•	R	S	R	R
Decrement	DEC	M − 1 → M	•	•	↕	↕	④	•
	DECA	A − 1 → A	•	•	↕	↕	④	•
	DECB	B − 1 → B	•	•	↕	↕	④	•
Increment	INC	M + 1 → M	•	•	↕	↕	⑤	•
	INCA	A + 1 → A	•	•	↕	↕	⑤	•
	INCB	B + 1 → B	•	•	↕	↕	⑤	•
Load Acmltr	LDAA	M → A	•	•	↕	↕	R	•
	LDAB	M → B	•	•	↕	↕	R	•
Push Data	PSHA	A → M_{SP}, SP − 1 → SP	•	•	•	•	•	•
	PSHB	B → M_{SP}, SP − 1 → SP	•	•	•	•	•	•
Pull Data	PULA	SP + 1 → SP, M_{SP} → A	•	•	•	•	•	•
	PULB	SP + 1 → SP, M_{SP} → B	•	•	•	•	•	•
Rotate Left	ROL	M	•	•	↕	↕	⑥	↕
	ROLA	A 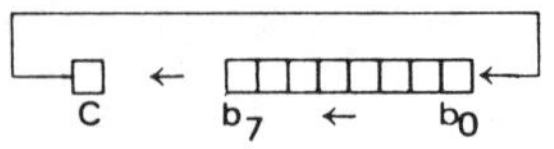 } C ← b_7 ← b_0	•	•	↕	↕	⑥	↕
	ROLB	B	•	•	↕	↕	⑥	↕
Rotate Right	ROR	M	•	•	↕	↕	⑥	↕
	RORA	A 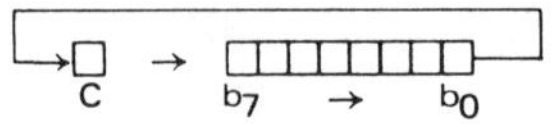 } C → b_7 → b_0	•	•	↕	↕	⑥	↕
	RORB	B	•	•	↕	↕	⑥	↕
Shift Left, Arithmetic	ASL	M	•	•	↕	↕	⑥	↕
	ASLA	A 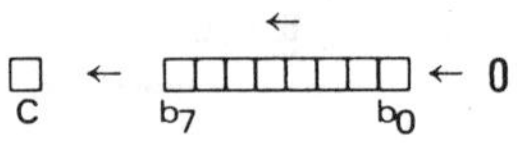 } C ← b_7 ← b_0 ← 0	•	•	↕	↕	⑥	↕
	ASLB	B	•	•	↕	↕	⑥	↕
Shift Right, Arithmetic	ASR	M	•	•	↕	↕	⑥	↕
	ASRA	A 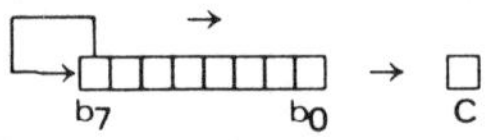 } b_7 → b_0 → C	•	•	↕	↕	⑥	↕
	ASRB	B	•	•	↕	↕	⑥	↕
Shift Right, Logic.	LSR	M	•	•	R	↕	⑥	↕
	LSRA	A 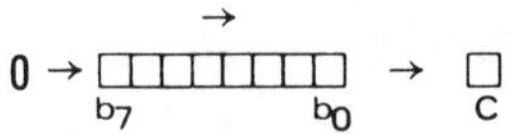 } 0 → b_7 → b_0 → C	•	•	R	↕	⑥	↕
	LSRB	B	•	•	R	↕	⑥	↕
Store Acmltr.	STAA	A → M	•	•	↕	↕	R	•
	STAB	B → M	•	•	↕	↕	R	•
Transfer Acmltrs	TAB	A → B	•	•	↕	↕	R	•
	TBA	B → A	•	•	↕	↕	R	•

④ (Bit V) Test: Operand = 10000000 prior to execution?

⑤ (Bit V) Test: Operand = 01111111 prior to execution?

⑥ (Bit V) Test: Set equal to result of $N \oplus C$ after shift has occurred.

Fig. 7-17 Data-handling instructions. (Courtesy of Motorola, Inc.)

INDEX REGISTER AND STACK POINTER OPERATIONS	MNEMONIC	BOOLEAN/ARITHMETIC OPERATION	5 H	4 I	3 N	2 Z	1 V	0 C
Compare Index Reg	CPX	$(X_H/X_L) - (M/M+1)$	•	•	①	↕	②	•
Decrement Index Reg	DEX	$X - 1 \rightarrow X$	•	•	•	↕	•	•
Decrement Stack Pntr	DES	$SP - 1 \rightarrow SP$	•	•	•	•	•	•
Increment Index Reg	INX	$X + 1 \rightarrow X$	•	•	•	↕	•	•
Increment Stack Pntr	INS	$SP + 1 \rightarrow SP$	•	•	•	•	•	•
Load Index Reg	LDX	$M \rightarrow X_H, (M+1) \rightarrow X_L$	•	•	③	↕	R	•
Load Stack Pntr	LDS	$M \rightarrow SP_H, (M+1) \rightarrow SP_L$	•	•	③	↕	R	•
Store Index Reg	STX	$X_H \rightarrow M, X_L \rightarrow (M+1)$	•	•	③	↕	R	•
Store Stack Pntr	STS	$SP_H \rightarrow M, SP_L \rightarrow (M+1)$	•	•	③	↕	R	•
Indx Reg → Stack Pntr	TXS	$X - 1 \rightarrow SP$	•	•	•	•	•	•
Stack Pntr → Indx Reg	TSX	$SP + 1 \rightarrow X$	•	•	•	•	•	•

① (Bit N) Test: Sign bit of most significant (MS) byte of result = 1?

② (Bit V) Test: 2's complement overflow from subtraction of LS bytes?

③ (Bit N) Test: Result less than zero? (Bit 15 = 1)

Fig. 7-18 Index register and stack pointer instructions. (Courtesy of Motorola, Inc.)

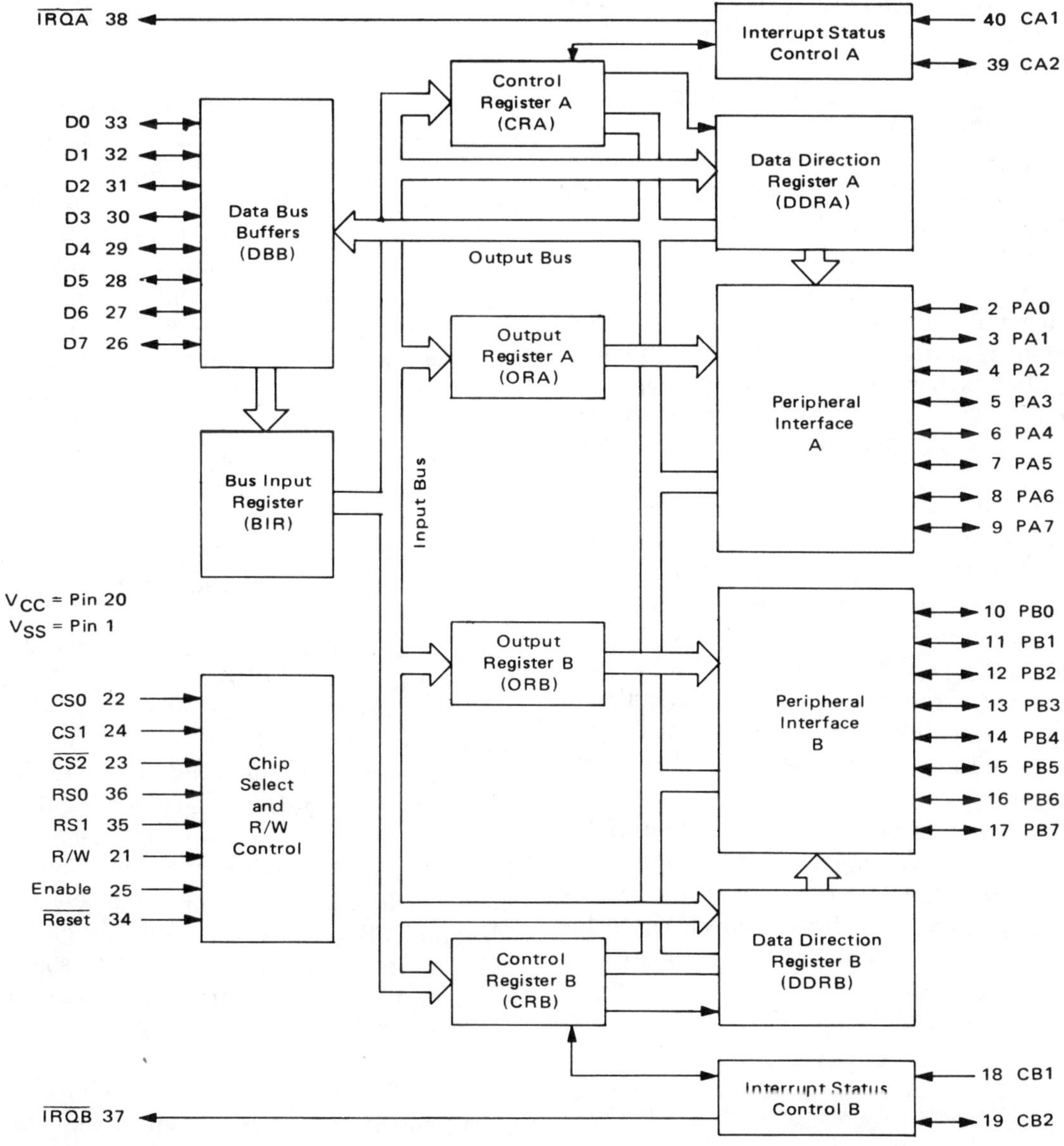

Fig. 7-19 Functional block diagram of the MC6820 Parallel Interface Adapter. (Courtesy of Motorola, Inc.)

The MC6820 Peripheral Interface Adapter

The MC6820 is one of two primary interface chips. The Peripheral Interface Adapter (PIA) permits the MPU to send and receive data from devices that accept and generate parallel bit data.[6] A block diagram of the MC6820 is shown in Fig. 7-19. Notice that the PIA has provision for two 8-bit parallel data connections. An 8-bit bidirectional data bus connects to the MPU data bus and to each peripheral device.

The data buses are shown as D0 through D7 for input data, PA0 through PA7 for peripheral bus A, and PB0 through PB7 for peripheral bus B.

The PIA has several programmable functions. The direction of data flow is determined by the status of the data direction register bits for the data lines to be controlled. A 0 in a data direction register bit causes the data line to act as an input. A 1 will cause the line to act as an output. When in the input mode, the data on the peripheral data lines will appear on the corresponding MPU data line. The input to the PIA imposes a load equal to a maximum of one standard TTL load.

In the output mode, the contents of the output register will appear on the data lines programmed to act as outputs. The two peripheral data bus output buses are not identical. Lines PA0–PA7 do not employ tri-state buffers. Lines PB0–PB7 do. This permits the output buffers of the B section to enter a high-impedance state when the peripheral data lines are used as inputs. Section B buffers are compatible with standard TTL and are capable of sourcing up to 1 mA at 1.5 V when directly driving a transistor switch.

Table 7-2 describes the control signal status that is required to address the programmable registers in the PIA. There are three registers for section A and three for register B. Primary selection is made using the two register select lines, RS0 and RS1. Addressing of the peripheral and data direction registers rely additionally on the status of the control register bit_2. Using the control bit patterns shown in Table 7-2, the six internal registers can be addressed.

The six registers control the operation of the PIA based on the programmed status of the register bits. The action of the data direction registers was discussed in connection with data lines. A 0 in the register sets the data line to the input mode and a 1 selects the output mode.

[6]The PIA can be programmed to transmit and receive serial data through only one input/output line, but it will be used in this volume for parallel bit data.

Table 7-2 Internal addressing. (Courtesy of Motorola, Inc.)

		Control Register Bit		
RS1	*RS0*	*CRA-2*	*CRB-2*	*Location Selected*
0	0	1	x	Peripheral Register A
0	0	0	x	Data Direction Register A
0	1	x	x	Control Register A
1	0	x	1	Peripheral Register B
1	0	x	0	Data Direction Register B
1	1	x	x	Control Register B

x = Don't Care

Table 7-3 Control word format. (Courtesy of Motorola, Inc.)

	7	6	5	4	3	2	1	0
CRA	IRQA1	IRQA2	CA2 Control			DDRA Access	CA1 Control	
	7	6	5	4	3	2	1	0
CRB	IRQB1	IRQB2	CB2 Control			DDRB Access	CB1 Control	

The control registers determine the operation of the peripheral control lines CA1, CA2, CB1, and CB2. CA1 and CB1 are interrupt input lines. They are used to set the interrupt flags of the control registers. Whether the low-to-high or the high-to-low transisition of the interrupt input signal is active also is subject to control by the control registers.

Peripheral controls CA2 and CB2 can be programmed to be either interrupt inputs or peripheral control outputs. The control lines present a high impedance when programmed as an input. As an output, the line is TTL compatible and is capable of sourcing up to 1 mA at 1.5 V.

Table 7-3 shows the format of the control register bits. The lowest 6 bits are accessible to the MPU for reading or writing when the proper chip and register select signals are applied. The last 2 bits are under the control of interrupts occurring on CA1, CA2, CB1, and CB2. These last 2 bits are read-only bits. The use of the control signals is shown in Tables 7-4 through 7-7.

A pin-out diagram of the MC6820 is included in Appendix C. It shows connections for the lines just discussed in addition to several lines that have not been discussed. Three of the signal lines that have not been discussed are the Chip Select lines CS0, CS1, and $\overline{\text{CS2}}$. These three signals connect to the MPU address bus and are used to select the PIA when required by program control. CS0 and CS1 will be high while $\overline{\text{CS2}}$ will be low when the PIA is selected.

The PIA Enable (E) and the PIA Read/Write (R/W) signals combine to control the "read to" or

Table 7-4 Control of interrupt inputs CA1 and CB1. (Courtesy of Motorola, Inc.)

CRA-1 (CRB-1)	CRA-0 (CRB-0)	Interrupt Input CA1 (CB1)	Interrupt Flag CRA-7 (CRB-7)	MPU Interrupt Request $\overline{IRQA}$ ($\overline{IRQB}$)
0	0	↓ Active	Set high on ↓ of CA1 (CB1)	Disabled—$\overline{IRQ}$ remains high
0	1	↓ Active	Set high on ↓ of CA1 (CB1)	Goes low when the interrupt flag bit CRA-7 (CRB-7) goes high
1	0	↑ Active	Set high on ↑ of CA1 (CB1)	Disabled—$\overline{IRQ}$ remains high
1	1	↑ Active	Set high on ↑ of CA1 (CB1)	Goes low when the interrupt flag bit CRA-7 (CRB-7) goes high

Notes: 1. ↑ indicates positive transition (low to high).
2. ↓ indicates negative transition (high to low).
3. The interrupt flag bit CRA-7 is cleared by an MPU Read of the A Data Register, and CRB-7 is cleared by an MPU Read of the B Data Register.
4. If CRA-0 (CRB-0) is low when an interrupt occurs (Interrupt disabled) and is later brought high, $\overline{IRQA}$ ($\overline{IRQB}$) occurs after CRA-0 (CRB-0) is written to a "one".

Table 7-5 Control of CA2 and CB2 as interrupt inputs; CRA5 (CRB5) is low. (Courtesy of Motorola, Inc.)

CRA-5 (CRB-5)	CRA-4 (CRB-4)	CRA-3 (CRB-3)	Interrupt Input CA2 (CB2)	Interrupt Flag CRA-6 (CRB-6)	MPU Interrupt Request $\overline{IRQA}$ ($\overline{IRQB}$)
0	0	0	↓ Active	Set high on ↓ of CA2 (CB2)	Disabled—$\overline{IRQ}$ remains high
0	0	1	↓ Active	Set high on ↓ of CA2 (CB2)	Goes low when the interrupt flag bit CRA-6 (CRB-6) goes high
0	1	0	↑ Active	Set high on ↑ of CA2 (CB2)	Disabled—$\overline{IRQ}$ remains high
0	1	1	↑ Active	Set high on ↑ of CA2 (CB2)	Goes low when the interrupt flag bit CRA-6 (CRB-6) goes high

Notes: 1. ↓ indicates positive transition (low to high).
2. ↑ indicates negative transition (high to low).
3. The interrupt flag bit CRA-6 is cleared by an MPU Read of the A Data Register, and CRB-6 is cleared by an MPU Read of the B Data Register.
4. If CRA-3 (CRB-3) is low when an interrupt occurs (Interrupt disabled) and is later brought high, $\overline{IRQA}$ ($\overline{IRQB}$) occurs after CRA-3 (CRB-3) is written to a "one".

Table 7-6 Control of CB2 as an output; CRB-5 is high. (Courtesy of Motorola, Inc.)

			CB2	
CRB-5	CRB-4	CRB-3	Cleared	Set
1	0	0	Low on the positive transition of the first E pulse following an MPU Write "B" Data Register operation.	High when the interrupt flag bit CRB-7 is set by an active transition of the CB1 signal.
1	0	1	Low on the positive transition of the first E pulse after an MPU Write "B" Data Register operation.	High on the positive edge of the first "E" pulse following an "E" pulse which occurred while the part was deselected.
1	1	0	Low when CRB-3 goes low as a result of an MPU Write in Control Register "B".	Always low as long as CRB-3 is low. Will go high on an MPU Write in Control Register "B" that changes CRB-3 to "one".
1	1	1	Always high as long as CRB-3 is high. Will be cleared when an MPU Write Control Register "B" results in clearing CRB-3 to "zero".	High when CRB-3 goes high as a result of an MPU Write into Control Register "B".

Table 7-7 Control of CA-2 as an output; CRA-5 is high. (Courtesy of Motorola, Inc.)

			CA2	
CRA-5	*CRA-4*	*CRA-3*	*Cleared*	*Set*
1	0	0	Low on negative transition of E after an MPU Read "A" Data operation.	High when the interrupt flag bit CRA-7 is set by an active transition of the CA1 signal.
1	0	1	Low on negative transition of E after an MPU Read "A" Data operation.	High on the negative edge of the first "E" pulse which occurs during a deselect.
1	1	0	Low when CRA-3 goes low as a result of an MPU Write to Control Register "A".	Always low as long as CRA-3 is low. Will go high on an MPU Write to Control Register "A" that changes CRA-3 to "one".
1	1	1	Always high as long as CRA-3 is high. Will be cleared on an MPU Write to Control Register "A" that clears CRA-3 to a "zero".	High when CRA-3 goes high as a result of an MPU Write to Control Register "A".

"write from" functions of the PIA. The E signal is derived from the Φ2 clock signal. A low on the R/W line of the PIA will cause data to be transferred to the PIA from the MPU on the E signal if the chip has been selected. A high on the R/W line causes the PIA to feed data back to the MPU bus on E if the chip is addressed.

The $\overline{\text{Interrupt Request}}$ ($\overline{\text{IRQA}}$ and $\overline{\text{IRQB}}$) signals interrupt MPU action based on internal interrupt enable bits. Interrupts are required when peripheral devices cannot respond as quickly as the MPU requires or when some external data needs to enter the processing. The interrupt request will be serviced by the MPU only after the cycle in progress is completed.

The last input to the PIA is the $\overline{\text{Reset}}$. When this line is brought low, all register bits in the PIA will be set to 0. This is usually done at power-on and may be incorporated into the program.

Figures 7-20 through 7-22 show some typical applications for PIAs. A PIA is used to interface a keyboard to the MPU system shown in Fig. 7-20. Notice that simple circuitry is included to decode some control signals.

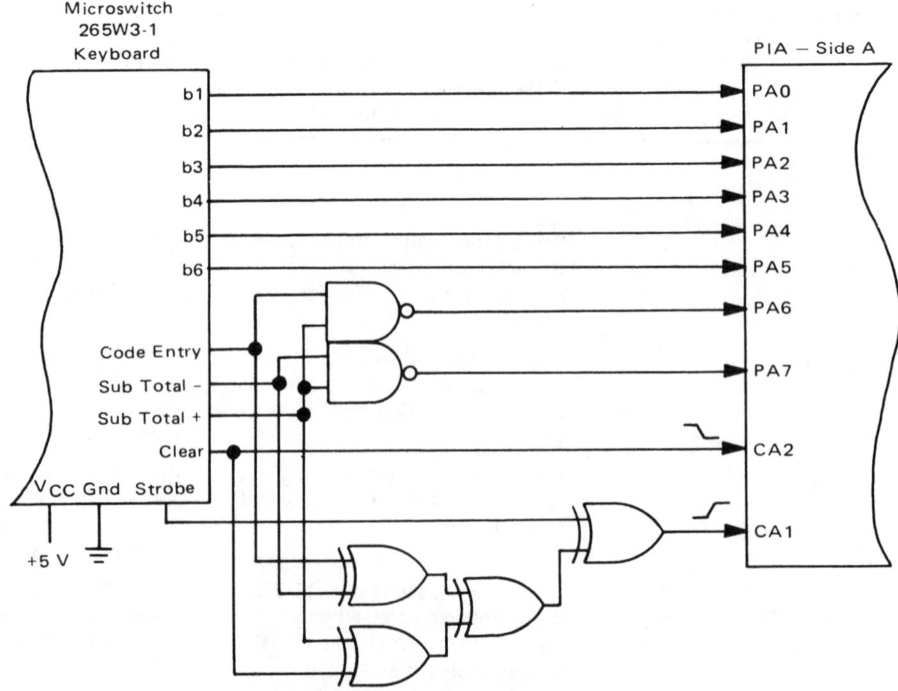

Fig. 7-20 Keyboard/PIA hardware interface. (Courtesy of Motorola, Inc.)

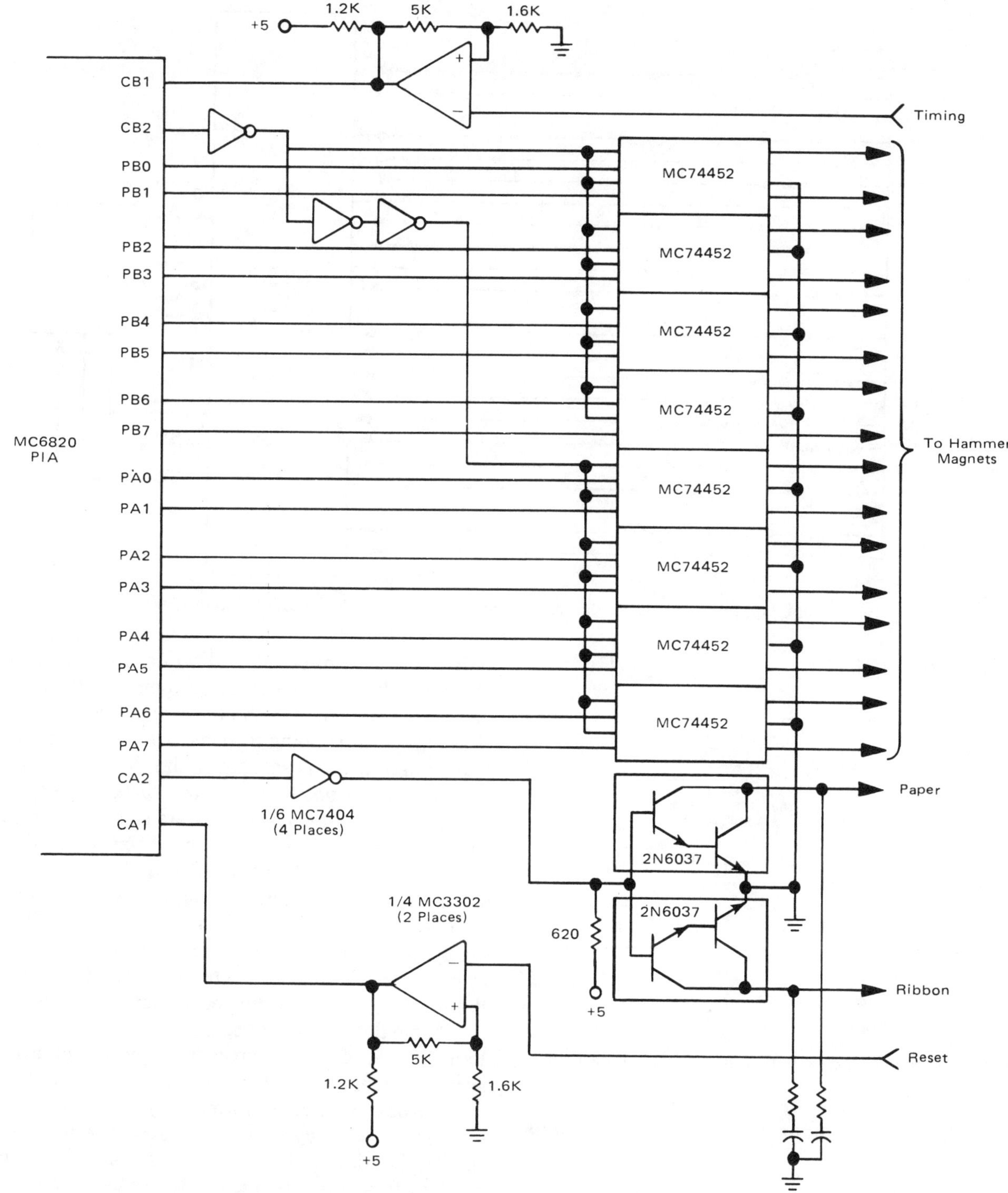

Fig. 7-21 SEIKO printer circuit requirements. (Courtesy of Motorola, Inc.)

Figure 7-21 shows the versatility of the PIA. In this application, the PIA is used to drive a printer. Notice that both A and B output lines are used.

The third example is a common use for a PIA in microcomputer systems. The PIA is used to interface and control a tape drive. In this application, data is handled in serial fashion, and the majority of the output lines are used for control functions.

A simple use of the PIA will be included in the example system described later in the chapter. The flexibility of the 6820 is clear. Its use is limited primarily by the imagination of the designer.

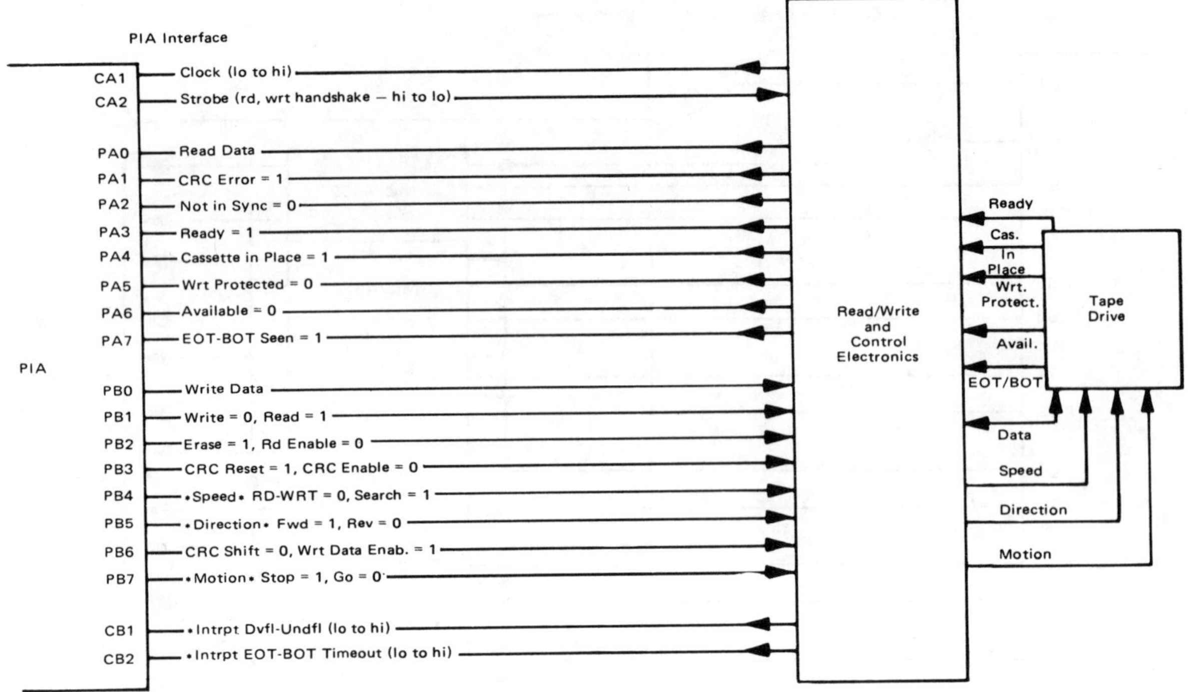

Fig. 7-22 PIA, tape drive, and read/write control electronics interface. (Courtesy of Motorola, Inc.)

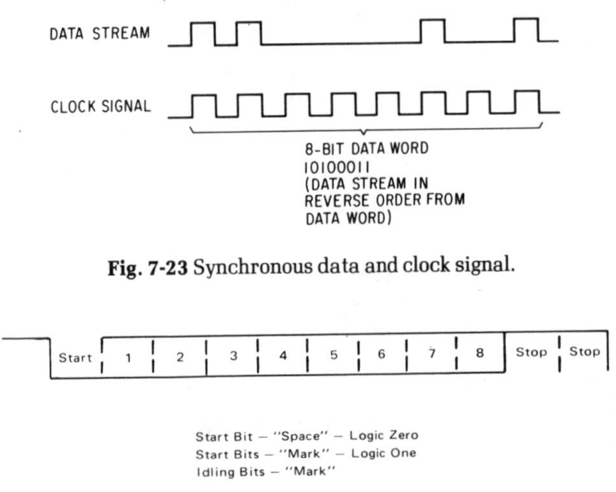

Fig. 7-23 Synchronous data and clock signal.

Fig. 7-24 Asynchronous data format. (Courtesy of Motorola, Inc.)

The MC6850 Asynchronous Communications Interface Adapter

The MC6850 Asynchronous Communications Interface Adapter (ACIA) is an interesting system chip that expands the capabilities of the MC6802 system. The 6850 permits asynchronous serial data to be used in the 8-bit parallel bus system of the 6802.

Synchronous data is data that is generated in relation to a clock. Each time the clock goes high, for example, a data bit is generated. Since a stream of serial data is a sequence of high data bits and low data bits, some method is needed to differentiate between a *low* data bit and *no* data bit. If a data stream is transmitted in synchronism with a clock signal, the presence of the clock signal will indicate that a data bit is being transmitted. If the data bit is high, a high will be recorded at the clocking signal. If the data bit is not high, a low will be recorded at clocking. Figure 7-23 shows a data stream compared to a clock signal. The data is synchronized so that a high clock signal coincides with data bits. To convert this serial signal to parallel data, it is necessary to simply count the clock pulses and make each eight clock pulses equal to one parallel word (in 8-bit word systems). Shift register techniques can be used for the serial-to-parallel data conversion.

Some data is not generated in sync with a clock. Entering data from a keyboard is slow and erratic. If serial conversion techniques were used, a lot of zeros would be placed in memory as the user was deciding which key to depress next. The answer to this kind of data entry is to identify each data word or bit with a start and stop code. As long as no data is being sent, no start code will instruct the system to accept data. When data is interrupted, a stop code will indicate end-of-data to the system. Figure 7-24 illustrates the asynchronous data format.

Figure 7-25 shows a block diagram of the 6850 ACIA. D0 through D7 constitute an 8-bit bidirectional data bus. The 8-bit data word from this data bus is transferred to the transmit data register, if the necessary read conditions are met. The data from the transmit data register is transferred to the

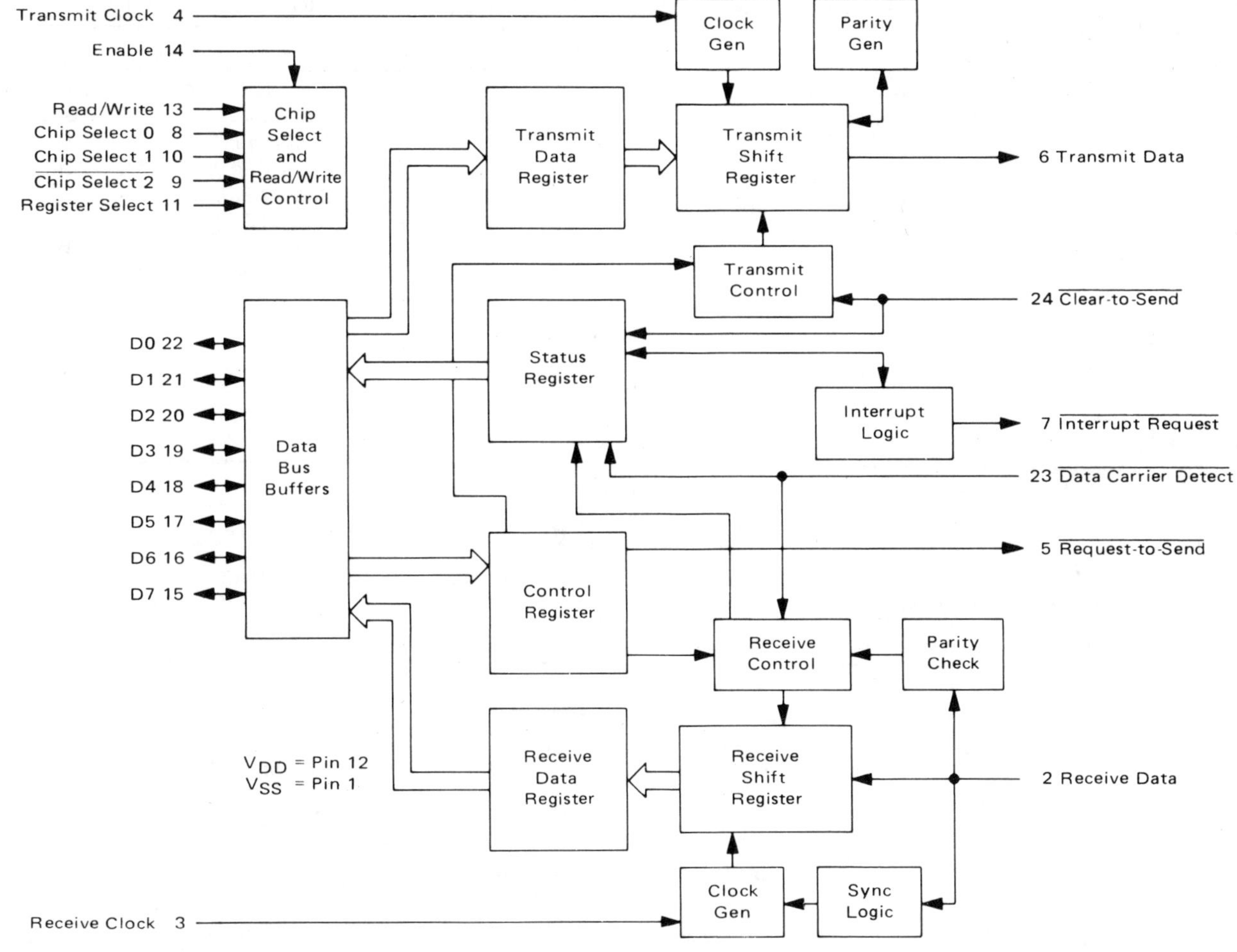

Fig. 7-25 Functional block diagram of the 6850 Asynchronous Communications Interface Adapter. (Courtesy of Motorola, Inc.)

transmit shift register where it is serialized. At the appropriate time as determined by the status of the Clear-to-Send and Interrupt Request, the serial data is clocked out of the shift register onto the data output line (pin 6). Prior to transmission, a parity bit can be added by the parity generator.

Receiving from the peripheral also is dependent on several status signals. When the conditions of these signals are correct, serial data will enter the ACIA at the receive data line (pin 2). As the data enters, parity is checked and the data word is clocked into the receive shift register. The shift register converts the serial data into parallel data, which is then transferred into the receive data register. When the read function is enabled, the data is passed to the MPU data bus.

Several of the signals and connections to the ACIA are similar to those of the PIA. The MPU data bus buffers perform the same function as those of the PIA. These buffers are tri-state. The Enable (E) signal is related to the $\Phi 2$ clock. It enables the I/O buffers and clocks data to and from the ACIA. The Read/Write (R/W) and Register Select (R/S) signal combine to determine whether the transmit/receive data register or the control/status registers will be read into or written from. A high R/S will select the data register, and a low will select the control/status register. The interrupt request signal operates as it did for the 6820. CS0, CS1, and CS2 are chip select signals.

Two clock inputs are available. The Receive Clock (R × Clk) and the Transmit Clock (T × Clk) operate independently. The clocks can be programmed for different modes of operation, as shown in Table 7-8 for bits CR0 and CR1.

Table 7-9 describes the meaning of the other bits of the control register as well. In like fashion, Table 7-10 describes the function of the bits in the status register. These two registers describe the bulk of the programming and operation of the ACIA.

The ACIA is designed to operate in transmission circuits. This usually requires the use of a data modem such as the MC6860. A typical transmission system is shown in Fig. 7-26. In this system, teletype

Table 7-8 Selection of ACIA clock mode and reset. (Courtesy of Motorola, Inc.)

CR1	*CR0*	*Function*
0	0	÷ 1
0	1	÷ 16
1	0	÷ 64
1	1	Master Reset

Table 7-9 Function of select bits. (Courtesy of Motorola, Inc.)

CR4	*CR3*	*CR2*	*Function*
0	0	0	7 Bits + Even Parity + 2 Stop Bits
0	0	1	7 Bits + Odd Parity + 2 Stop Bits
0	1	0	7 Bits + Even Parity + 1 Stop Bit
0	1	1	7 Bits + Odd Parity + 1 Stop Bit
1	0	0	8 Bits + 2 Stop Bits
1	0	1	8 Bits + 1 Stop Bit
1	1	0	8 Bits + Even Parity + 1 Stop Bit
1	1	1	8 Bits + Odd Parity + 1 Stop Bit

Table 7-10 Definition of ACIA register contents. (Courtesy of Motorola, Inc.)

	Buffer Address			
Data Bus Line Number	*RS • $\overline{R/W}$ Transmit Data Register*	*RS • R/W Receive Data Register*	*$\overline{RS}$ • $\overline{R/W}$ Control Register*	*$\overline{RS}$ • R/W Status Register*
	(Write Only)	*(Read Only)*	*(Write Only)*	*(Read Only)*
0	Data Bit 0*	Data Bit 0	Counter Divide Select 1 (CR0)	Receive Data Register Full (RDRF)
1	Data Bit 1	Data Bit 1	Counter Divide Select 2 (CR1)	Transmit Data Register Empty (TDRE)
2	Data Bit 2	Data Bit 2	Word Select 1 (CR2)	Data Carrier Detect ($\overline{DCD}$)
3	Data Bit 3	Data Bit 3	Word Select 2 (CR3)	Clear-to-Send ($\overline{CTS}$)
4	Data Bit 4	Data Bit 4	Word Select 3 (CR4)	Framing Error (FE)
5	Data Bit 5	Data Bit 5	Transmit Control 1 (CR5)	Receiver Overrun (OVRN)
6	Data Bit 6	Data Bit 6	Transmit Control 2 (CR6)	Parity Error (PE)
7	Data Bit 7***	Data Bit 7**	Receive Interrupt Enable (CR7)	Interrupt Request (IRQ)

*Leading bit = LSB = Bit 0
**Data bit will be zero in 7-bit plus parity modes.
***Data bit is "don't care" in 7-bit plus parity modes.

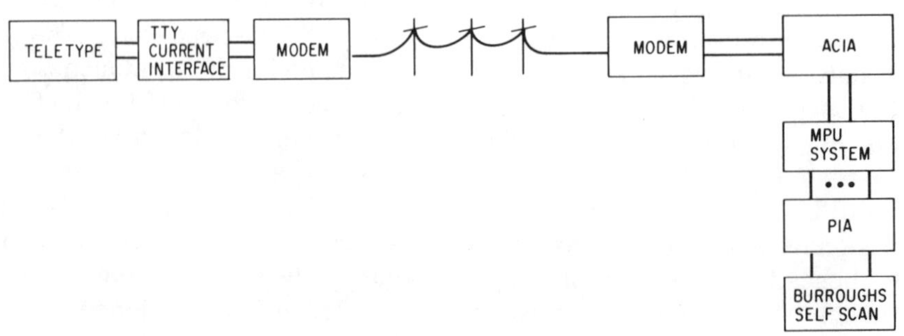

Fig. 7-26 Teletype system interconnections. (Courtesy of Motorola, Inc.)

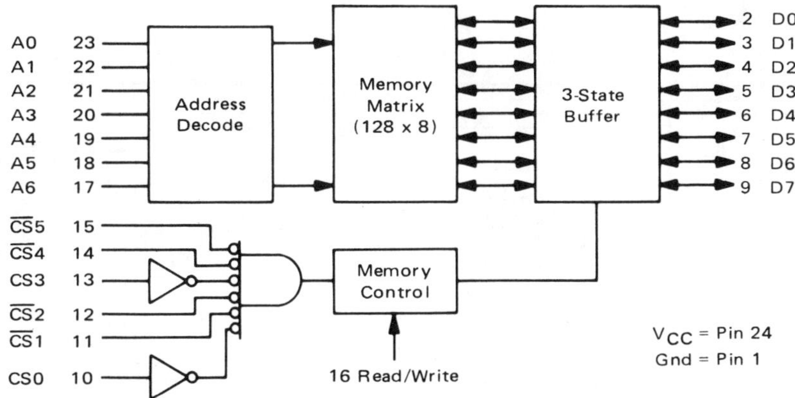

Fig. 7-27 Functional block diagram and pin-outs of the 6810 RAM. (Courtesy of Motorola, Inc.)

(TTY) signals are carried via a telephone system to the computer location. The modems[7] convert binary data into audio tones that can be transmitted over the telephone wire. The modem does this by changing the data into a frequency shift keying (FSK) signal. The ACIA interfaces the modem with the MPU, and the system can operate via the telephone circuit.

The MCM6810A Random Access Memory

The 6802 MPU incorporates 128 × 8 bits of Random Access Memory (RAM) on the MPU chip. For most simple breadboard experiments, this amount of RAM will be adequate. When additional memory is needed, however, the MCM6810 is directly compatible with the 6802. The MCM6810 is a static device. It requires no refreshing. The contents of the 6810 are volatile. When power is removed from the chip, the contents of memory are lost.

The block diagram of the MCM6810 is shown in Fig. 7-27. Data flows on eight bidirectional data lines that connect to a tri-state buffer. When unselected, the eight data lines are in the third or high-impedance state. The read operation is initiated when the device is chosen by the six chip select lines and when the R/W control line is high. The write operation requires chip selection and a low R/W line. The six chip select lines provide for extensive memory selection. The 128 words of memory can be individually addressed using the seven address input lines. These seven inputs are decoded and the result is passed to the memory matrix, which, in turn, connects to the tri-state buffers. Read and write operations for any specific word are accomplished using the decoder.

Information about the 6810 or any of the ROM or EPROM chips can be obtained from Motorola. Memory must be chosen based on the purposes of the designer. If he desires permanent memory, ROM is best. If he desires temporary memory, EPROM and RAM will be the best choices. In the experimental system, RAM has been chosen to permit rapid programming change.

The 6802 Demonstration System

It was indicated earlier that construction of a full-scale microcomputer was not the objective of this chapter. By using the MC6802, a simplified processing system that possesses great power can be constructed on a breadboard. Using Motorola's system chips, the processor shown in Fig. 7-28, which uses only three main chips, can be easily assembled.

In addition to the system chips, the reset, single step, and I/O circuitry is required. The system shown is designed to accept 8-bit words using breadboard latches or simple switches for entry. The entered number will be displayed on eight LEDs. Circuitry is provided to initiate MPU operation when a "data ready" signal is received.

Data input uses the "A" sections of the PIA. The output data uses the "B" section of the PIA. LEDs are provided to display the data until cleared or until new data is received. Tandem inverters drive eight LEDs to display the binary output.

It will be noted that the 6802 does not operate in the same manner as the 14500B. The 14500B executed one instruction per machine cycle. Inspection of the 6802 instruction set shown in Fig. 7-11 will show that multiple MPU cycles are required for completion of various instructions. The number of cycles required are listed in the column marked with the symbol " ~ " and range from two cycles, for instructions such as the ADDA, to eight cycles for the BSR instruction.

The reset circuitry is designed to place the $\overline{\text{Reset}}$ pulse on the $\overline{\text{Reset}}$ pins of each system chip

[7]Modem stands for modulator/demodulator.

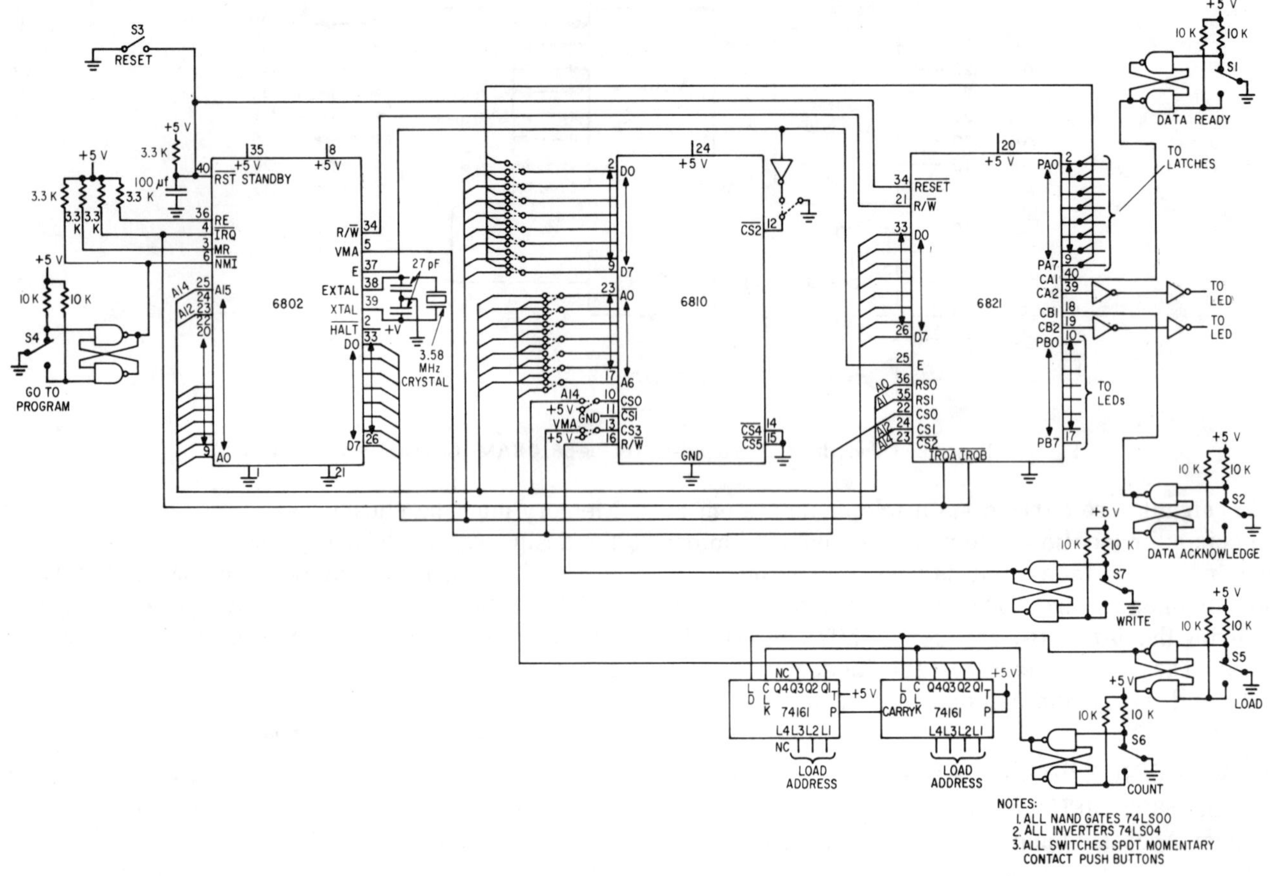

Fig. 7-28 Breadboard microcomputer.

that requires it. The circuit will apply the $\overline{\text{Reset}}$ pulse for a minimum of eight MPU cycles. The pulse must be applied for at least eight cycles for the reset process to be completed.

While the I/O circuitry is minimal and limiting, the system can still carry out high level and sophisticated processing. For this to be accomplished, data will have to be entered one word at a time into memory, and the program will recall the data from memory for processing. The results of processing will be placed in memory, if more than one result or more than 8-bit words are produced. The stored results will then be called from memory and displayed via the LEDs one 8-bit word at a time.

Programming the 6802 MPU System

The 6802 system is versatile in its use of numerical systems. Four different number systems can be used with the system. The arithmetic logic unit of the MPU uses the 2s complement number system for its calculations. Each byte will be interpreted as a signed 2s complement number in the range from − 128 to + 127. Under this system, the MSB (b_7) is the number sign with "1" representing minus and "0" representing plus. The following illustrates the signed 2s complement numbering system.

±	2^6	2^5	2^4	2^3	2^2	2^1	2^0	*Binary Value*
b_7	b_6	b_5	b_4	b_3	b_2	b_1	b_0	*Bit Number*
1	0	0	0	0	0	0	0	(− 128 in 2s complement)
1	1	1	1	1	1	1	1	(− 1 in 2s complement)
0	0	0	0	0	0	0	0	(0 in 2s complement)
0	0	0	0	0	0	0	1	(+ 1 in 2s complement)
0	1	1	1	1	1	1	1	(+ 127 in 2s complement)

A second number system interprets each byte as an unsigned binary number in the range from 0 to 255. Examples of this system are

2^7	2^6	2^5	2^4	2^3	2^2	2^1	2^0	*Binary Value*
b_7	b_6	b_5	b_4	b_3	b_2	b_1	b_0	*Bit Number*
0	0	0	0	0	0	0	0	(0 in unsigned binary)
1	1	1	1	1	1	1	1	(255 in unsigned binary)

A third number system is BCD. In this system, each byte contains one 4-bit BCD number in the four LS bits. The four MS bits will always be zero. This number system is referred to as *unpacked BCD*. Examples of this system are

2^7	2^6	2^5	2^4	2^3	2^2	2^1	2^0	*Binary Value*
b^7	b^6	b^5	b^4	b^3	b^2	b^1	b^0	*Bit Number*
0	0	0	0	0	0	0	0	(BCD 0)
0	0	0	0	0	1	0	1	(BCD 5)
0	0	0	0	1	0	0	1	(BCD 9)
always 0								

The fourth number system uses two 4-bit BCD numbers in each byte. Under this system, each byte can represent numbers in the range from 0 to 99. Examples of this system are

2^3	2^2	2^1	2^0	2^3	2^2	2^1	2^0	*Binary Value*
b^7	b^6	b^5	b^4	b^3	b^2	b^1	b^0	*Bit Number*
0	0	0	0	0	0	0	0	(BCD 00)
0	0	0	1	0	0	0	0	(BCD 10)
1	0	0	0	1	0	0	0	(BCD 88)

As indicated earlier, the MPU operates internally using the 2s complement numbering system for some operations. When the various numbering systems are used with the 6802, the program coding must be written to ensure proper results for the numbering system used. The program can be specific to the number system or it can convert all of the numbers to a common base prior to calculations. As an example of this, consider the following routines for addition. The first procedure adds two binary numbers to obtain a binary result. The second procedure adds two BCD numbers to obtain a BCD result.

Operator	*Operand*	*Process Description*
LDAA	A001	Load accumulator A with contents of memory location A001
ADDA	A002	Add contents of A002 to contents of accumulator A
STAA	A003	Store contents of accumulator A in memory location A003

As an example of this program in operation consider the following:

Step	*Value of Accumulator A*	*Memory Location/Value*
—	0000 0000	A001/00001011
LDAA/A001	0000 1011	A002/00001001
ADDA/A002	0001 0100	A003/00000000
STAA/A003	—	A003/00010100

If this same program is used with BCD addition, however, problems result. The BCD addition would be as follows:

Step	*Value of Accumulator A*	*Memory Location/Value*
—	0000 0000	A001/0001 0001 (BCD)
LDAA/A001	0001 0001	A002/0000 1001 (BCD)
ADDA/A002	0001 1010	A003/0000 0000
STAA/A003	—	A003/0001 1010 (BCD)

The error that results for BCD addition is clear. The BCD values have the decimal equivalents of 11 and 9. When added, the sum should be 20. The BCD value of the actual sum is MSB (0001) = 10, LSB (1010) = 10. A BCD number cannot have greater value than decimal 9, however. The LSB number (10), is therefore not correct. To solve the problem, the decimal number 6 needs to be added to the sum. (This process was explained in the description of the binary-to-BCD converter in Chap. 2.) The 6802 instruction set contains an instruction to do this. The following program shows how simple the solution to the BCD problem can be.

Operator	*Operand*	*Process Description*
LDAA	A001	Load A001 into accumulator A
ADDA	A002	Add above with A002
DAA		Add 6_{10} to the sum
STAA	A003	Store result in A003

Using the numbers from the prior example, the problem is

Step	*Value of Accumulator A*	*Memory Location/Value*
—	0000 0000	A001/0001 0001
LDAA/A001	0001 0001	A002/0000 1001
ADDA/A002	0001 1010	A003/0000 0000
DAA	0110	
STAA/A003	0010 0000	A003/0010 0000

The result is now correct.

Numbers that are composed of more than 1 byte require the use of a technique called *multiple precision*. Multiple precision operates by dividing the numbers into segments equal in length to the system's data word. If 16-bit binary numbers are included, the numbers are broken into two 8-bit numbers, and the procedure is referred to as *double precision*. The following binary subtraction of two 16-bit numbers is an example of double precision:

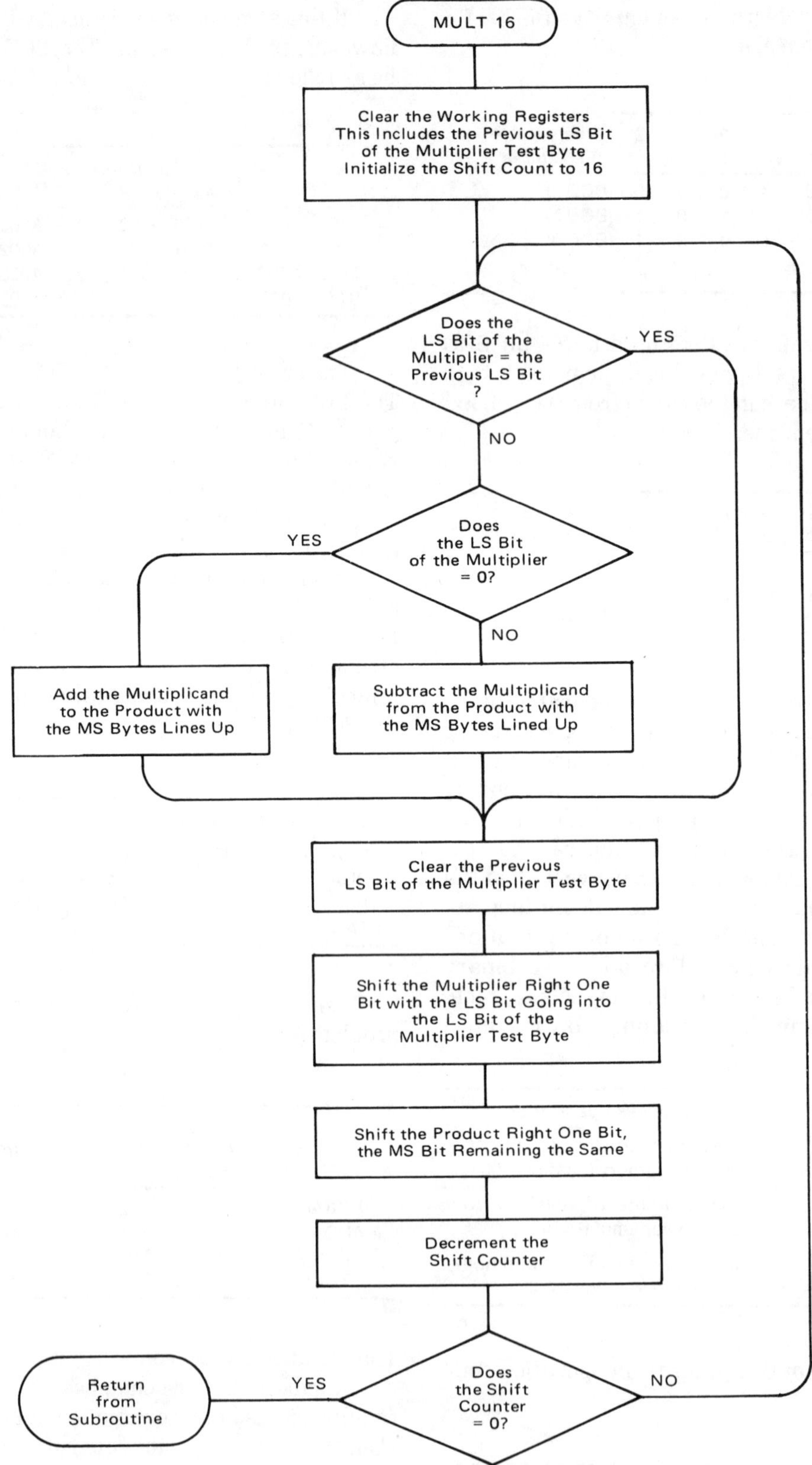

Fig. 7-29 Flowchart for Booth's algorithm. (Courtesy of Motorola, Inc.)

Operator	*Operand*	*Process Description*
LDAA	A001	Load LS byte of first number
LDAB	A002	Load MS byte of first number
SUBA	B001	Subtract LS bytes of the two numbers
SBCB	B002	Subtract MS bytes with borrow from LS bytes
STAA	C001	Store result LS byte in C001
STAB	C002	Store result MS byte in C002

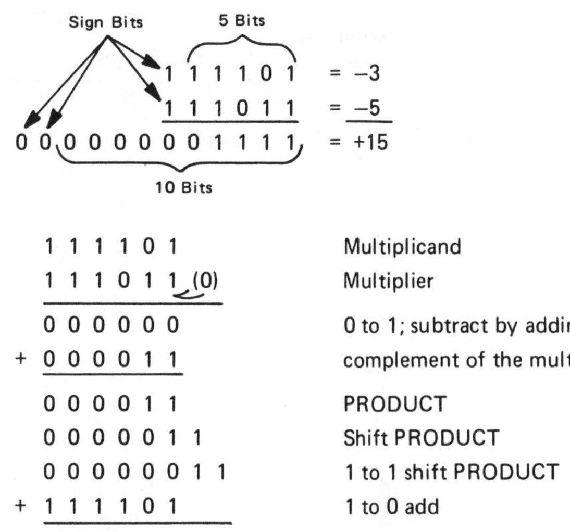

(A)

1. Test the transition of the multiplier bits from right to left assuming an imaginary 0 bit to the immediate right of the multiplier.
2. If the bits in question are equal, then 5.
3. If there is a 0 to 1 transition, the multiplicand is subtracted from the product, then 5.
4. If there is a 1 to 0 transition, the multiplicand is added to the product, then 5.
5. Shift the product right one bit with the MSBit remaining the same.
6. Go to 1 to test the next transition of the multiplier.

(B)

Fig. 7-30 Multiplication using Booth's algorithm: (a) function outline and (b) process statement. (Courtesy of Motorola, Inc.)

This program approaches the binary subtraction in two steps. The least significant (LS) bytes are subtracted first. Next, the MS bytes are subtracted. The MS bytes will have input from the LS byte subtraction through the condition code register. The results of the MS byte subtraction are stored adjacent to the LS byte subtraction result. The 16-bit result can then be called from memory 1 byte at a time and reassembled into the full 16-bit word.

Multiplication represents a more complicated programming task. The 6802 instruction set includes instructions that accommodate the processes of multiplication. Even with these special instructions, however, programming is still relatively complicated. To make allowances for negative numbers, a special routine is required. This routine has been named *Booth's algorithm.* Figure 7-29 shows a flowchart for Booth's algorithm. The increase in program sophistication is evident. Figure 7-30a lists the programming decisions that are made within the flowchart. Figure 7-30b illustrates the process. The MPU instructions that permit this multiplication process are the rotate and shift instructions. The programmer must devise the test routine that determines when the shift is required. In this example, the test is whether or not the multiplier bit is equal to the previous multiplier bit. Based on that test, one of three actions is taken:

1. If the two bits are equal, the product is shifted right one bit.
2. If the previous bit is "0" and the next is "1," the multiplicand is subtracted from the product and that result is then shifted.
3. If the previous bit is "1" and the next is "0," the multiplicand is added to the product and that result is then shifted.

These options are accommodated using branching instruction. Consider the program listing in Fig. 7-31. Statement 5 uses an EOR (exclusive-OR) instruction to test the LS bit of the multiplier against a test bit that is initialized to zero and will have the successive LS bit of the multiplier placed into it by the program. Based on the result of the EOR test (1 or 0), the next instruction will determine program flow. If the EOR test results in a 0, the BEQ (branch if equal to zero) instruction will direct the program to branch to the instruction labeled SHIFT (statement 22). The intervening statements will be bypassed in this case. Statements 23–30 are equivalent to step 5 in Fig. 7-30b.

If statement 6 indicates that the 2 bits are not equal, the program moves on to the test of statement 7. Does the LS bit of the multiplier equal zero? If it does, statement 8 sends the program to the statement labeled ADD. If it does not, the subtract routine of statements 10–14 is accomplished. Both the add

Statement No.	Label	Operator	Operand	Process Description
1		LDX	#16	Initialize shift counter to 16
2	BEGIN	LDAA	Y + 1	Put LSB of multiplier in ACCA*
3	LP2	ANDA	#1	
4		TAB		Put value of accumulator A in ACCB
5		EORA	FF	Test ACCA and FF (FF = 0) (Does ACCA = FF?)
6		BEQ	SHIFT	If yes, go to SHIFT
7		TST B		IF no, compart ACCA and zero (does ACCA = 0?)
8		BEQ	ADD	If yes, go to ADD
9		LDAA	U + 1	
10		LDAB	U	If no, put ACCA in location U, subtract MSB, and subtract LSB with carry. Store result in U
11		SUBA	XX + 1	
12		SBCB	XX	
13		STAA	U + 1	
14		STAB	U	
15		BRA	SHIFT	
16	ADD	LDAA	U + 1	
17		LDAB	U	Go to SHIFT routine. Add multiplicand to product with MSBytes lined up
18		ADDA	XX + 1	
19		ADCB	XX	
20		STAA	U + 1	
21		STAB	U	
22	SHIFT	CLR	FF	Clear test byte (FF)
23		ROR	Y	Shift the multiplier right one bit with LSB into LSB
24		ROR	Y + 1	
25				
26		ROL	FF	
27		ASR	U	Of FF, shift product right one bit, the MSB remaining the same
28		ROR	U + 1	
29		ROR	U + 2	
30		ROR	U + 3	
31		DEX		Decrement shift counter
32		BNE	LP2	Test shift counter; if ≠ 0, go to LP2

*This program presupposes that the multiplier has been loaded in memory location Y and the multiplicand in XX (Y = MSB, Y + 1 = LSB, XX = MSB, XX + 1 = LSB). Product will be in U (U = MSB, U + 1 = LSB).

Fig. 7-31 Program for multiplication.

and subtract routines conclude by entering the shift routine. The process that aligns the products as required for proper multiplication begins at statement 22. Statement 30 decrements the shift counter, which is tested for zero by statement 31. If a 0 is detected, it is concluded that all 16 bits of the 16-bit multiplier have been used and the program will end. If the shift counter is not equal to zero, the BNE instruction of statement 31 will send the program back to the start of the process at statement 2.

Evaluation of the programming steps required to accomplish binary multiplication illustrates the importance of formulating a usable algorithm. The process used to accomplish multiplication has less similarity to standard arithmetic processes than to observation. The program duplicates the manipulations that are observed to be used in the multiplication. Programs are often developed in this manner. By reducing the observed process to discrete steps, the instruction-set elements can be combined to accomplish each step.

The branch instruction used in the multiplication process is a type of subroutine instruction. It directs the program flow to another part of the pro-

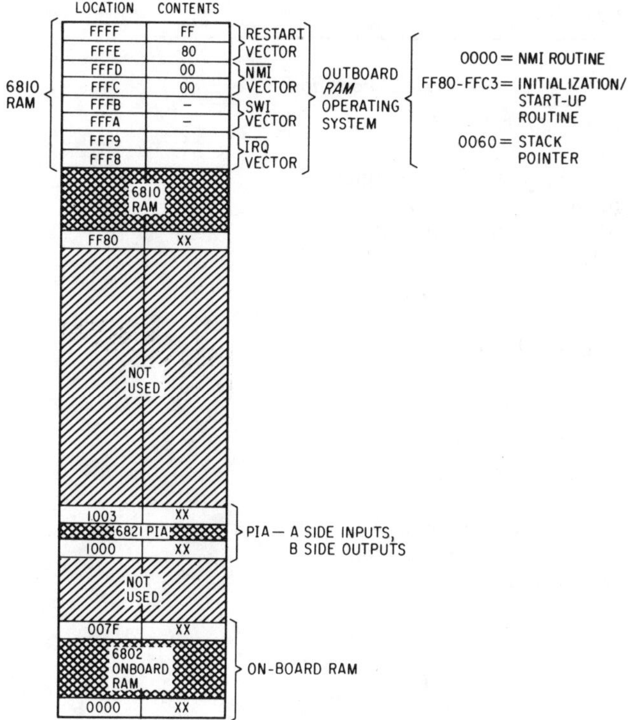

Fig. 7-32 Memory map of the 6802 breadboard computer.

gram. In the multiplication example, the SHIFT routine was entered from several decisions within the program flow. This one block of code was then used repeatedly. This is the basic idea of the subroutine. It is a self-contained block of instructions that accomplishes some task that is used by several parts of the program. The block could be removed intact to be used in other programs. An actual subroutine would normally be constructed outside the main program flow and would be entered by appropriate statements in the main program. After the subroutine completes its processing, the results would be returned to the main program for completion of processing. The decision to use branch rather than subroutine instruction is a question of efficiency.

Experiment 7-1: The Addition Program

The demonstration system shown in Fig. 7-28 will not use nonvolatile memory. In most systems, nonvolatile memory would be used to direct MPU during its initialization routines. If the MPU is used to interpret keyboard closures (debouncing and keyboard encoding can be accomplished with software rather than hardware) these routines will be included in ROM or PROM. Since the demonstration system does not include PROM or EPROM, an initialization or "front panel" routine will have to be reentered each time the system is powered up. The only exception to this is the information that is placed in the 32 bytes of retainable RAM within the 6802. (A battery must be included to permit this retainable RAM function to be used.)

The controller of the system shown in Fig. 7-28 is, of course, the 6802 MPU. For the 6802 to know what function it is to perform, instructions must be obtained. These instructions are stored in memory and constitute the program. The MPU obtains the instructions by placing an address on the 16 address lines that enable the addressed memory. The 8-bit data word stored in the addressed memory space will be placed on the data bus to be received by the MPU. Figure 7-32 is a memory map for the system. The 6802 is capable of addressing over 65,000 bytes of memory. The breadboard system uses only a few of these. Figure 7-28 indicates that all 16 address lines are not wired into the system. If the chip select inputs to the various devices are listed, it will be noted that the system uses only the address lines A0–A6, A12, and A14. Since the number of address locations is small, these few lines are adequate to provide unique addressing for each memory word and each system chip.

Construct the breadboard computer shown in Fig. 7-28. Construction will be easiest if multiconductor ribbon cable is used for data and address buses where several parallel lines are needed. To facilitate changing the jumpers from program load to execute mode, DIP plugs or headers can be connected to the ends of the ribbon cable that connects to the 6810 RAM. If the chips are arranged to reduce crossed wires as much as possible, construction and debugging will be easier. Also, if the two sets of lines between which the RAM jumpers must be alternated are placed on opposite sides of a breadboard strip, it will be a simple matter to move the three DIP headers from one set of connections to the other.

After the breadboard computer has been constructed, a simple program can be loaded to check the system for proper operation. For all programs, it is necessary to use the initialization routine shown in Fig. 7-33a.

The initialization routine for the breadboard computer is loaded into the 6810 by changing the address, data, and control jumpers to the load positions. Data can then be placed in the data inputs and will be written into memory by depressing the *write* switch, S7. Switch S6, the *count* switch, will be depressed after each byte of data is entered to increment the counter circuit. If a specific memory location is desired, it can be placed on the load inputs, with S5, the *load* switch, depressed. This will transfer the location at the load inputs to the outputs and simultaneously to the 6810 address inputs. If LEDs are placed on each of the counter output lines, it will be easy to determine what address is being

(A) Initialization Routine

Memory Location	*Instruction/ Data*	*Mnemonic*	*Addressing Mode*	*Process Description*
FF80	86	LDAA	I	Load 32_{16} in ACC A
FF81	32			
FF82	B7	STAA	E	Access data direction register A (DDRA) of PIA by storing 32_{16} in control register A (CRA)
FF83	10			
FF84	01			
FF85	B7	STAA	E	Access DDRB by storing 32_{16} in CRB
FF86	10			
FF87	03			
FF88	86	LDAA	I	Load 00 in ACC A
FF89	00			
FF8A	B7	STAA	E	Set peripheral register A (PRA) as inputs by loading lows in DDRA
FF8B	10			
FF8C	00			
FF8D	86	LDAA	I	Load FF_{16} in ACC A
FF8E	FF			
FF8F	B7	STAA	E	Set PRB as outputs by loading highs in DDRB
FF90	10			
FF91	02			
FF92	86	LDAA	I	Load 36_{16} in ACC A
FF93	36			
FF94	B7	STAA	E	Access peripheral register A (PRA) by loading 36_{16} in CRA
FF95	10			
FF96	01			
FF97	B7	STAA	E	Access PRB by loading 36_{16} in CRB
FF98	10			
FF99	03			
FF9A	8E	LDS	I	Set the stack pointer to memory location 0060
FF9B	00			
FF9C	60			
FF9D	CE	LDX	I	Load the index register with memory location 0000
FF9E	00			
FF9F	00			

Note: Initialization will be completed after the test program or monitor routine is loaded by setting the proper interrupt vectors as follows:

FFFC	00	MNI vector address—program start
FFFD	00	
FFFE	FF	Restart vector address—bottom of initialization routine
FFFF	80	

Note: Addressing mode abbreviations
N = inherent
I = immediate
D = direct
E = extended
R = relative
X = indexed

Fig. 7-33 (a) Initialization routine.

(B) Test Program

Memory Location	Instruction/ Data	Mnemonic	Addressing Mode	Process Description
FFA0	B6	LDAA	E	Polling loop—load CRA byte in ACC A
FFA1	10			
FFA2	01			
FFA3	81	CMPA		Compare CRA byte with $B6_{16}$
FFA4	B6			
FFA5	26	BNE	R	If CA1 not set (S1 has not been depressed) branch back to FFA0 (CRA byte $\neq B6_{16}$)
FFA6	F9*			
FFA7	B6	LDAA	E	If switch has been pressed, load data from PRA into ACC A
FFA8	10			
FFA9	00			
FFAA	B7	STAA	E	Output data to PRB LEDs
FFAB	10			
FFAC	02			
FFAD	20	BRA	R	Branch back to program start
FFAE	F1			

Note: Load interrupt vectors shown at bottom of initialization routine.

*Branch offset for backward branching uses 2s complement number, i.e., backward branch of 7 steps = 10000111 in binary MSB 1 = backward branch; 1s complement = 11111000, 2s complement = 11111001 = F9.

Note: Addressing mode abbreviations
N = inherent
I = immediate
D = direct
E = extended
R = relative
X = indexed

Fig. 7-33 (b) Test program.

acted upon. If the Smitty breadboard is used, the data latches and the switch-conditioning circuits will be available on the breadboard. The 6810 RAM must be connected to the system power supply during the loading process and must remain connected thereafter so that its memory contents will be retained when the jumpers are returned to the normal positions. At this time, load the initialization routine and the test program shown in Fig. 7-33b.

After the RAM is loaded and returned to the system buses, the $\overline{\text{reset}}$ switch S3 must be depressed. This will clear the MPU registers and will set all of the PIA registers to zero. It also will cause the MPU to address the vector memory locations FFFF and FFFE. These addresses will reach the outboard RAM and will place address FF80 on the data bus. This is the starting address of the initialization routine shown in Fig. 7-33a. As indicated in the figure, the initialization routine will establish the A side peripheral data lines of the 6821 PIA as inputs, the B lines as outputs, CA1 as a data ready input, CA2 as output, CB1 as data received input, and CB2 as output. The stack pointer is initialized at hex address 0060 and the index register at 0000. The initialization routine is followed by the test program. The first element of the test program is a "polling" loop. This polling loop will test the IRQA1 flag for presence of a high. If the flag is not high, the MPU will loop repeating steps FFA0 through FFA5 of the program and again polling the IRQA1 flag. This process will continue until S1 (data ready) is pressed, conveying a low-to-high transition to CA1. CA1 causes the flag to go high, indicating that data is available on the A peripheral data lines of the PIA. When the flag goes high, the program will continue with the next instruction and will ignore the Branch if Not Equal (BNE) instruction. The data loaded on the PIA A peripheral data lines will be loaded into accumulator A by the LDAA (B6) instruction. This byte of data will then be stored at the PIA B peripheral data lines by the STAA (B7) instruction. The LEDs connected to the PIA B peripheral lines will indicate the same sequence of highs and lows as were placed on the A-side input. The program will now loop back to the start of the program at FFA0 in response to the Branch Always (BRA) instruction. New bytes can now be loaded and displayed as desired by loading the PIA A-side inputs and depressing S1.

(C) Monitor Program

Memory Location	*Instruction/ Data*	*Mnemonic*	*Addressing Mode*	*Process Description*
			Program Load Routine	
FFA0	86	LDAA	I	Load $3E_{16}$ in ACC A
FFA1	3E			
FFA2	B7	STAA	E	Turn on CA2 LED by loading $3E_{16}$ in CRA
FFA3	10			
FFA4	01			
FFA5	B6	LDAA	E	Polling loop—load contents of CRA in ACC A
FFA6	10			
FFA7	01			
FFA8	81	CMPA	I	Compare CRA byte with BE_{16}
FFA9	BE			
FFAA	26	BNE	R	Return to FFA5 if CRA byte $\neq BE_{16}$
FFAB	F9			
FFAC	B6	LDAA	E	Load the data from PRA in ACC A
FFAD	10			
FFAE	00			
FFAF	A7	STAA	X	Store data at the memory location = index reg + 0 offset
FFB0	00			
FFB1	08	INX	N	Increment index register
FFB2	86	LDAA	I	Load 36_{16} in ACC A
FFB3	36			
FFB4	B7	STAA	E	Clear CA1 and turn the CA2 LED off by loading 36_{16} in CRA
FFB5	10			
FFB6	01			
FFB7	86	LDAA	I	Time delay—load FF_{16} in ACC A
FFB8	FF			
FFB9	C6	LDAB	I	Load FF_{16} in ACC B
FFBA	FF			
FFBB	5A	DECB	N	Decrement B
FFBC	26	BNE	R	If ACC B $\neq$ 0, branch back to FFBB
FFBD	FD			
FFBE	4A	DECA	N	Decrement A
FFBF	26	BNE	R	If ACC A 0, branch back to FFB9
FFC0	F8			Time delay = time required for MPU to count down to zero from 255, 255 times—LED will stay off long enough to see blink
FFC1	B6	LDAA	E	End of program polling loop—load CRB byte in ACC A
FFC2	10			
FFC3	03			
FFC4	81	CMPA	I	Compare CRB byte with B6
FFC5	B6			
FFC6	26	BNE		If CRB byte = 0, S2 not depressed; branch back to FFA0
FFC7	D8			

Fig. 7-33 (c) Monitor program.

Memory Examine Routine				
FFC8	CE	LDX	I	Load the index register with the location of program start (0000)
FFC9	00			
FFCA	00			
FFCB	86	LDAA	I	Load $3E_{16}$ in ACC A
FFCC	3E			
FFCD	B7	STAA	E	Turn on the CB2 LED by storing $3E_{16}$ in CRB
FFCE	10			
FFCF	03			
FFD0	A6	LDAA	X	Load the data at memory location = index register + 0 offset in ACC A
FFD1	00			
FFD2	B7	STAA	E	Store data at the PIA B-side LEDs
FFD3	10			
FFD4	02			
FFD5	B6	LDAA	E	Data acknowledge polling loop—load CRB byte in ACC A
FFD6	10			
FFD7	03			
FFD8	81	CMPA	I	Compare CRB byte with BE_{16}—checks for closure of S2 (data acknowledge)
FFD9	BE			
FFDA	26	BNE	R	If CRB byte $\neq$ BE_{16}, loop to FFD5
FFDB	F9			
FFDC	08	INX	N	Increment index register
FFDD	86	LDAA	I	Load $3E_{16}$ in ACC A
FFDE	3E			
FFDF	B7	STAA	E	Clear CB1 while leaving CB2 LED lit by storing $3E_{16}$ in CRB
FFE0	10			
FFE1	03			
FFE2	20	BRA	R	Branch back to start of memory examine routine (FFD0)
FFE3	EC			
FFE4				
— — — —	These memory locations not used by initialization routine and monitor program.			
FFFB				
FFFC	00			NMI Vector (program start)
FFFD	00			
FFFE	FF			Restart vector (bottom of initialization routine)
FFFF	80			

Note: Addressing mode abbreviations
N = inherent
I = immediate
D = direct
E = extended
R = relative
X = indexed

Fig. 7-33 (c) cont'd Monitor program.

If this program operates properly, it can be assumed that the system is operating properly. If it does not, the reader should check his wiring and programming for errors. The program can be checked by moving the jumpers to the load positions and replacing the data input lines with LEDs. As the count switch (S6) is depressed, the contents of each memory location in turn will be displayed. If errors are noted, the input lines can be reinserted, the proper data set at the inputs, and S7 (write) depressed. Removal of the input lines will now permit display of the memory contents. When the system is determined to be operating properly, loading of the monitor program in Fig. 7-33c can be accomplished. The monitor program will be loaded immediately after the initialization routine in place of the test program.

Since no ROM or EPROM is used in the system, the 6810 RAM must be loaded with the monitor program. A *monitor program* is a group of instructions that direct the MPU to perform the steps necessary to load program and data information and to execute it. The monitor program will accept data from the PIA and put it in memory at a location determined by the value in the index register. Another polling loop will then be encountered that will loop back to the above polling loop unless CB1 causes the IRQB1 flag to go high, indicating that data is loaded and memory examination is desired.

If the first loop is reentered, the index register will be incremented by one for each time the program passes through the loop. If memory examine is chosen,[8] the index register will be returned to program start, and each memory location will be displayed in sequence at the PIA B outputs. The *data acknowledge* switch S2 must be depressed after each memory location is displayed in order for the next location to be routed to the output LEDs.

As data is loaded during the first portion of the monitor program, the CA2 LED will signal that the load process is progressing properly. Following initialization, the CA2 LED will be lit. Each time the *data ready* switch (S1) is depressed the CA2 LED will blink, indicating that the data has been loaded. When the memory examine routine is accessed, the CB2 LED will be lit.

The above actions can be terminated by depressing S4, the *go to program* switch. When this is done, the MPU will vector to the address stored in the NMI vector addresses FFFD and FFFC. This is, of course, the location of the start of the program. If mistakes are made in programming, the program will have to be reentered. This is easily done by depressing the *restart* switch S3 and reentering the program.

The process flow of the monitor program is informative. Follow the program step-by-step and notice how the op code and addressing mode are manipulated to process the data. The process descriptions should assist in understanding the program flow.

Now that the monitor program is loaded, the addition program shown in Fig. 7-34a can be loaded and executed. When the monitor program is loaded, the *restart* switch should be depressed. This will send the MPU back to FF80, the start of the initialization routine. This will be processed up to the first polling loop. At this point, the MPU will wait until program data is placed on the PIA A inputs and S1, *data ready*, is depressed. This data will be entered and the MPU will loop back to get the next data byte. This will continue until the program is loaded and the *examine memory* switch[9] or the *go to program* switch is activated, and the program will execute, resulting in the sum of the two numbers being displayed at the PIA B outputs.

New numbers can be added by entering them and depressing the *data ready* switch. After each two numbers are added, the sum will be displayed. For program simplicity, numbers whose sum is no more than 8 bits (255_{10}) should be used. The program can be expanded to display 2-byte sums, if desired.

Enter the numbers listed in Fig. 7-34b. Record the sums and check them for accuracy. This experiment illustrates a stored program sequence and keyboard modification techniques. Notice the use of the various address modes in this program. The variety of addressing modes used in this simple program demonstrates the flexibility and power of the 6802 instruction set. Choice of the appropriate addressing mode is as important as choice of the appropriate operation code.

The sequence of instructions in the polling loop is interesting. The function of this sequence is to check bit 7 of the PIA A-side control register for presence of a high. If the bit is high, it indicates that switch S1 has been depressed to show that data is ready on the input lines. To check bit 7, the contents of the control register are loaded into accumulator A. Comparison is now made of the byte of information loaded from the A-side control register with the hexadecimal value B6. The control register byte will be 36_{16} until a positive transition is received at the CA1 input. The 36_{16} will then become $B6_{16}$, and the two bytes will match. The Branch if Not Equal (BNE) instruction will branch back to the top of the polling

[8]To initiate memory examine, switch S2 will be depressed, and then switch S1 will be depressed.

[9]*Examine memory* switch = *data acknowledge* switch.

(A) Addition Program

Memory Location	*Instruction/ Data*	*Mnemonic*	*Addressing Mode*	*Process Description*
1	86	LDAA	I	Load 36_{16} in ACC A
2	36			
3	B7	STAA	E	Clear CA1 by storing 36_{16} in CRA
4	10			
5	01			
6	B6	LDAA	E	Polling loop—load CRA byte in ACC A
7	10			
8	01			
9	81	CMPA	I	Compare CRA byte with $B6_{16}$
10	B6			
11	26	BNE	R	If CRA byte $\neq$ $B6_{16}$, branch to top of loop
12	F9			
13	B6	LDAA	E	Load first number from PRA into ACC A
14	10			
15	00			
16	C6	LDAB	I	Load 36_{16} in ACC B
17	36			
18	F7	STAB	E	Clear CA1 by storing 36_{16} in CRA
19	10			
20	01			
21	F6	LDAB	E	Polling loop—load CRA byte in ACC B
22	10			
23	01			
24	C1	CMPB	I	Compare CRA byte with $B6_{16}$
25	B6			
26	26	BNE	R	If CRA byte $\neq$ $B6_{16}$, branch to top of loop
27	F9			
28	F6	LDAB	E	Load second number from PRA into ACC B
29	10			
30	00			
31	1B	ABA	N	Add ACC B to ACC A and store result in ACC A
32	B7	STAA	E	Store sum at PRB LEDs
33	10			
34	02			
35	7E	JMP	E	Jump (loop) to program start (0000)
36	00			
37	00			

Note: Addressing mode abbreviations
N = inherent
I = immediate
D = direct
E = extended
R = relative
X = indexed

Fig. 7-34 (a) Addition program.

(B) Test Numbers for Addition Program

	Number A		Number B	
	Decimal	*Hex*	*Decimal*	*Hex*
1.	12	00001100	10	00001010
2.	52	00110100	44	00101100
3.	100	01100100	7	00000111
4.	91	01011011	25	00011001
5.	36	00100100	36	00100100

Fig. 7-34 (b) Test numbers for addition program.

loop until the two bytes match. Then the BNE instruction will be bypassed, and program execution will continue with the step below the BNE offset byte. Until the bytes match, the BNE instruction will continue branching to the top of the loop each time it is executed. The branch will be to the address location equal to the current location plus two plus the value of the operand. In this case, the operand is – 7, and the program branches back to the start of the sequence.

Experiment 7-2: The Subtraction Program

Write a subtraction program for 8-bit binary numbers. Compare the program with Fig. 7-35. Load and execute the program and confirm that the program action is correct.

Experiment 7-3: BCD Subtraction

Design a program for BCD subtraction. Prepare a flowchart and a program listing. Reference to the BCD addition routine (see Fig. 7-36a) will help in writing the program. Figure 7-36b can be used to check your program. Try running your program even if it is different from the one shown in Fig. 7-36b. There are numerous ways to program most procedures. If your program is different but obtains correct results, use the instruction set chart to count the number of MPU cycles required for each program. The program that requires the least number of MPU cycles is the most time-efficient.

Experiment 7-4: Logic Programs

The 6802 system also can perform logic functions. The basic AND, OR, and EXOR functions are included in the instruction set. Other logic functions require some programming. Figure 7-37 includes the programming listings for several simple logic operations. Notice that the logic function covers all 8 bits.

The programmable nature of the 6802 system makes it possible for the MPU to simulate the function of other logic blocks. Even sophisticated logic functions can be programmed. Figures 7-38 through 7-40 show logic diagrams for several more complex logic circuits. A program listing is given for the first (see Fig. 7-38). Write programs for the other two. Load and run these programs to gain an understanding of them.

Experiment 7-5: The Look-up Table and the Subroutine

The 6802 is capable of handling alphabetic as well as numeric data. It treats the alphabetic data as numbers as far as MPU operation is concerned, but through proper programming, alphabetic manipulations can be performed. A partial listing of the ASCII code is shown in Fig. 7-41. (ASCII stands for American Standard Computer Information Interchange). A seven-digit code is used to represent the ten numbers of the decimal system and the letters of the alphabet. A typical alphabetic operation would be a search routine to identify a letter or word within a file of information stored in memory. If, for instance, a department store listed in its computer memory the names of all persons who had given bad checks, the store could then check each check writer against this list to prevent acceptance of additional bad checks from these people. A simple example of this type of operation is shown in Fig. 7-42. This program is designed to compare a letter with those letters placed in the memory listing. If the letter is not in the file, an N (ASCII 1001110) for "no match" is outputted. If the letter is included in the file, an M (ASCII 1001101) for "match" is outputted. In this case, the listing was predetermined at the time the program was loaded. This type of file is sometimes referred to as a look-up table. Logarithm tables, conversion tables, and similar lists often take the form of look-up tables in computer systems.

Load the program shown in Fig. 7-42 and input in turn each letter of the alphabet. If a record is kept of which letters indicate a match in the list, the list in the computer can be determined.

Notice that the MPU deals with the ASCII character as if it were a binary number. At each comparison, the MPU is actually asking the question, "Does subtraction of these two numbers result in zero?" The program, however, interprets the mathematical actions by saying, "If the result is zero, the two letters are the same and if the result is not zero, the letters are not the same." It should be understood that if two ASCII letters are treated mathematically (added, subtracted, etc.), the result will be undefined. It is the programmer's job to direct the MPU to deal with alphabetic data appropriately.

Alphabetic data is subject to only a few computer-controlled actions. Typical computer manipu-

Subtraction Program

Step	*Instruction/ Data*	*Mnemonic*	*Addressing Mode*	*Process Description*
1	86	LDAA	I	Load 36_{16} in ACC A
2	36			
3	B7	STAA	E	Reset CA1 by storing 36_{16} in CRA
4	10			
5	01			
6	B6	LDAA	E	Polling loop—load CRA
7	10			
8	01			
9	81	CMPA	I	Compare CRA byte and $B6_{16}$
10	B6			
11	26	BNE	R	Branch if not equal
12	F9			
13	B6	LDAA	E	Load data byte from PRA into ACC A
14	10			
15	00			
16	C6	LDAB	I	Load 36_{16} into ACC B
17	36			
18	F7	STAA	E	Clear CA1 by storing 36_{16} into CRA
19	10			
20	01			
21	F6	LDAB	E	Polling loop—load CRA byte into ACC B
22	10			
23	01			
24	C1	CMPB	I	Compare CRA byte with $B6_{16}$
25	B6			
26	26	BNE	R	Branch if not equal
27	F9			
28	F6	LDAB	E	Load data byte from PRA into ACC B
29	10			
30	00			
31	10	SBA	N	Subtract ACC B from ACC A and store results in ACC A
32	B7	STAA	E	Store results at PRB LEDs
33	10			
34	02			
35	7E	JMP	E	Jump (loop) to top of program
36	00			
37	00			

Note: Addressing mode abbreviations
N = inherent
I = immediate
D = direct
E = extended
R = relative
X = indexed

Fig. 7-35 Subtraction program.

(A) BCD Addition

Step	Instruction/ Data	Mnemonic	Addressing Mode	Process Description
1				Reset CA1 (steps 1-5 in Fig. 7-35)
2				Polling loop—use ACC A (steps 6-12 in Fig. 7-35)
3	B6	LDAA	E	Load first number in ACC A
4	10			
5	00			
6				Clear CA1 and polling loop—use ACC B (steps 16-27 in Fig. 7-35)
7	BB	ADDA	E	Add number at PRA to number in ACC A and store sum in ACC A
8	10			
9	00			
10	19	DAA	N	Decimal adjustment—add the appropriate number to binary sum to provide BCD number
11	B7	STAA	E	Store BCD sum at PRA LEDs
12	10			
13	02			
14	7E	JMP	E	Loop to start of program
15	00			
16	00			

Note: Addressing mode abbreviations N = inherent
I = immediate
D = direct
E = extended
R = relative
X = indexed

Fig. 7-36 (a) BCD addition program.

lations of alphabetic data include (1) storage, (2) sorting, and (3) transfer between memory locations.

Due to the amazing speed of the computer, these simple functions can be assembled into powerful and productive programs. For example, a computer can sort through lengthy lists in a fraction of the time that would be required for manual sorting.

The program shown in Fig. 7-43 illustrates a file search and retrieval routine. The program builds a file of family names for four family members. The program uses four-letter names. Inputting the title of the family member (i.e., Dad = D, Mom = M, Son = S, Daughter = D) will result in the appropriate name being provided to the output as successive ASCII letters. Since the MPU would output these letters so rapidly that the LEDs would not display them, recording each successive letter would not be possible. To correct this problem, a program sequence is included that will require the MPU to enter and remain in a loop for 65,535 times. Each time a letter is outputted, the MPU will enter and complete this looping sequence prior to outputting the next letter. The result of this process will be that the letters will remain on the display for the length of this looping sequence.

The looping sequence is handled as a subroutine. Instead of the program code having to be entered each time the program outputs a letter, the looping subroutine is accessed instead. This represents a great saving in program memory space. The breadboard computer does not have adequate memory for this program. It is included as an example but is too lengthy for breadboard entry.

Interrupts

The 6802 provides for several kinds of interrupt routines. The Hardware Interrupt (IRQ) is usually generated by a peripheral device that desires to be serviced by the MPU. The PIA can be programmed to provide the interrupt request. It is not used in this system since only one peripheral is included. If several peripheral devices were connected to the MPU, they would request service by bringing the IRQ line low. The MPU would then go to interrupt vector FFF9 FFF8. A memory location would be

(B) BCD Subtraction

Step	*Instruction/ Data*	*Mnemonic*	*Addressing Mode*	*Process Description*
1				Reset CA1 (steps 1–5 in Fig. 7-35)
2	86	LDAA	I	Load 99_{10} in ACC A
3	99			
4				Polling loop (steps 6–12 in Fig. 7-35)
5	B0	SUBA	E	Subtract subtrahend from ACC A (99_{10})
6	10			Subtrahend input at PRA
7	00			
8	0D	SEC	N	Set carry bit
9				CA1 reset and polling loop (steps 16–27 in Fig. 7-35)
10	B9	ADCA	E	Add minuend with carry to ACC A value minuend input at PRA
11	10			
12	00			
13	19	DAA	N	Adjust for BCD
14	B7	STAA	E	Store BCD result at PRA LEDs
15	10			
16	02			
17	7E	JMP	E	Loop to start of program
18	00			
19	00			

Note A: Addressing mode abbreviations
N = inherent
I = immediate
D = direct
E = extended
R = relative
X = indexed

Note B: This program implements the following formula:

Minuend + (99 − subtrahend + 1) − 100 = result

The 9s complement of the subtrahend has 1 added to obtain the 10s complement. The DAA instruction adjusts the decimal point, effectively subtracting 100, and provides the proper BCD result.

Fig. 7-36 (b) BCD subtraction program.

loaded in this location, which would direct the MPU to service the interrupt request. If a system included several peripherals, the program would check each peripheral to determine which had initiated the interrupt request. The program also might include a priority routine that would service the various peripherals on a basis of priority. This would be needed if several peripherals requested an interrupt at the same time.

A second interrupt is the Software Interrupt (SWI). This is an interrupt programmed into the program flow. It might be used to permit processing of a second phase of a task, which would then provide results for the program sequence. The nature of interrupt action has benefits for the programmer. As shown in Fig. 7-44, an interrupt results in the storage of the active registers in the memory stack. When the interrupt is completed, a Return from Interrupt (RTI) instruction will reload the stored information in the MPU registers, and the processing can continue. The Wait for Interrupt (WAI) instruction also is available. It stacks the registers and places the MPU in a wait sequence until an interrupt is received. This procedure can speed up servicing of interrupts if rapid processing is required. The Software Interrupt (SWI) directs the MPU to vector addresses FFFA and FFFB. As with the $\overline{IRQ}$, an address will be stored in these locations, which will direct the MPU to the proper interrupt routine.

Step	*Instruction/ Data*	*Mnemonic*
AND		
1	B6	LDAA
2	10	
3	00	
4	B4	ANDA
5	10	
6	00	
7	B7	STAA
8	10	
9	02	
NAND		
1-6	Same as AND above	
7	C6	LDAB
8	FF	
9	D7	STAB
10	7E	
11	98	EORA
12	7E	
13	B7	STAA
14	10	
15	02	
OR		
1	B6	LDAA
2	10	
3	00	
4	BA	ORA
5	10	
6	00	
7	B7	STAA
8	10	
9	02	

Note: These programs do not include initialization and polling routines. Eight gates are simulated simultaneously. *Go to program* switch must be depressed to execute programs after entry.

Step	*Instruction/ Data*	*Mnemonic*
NOR		
1-6	Same as OR above	
7	C6	LDAB
8	FF	
9	D7	STAB
10	7E	
11	98	EORA
12	7E	
13	B7	STAA
14	10	
15	02	
XOR		
1	B6	LDAA
2	10	
3	00	
4	B8	EORA
5	10	
6	00	
7	B7	STAA
8	10	
9	02	
XNOR		
1-6	Same as XOR above	
7	C6	LDAB
8	FF	
9	D7	STAB
10	7E	
11	98	EORA
12	7E	
13	B7	STAA
14	10	
15	02	

Fig. 7-37 Programs for simple logic functions.

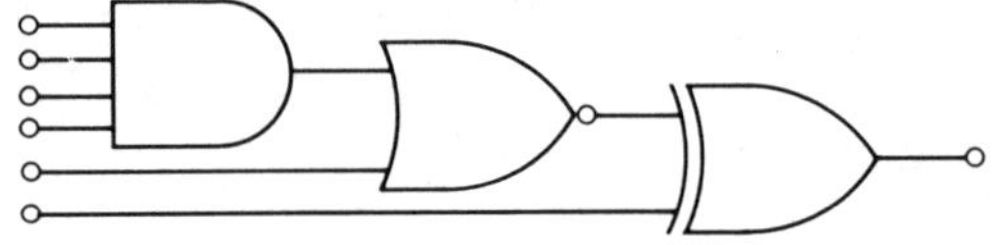

Step	*Instruction/ Data*	*Mnemonic*
1	B6	LDAA
2	10	
3	00	
4	B4	ANDA
5	10	
6	00	
7	B4	ANDA
8	10	
9	00	
10	B4	ANDA
11	10	
12	00	
13	BA	ORA
14	10	
15	00	
16	C6	LDAB
17	FF	
18	D7	STAB
19	7E	
20	98	EORA
21	7E	
22	B8	EORA
23	10	
24	00	
25	B7	STAA
26	10	
27	02	

Include polling routines
Initialization assumed
Depress *go to program* to execute

Fig. 7-38 Logic diagram and program.

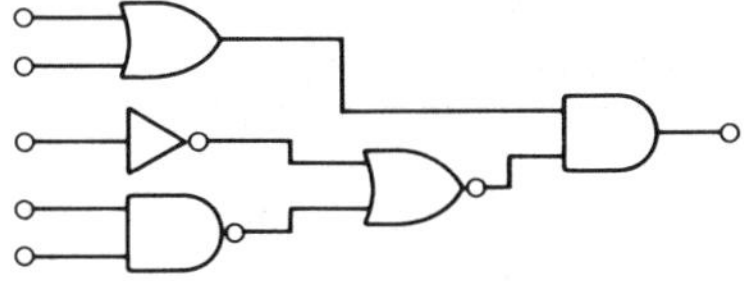

Fig. 7-39 Logic diagram.

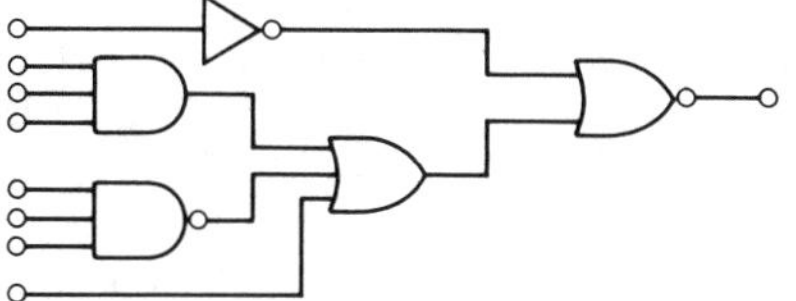

Fig. 7-40 Logic diagram.

Character	*ASCII Code*	*Character*	*ASCII Code*
0	0110000	I	1001001
1	0110001	J	1001010
2	0110010	K	1001011
3	0110011	L	1001100
4	0110100	M	1001101
5	0110101	N	1001110
6	0110110	O	1001111
7	0110111	P	1010000
8	0111000	Q	1010001
9	0111001	R	1010010
A	1000001	S	1010011
B	1000010	T	1010100
C	1000011	U	1010101
D	1000100	V	1010110
E	1000101	W	1010111
F	1000110	X	1011000
G	1000111	Y	1011001
H	1001000	Z	1011010

Fig. 7-41 ACSII computer code (partial listing).

Step	*Instruction/ Data*	*Mnemonic*	*Addressing Mode*	*Process Description*
1				Reset CA1 (steps 1–5 in Fig. 7-35)
2	86	LDAA	I	Load first letter in ACC A
3	44 (ASCII D)			
4	B7	STAA	E	Store letter at 0070
5	00			
6	70			
7	86	LDAA	I	Load second letter in ACC A
8	48 (ASCII H)			
9	B7	STAA	E	Store second letter at 0071
10	00			
11	71			
12	86	LDAA	I	Load third letter in ACC A
13	4C (ASCII L)			
14	B7	STAA	E	Store third letter at 0072
15	00			
16	72			
17	86	LDAA	I	Load fourth letter in ACC A
18	50 (ASCII P)			
19	B7	STAA	E	Store third letter at 0073
20	00			
21	73			
22	CE	LDX	I	Load index register with bottom memory location of look up table

Fig. 7-42 Program for look-up table.

Step	Instruction/ Data	Mnemonic	Addressing Mode	Process Description
23	00			
24	70			
25	C6	LDAB	I	Load 04_{16} in ACC B—(number of letters in look-up table)
26	04			
27				Polling loop (steps 6–12 in Fig. 7-35)
28	B6	LDAA	E	Load letter to be compared with look-up table from PRA
29	10			
30	00			
31	A1	CMPA	X	Compare letter in ACC A with letter at memory location = index register + 0 offset
32	00			
33	27	BEQ	R	If letters are the same, branch to match routine (step 46)
34	0C			
35	08	INX	N	Increment index register to next letter of look-up table
36	5A	DECB	N	Decrement ACC B
37	2E	BGT	R	If ACC B is greater than 0, loop back to step 31
38	F8			
39	86	LDAA	I	Load ACC A with letter *N* (ASCII—1001110)
40	4E			
41	B7	STAA	E	Store letter *N* at PRB LEDs
42	10			
43	02			
44	20	BRA	R	Branch to step 22
45	E2			
46	86	LDAA	I	Load ACC A with letter *M* (ASCII—1001101)
47	4D			
48	B7	STAA	E	Store letter *M* at PRB LEDs
49	10			
50	02			
51	20	BRA	R	Branch back to step 22
52	DB			

Note: Addressing mode abbreviations N = inherent
I = immediate
D = direct
E = extended
R = relative
X = indexed

Fig. 7-42 cont'd Program for look-up table.

Step	Instruction/ Data	Mnemonic	Addressing Mode	Process Description
		File Load Routine		
1				Clear CA1 (steps 1-5 in Fig. 7-35)
2	CE	LDX	I	Load index register with first look-up table memory location (XXXX)
3	XX			
4	XX			
5				Polling loop (steps 6-12 in Fig. 7-35)
6	B6	LDAA	E	Load ID letter; i.e., D = DAD, etc.
7	10			
8	00			
9	A7	STAA	X	Store ID letter at memory location = index register + 0
10	00			
11	C6	LDAB	I	Load ACC B with 4
12	04			
13	08	INX	N	Increment index register
14				Clear CA1 (step 1)
15				Polling loop (step 5)
16	B6	LDAA	E	Load letter of name in ACC A
17	10			
18	00			
19	A7	STAA	X	Store letter in memory location of previous letter + 1
20	00			
21	5A	DECB	N	Decrement ACC B
22	2E	BGT	R	Branch back to step 13 if ACC B $\neq$ 0
23	YY			
24				Polling routine—use data acknowledge CRB—additional file entry?
25	B6	LDAA	E	Load branch code—11_{16} = additional file entry; anything else = no more
26	10			
27	00			
28	81	CMPA	I	Compare PRA byte and 11_{16}
29	11			
30	27	BEQ	R	If PRA byte = 11_{16}, branch back to step 6
31	YY			

Fig. 7-43 File search and retrieval subroutine program.

Step	Instruction/ Data	*Mnemonic*	*Addressing Mode*	*Process Description*
		File Search Routine		
32	FF	STX	E	Store index register at memory location ZZZZ
33	ZZ			
34	ZZ			
35				Reset CA1 and polling routine (see step 1)
36	B6	LDAA	E	Load ID Letter of name in ACC A
37	10			
38	00			
39	CE	LDX	E	Set index register to start of file (XXXX)
40	XX			
41	XX			
42	E6	LDAB	X	Load ACC B with letter at memory location = index register + 0
43	00			
44	11	CBA	N	Compare accumulators
45	27	BEQ	R	If accumulators contain same letter, branch to output routine
46	YY			
47	08	INX	N	Increment index register to next ID letter
48	08	INX	N	
49	08	INX	N	
50	08	INX	N	
51	8C	CPX	E	Compare index register with contents of memory location ZZZZ (top of file)
52	ZZ			
53	ZZ			
54	2D	BLT	R	Branch back to step 40 if not at top of file
55	YY			
56	86	LDAA	I	Load no match code (FF) in ACC A
57	FF			
58	B7	STAA	E	Store no match code at PRB LEDs
59	10			
60	02			
61	20	BRA	R	Branch back to step 35
62	YY			

Fig. 7-43 cont'd File search and retrieval subroutine program.

Step	*Instruction Data*	*Mnemonic*	*Addressing Mode*	*Process Description*
		Output Routine		
63	C6	LDAA	I	Load ACC A with 04_{16}
64	04			
65	08	INX	N	Increment index register
66	A6	LDAA	X	Load ACC A with letter at memory location [7] index register + 0
67	00			
68	B7	STAA	E	Store letter at PRB LEDs
69	10			
70	02			
71	BD	JSR	D	Jump to time delay subroutine
72	XX			
73	5A	DECB	N	Decrement ACC B
74	2E	BGT	R	Branch back to step 63 if ACC B greater than 0
75	YY			
76	20	BRA	R	Branch back to step 35
77	YY			

XXXX represents unidentified memory locations
YY represents unidentified relative address offset
ZZZZ represents memory location for storage at top of file location

Step	*Instruction Data*	*Mnemonic*	*Addressing Mode*	*Process Description*
		Time Delay Subroutine		
1	B7	STAA	E	Display letter
2	10			
3	03			
4	86	LDAA	I	Store $65,536_{10}$ in locations
5	FF			
6	97	STAA	D	72 and 73
7	72			
8	97	STAA	D	
9	73			
10	08	INX	N	Increment X
11	BC	CPX	E	Compare memory locations and index register.
12	00			
13	72			
14	26	BNE	R	Loop in not =
15	87			
16	39	RTS	N	Return to main program

Note A: This program will require additional memory. Modification of the stack pointer locations also is needed. The stack should be initialized to an unused memory location. The subroutine will be placed in 6810 memory starting at location FFDF. The JSR instruction is used to reach the subroutine since branch instructions are limited to − 125 or + 127 bytes from the program counter location. This program is illustrative only and has not been debugged on the breadboard computer.

Note B: Addressing mode abbreviations
N = inherent
I = immediate
D = direct
E = extended
R = relative
X = indexed

Fig. 7-43 cont'd File search and retrieval subroutine program.

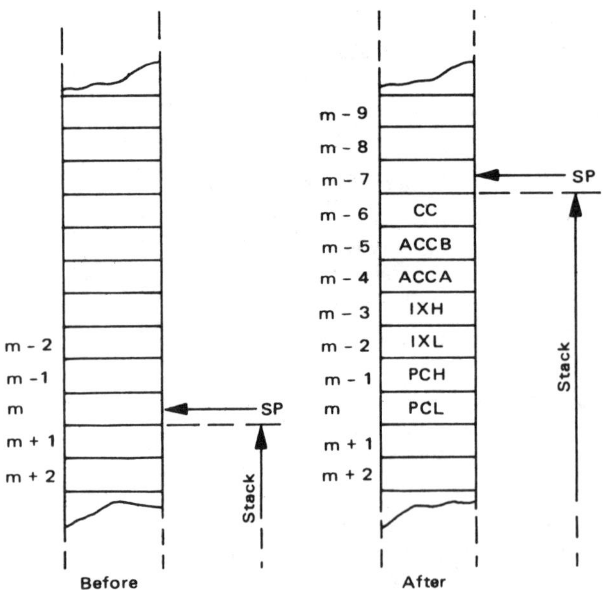

SP = Stack Pointer
CC = Condition Codes (Also called the Processor Status Byte)
ACCB = Accumulator B
ACCA = Accumulator A
IXH = Index Register, Higher Order 8 Bits
IXL = Index Register, Lower Order 8 Bits
PCH = Program Counter, Higher Order 8 Bits
PCL = Program Counter, Lower Order 8 Bits

Fig. 7-44 Stacking process of interrupt routine. (Courtesy of Motorola, Inc.)

The only interrupt that is designed into the breadboard system is the Nonmaskable Interrupt (NMI). The name for this interrupt is derived from the MPU interrupt mask bit. If this bit is set, IRQ and SWI will be prohibited. The NMI is not subject to the status of the mask bit, however. Therefore, the NMI will be serviced any time it is called for. In the breadboard system, the NMI is used for the *Go to Program* function. Depressing S4 will result in the MPU entering the NMI routine, which will stack the registers and direct the MPU to vector address FFFC/FFFD. For the breadboard system, the NMI routine will send the MPU to the start of the program for execution.

Summary of MPU Operation

The programming of the 6802 that has been demonstrated does not tap the power of the unit. It does, however, illustrate basic MPU operations. Memory storage and retrieval has been included. In larger and more complex systems, the memory manipulation will include larger memories, but the addressing and data flow will be accomplished in the same way as was used in this system.

Input/output for the demonstration system is very basic. Input/output for more advanced systems will take several forms. Some I/O devices will use a parallel data format and will be accessed via PIAs. Other I/O devices will use a serial data format. For some of these, the PIA will still be used. For devices located at a distance from the MPU, an ACIA and modems may be used. Cassette recorders, floppy disks, paper-tape punches, keyboards, video displays, and similar devices can be used with the MPU in advanced systems. Most of these will operate at slower speeds than the MPU and will use handshake controls, halts, and interrupts to interface with the system. For some applications, the 6840 timer will be used to generate time control sequences to permit data to be transmitted to or received from slower devices.

The programming examples in this chapter were limited to binary I/O encoding. The 6802 system is capable of handling more sophisticated data formats as well. This is usually accomplished using a special encoding such as ASCII. The MPU will interpret this code by using an intermediate program to obtain the binary machine codes that will direct the actual operation of the MPU. Expansion of the number of interfaces and the provision of alphanumeric keyboards and displays will make alphabetic data available to the 6802 system. Additional memory will, of course, be required.

Machine-level programming is tedious and time consuming. Use of "high-level" languages is much easier. If memory space is adequate, a language such as BASIC can be used with 6802. BASIC uses "plain English" commands such as "go to" and if-then" to direct the computer operation. The BASIC programming is loaded into the system, and a stored conversion program changes each BASIC statement into a series of machine-level codes. This greatly simplifies the programming task.

The purpose of this chapter was to present the rudiments of MPU construction and operation. Detailed programming must be left to more specialized volumes. However, since MPUs are being increasingly used to accomplish simple control tasks and dedicated functions, the kind of programming discussed here may serve the student's needs until he progresses into a fuller study of microcomputers. When that study is undertaken, this book will serve as a foundation that will permit the student to understand the hardware actions that underlie the programming statements. This is the type of knowledge that can mean the difference between excellence and mediocrity of programming skill.

APPENDIX A

The Melton Special Breadboard

The Melton Special is a simple but serviceable breadboard. Construction details are shown in Figs. A-1 through A-3. The unit is constructed on a 10′ × 14′ × 3′ metal chassis. The breadboard sockets, LEDs, and all panel controls are mounted on the top of the chassis, as shown in Fig. A-1. Dimensions can be altered to accommodate materials available to the reader. The custom connection strip can use pin jacks or a small breadboard socket. The power supplies, pulse generators, and LED circuits shown in Figs. A-2 and A-3 are mounted inside the chassis. The size of the transformer should be appropriate for mounting inside the chassis. The regulators are mounted on heat sinks attached to the rear of the chassis. If the variable supply is to be used primarily with CMOS, the regulator can be mounted on the chassis rather than a heat sink. Heat sink compound and appropriate insulators should be used.

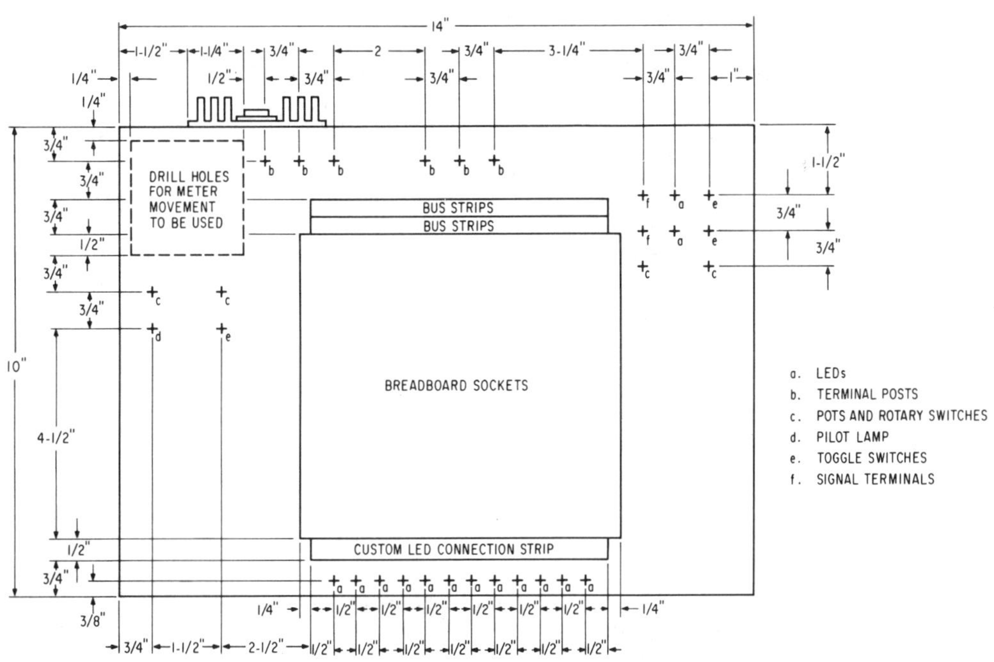

Fig. A-1 Dimensioning diagram for construction of breadboard chassis.

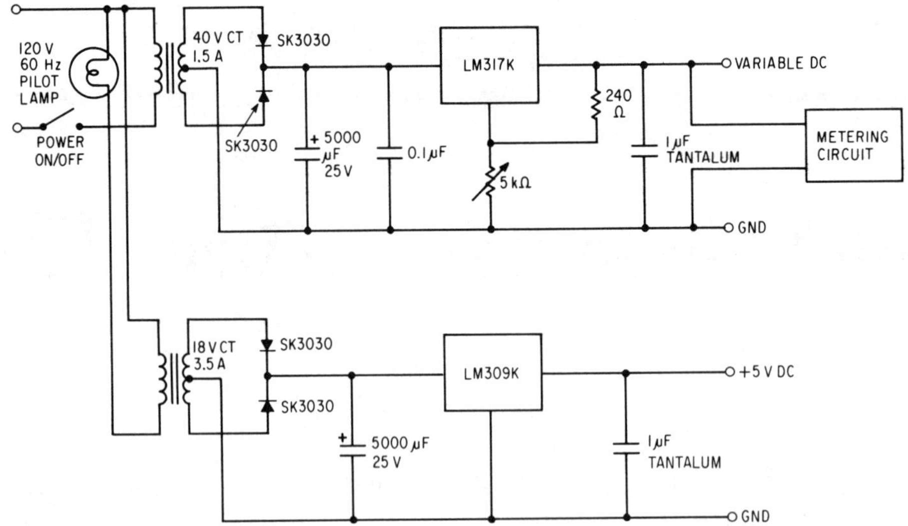

Fig. A-2 Dual power supply for breadboard. Regulated 5-V section seen below variable output section (shown with meter connection).

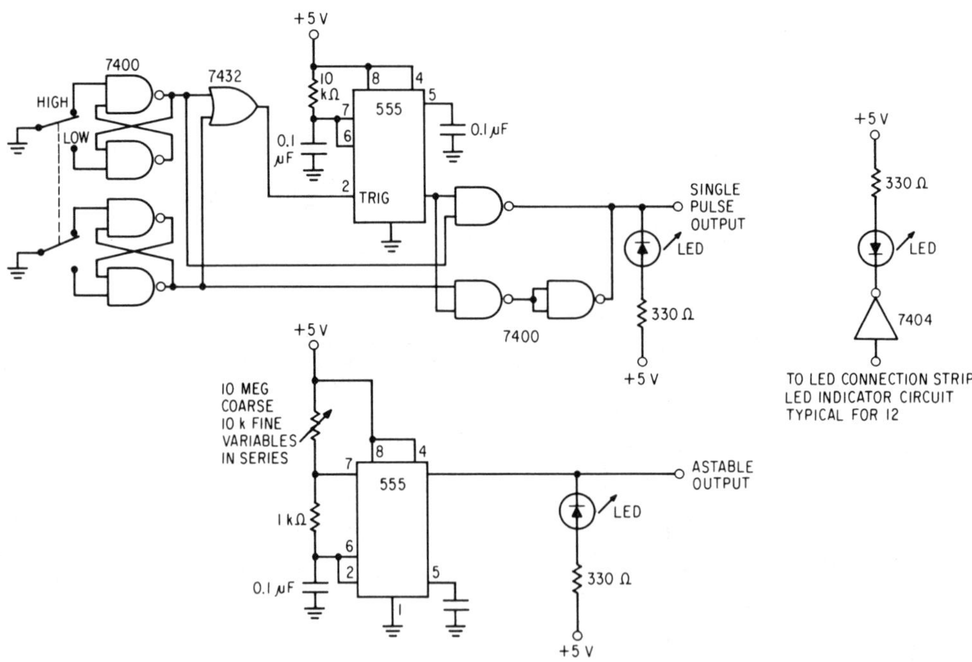

Fig. A-3 The two pulse generators utilized on the breadboard.

APPENDIX B

IC Type Number Identification

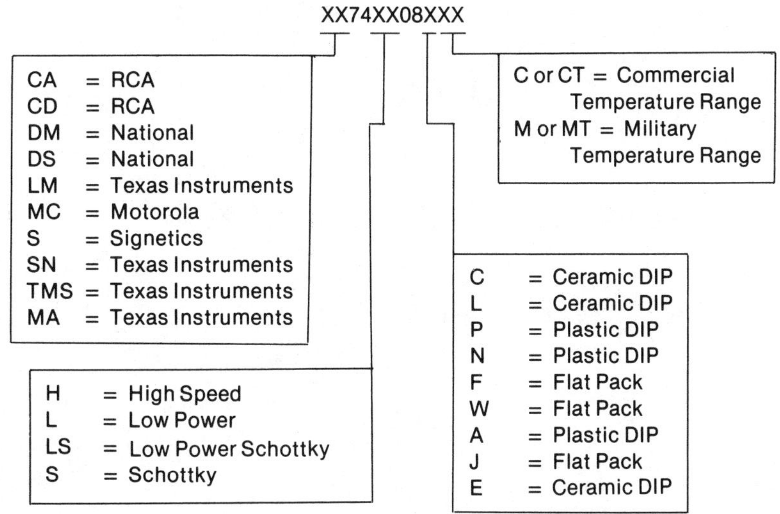

APPENDIX C

Pin-Out Diagrams

This appendix contains pin-out diagrams and a data sheet that are reprinted through the courtesy of Texas Instruments, Inc., Motorola, Inc., National Semiconductor Corp., and Nippon Electric Co., Ltd.

TTL Pin-Outs

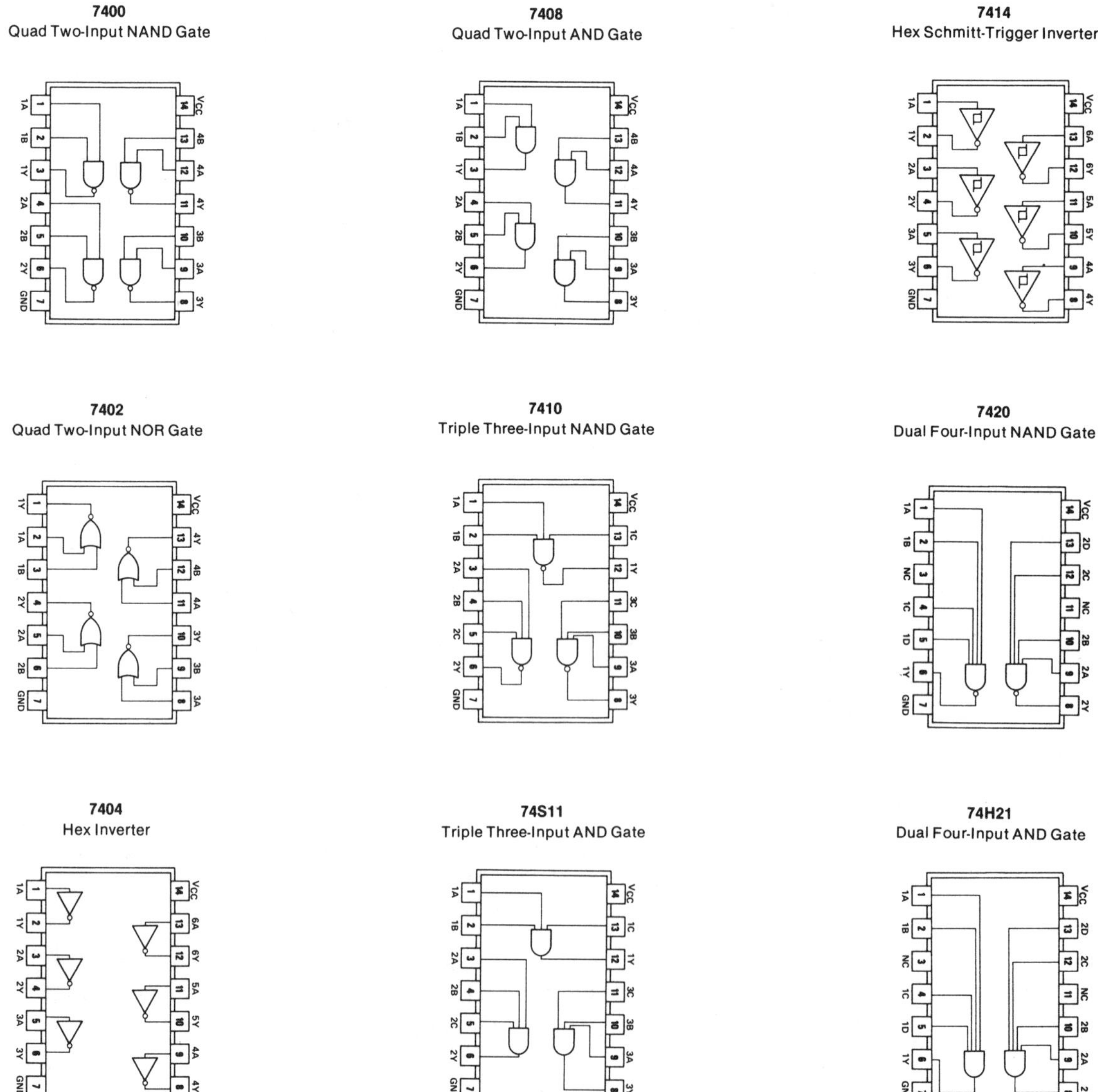

7430
Eight-Input NAND Gate

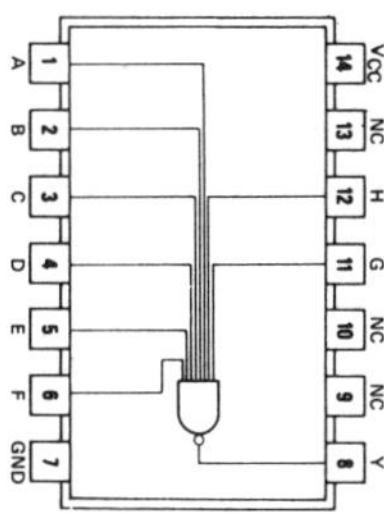

7432
Quad Two-Input OR Gate

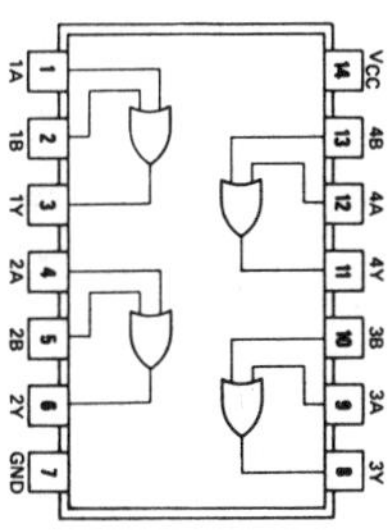

7437
Quad Two-Input NAND Buffer

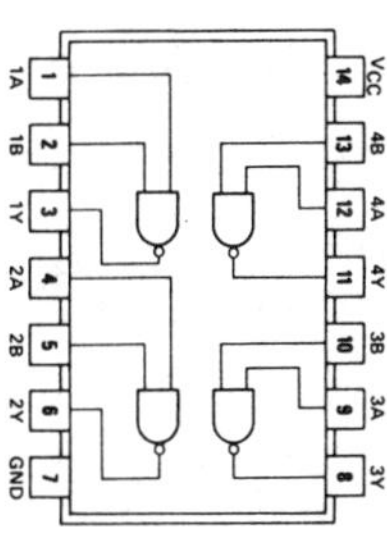

7440
Dual Four-Input NAND Buffer

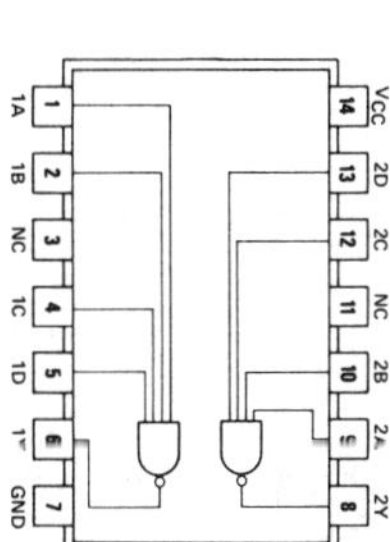

7442
BCD-to-Decimal Decoder

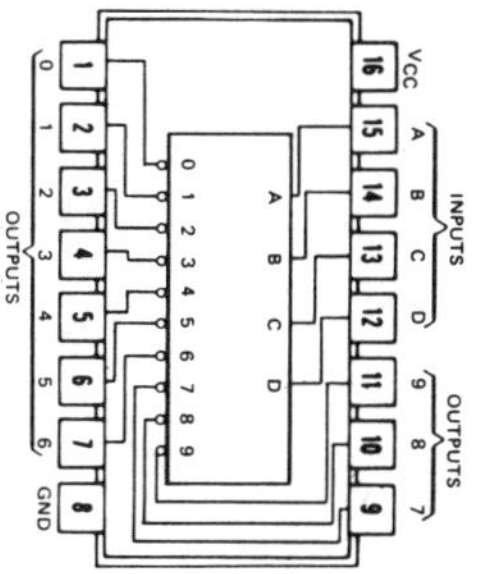

7445
BCD-to-Decimal Decoder/Driver

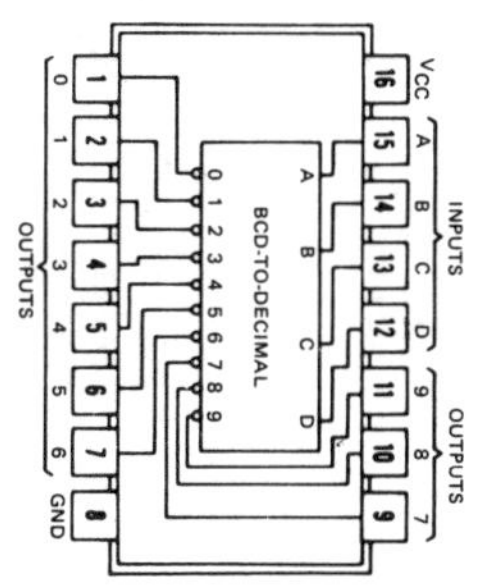

7447
BCD-to-Seven-Segment Decoder/Driver

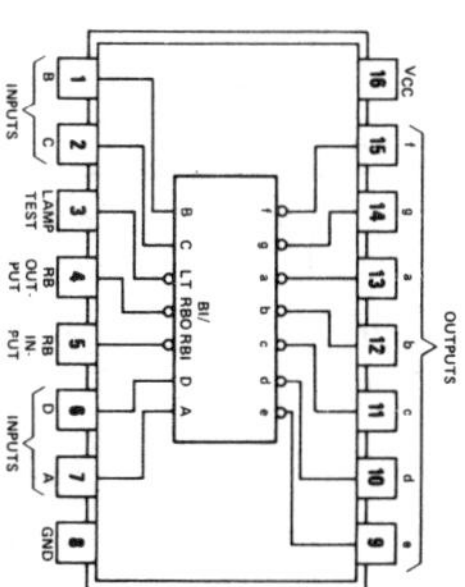

7470
AND-Gated J-K Flip-Flop

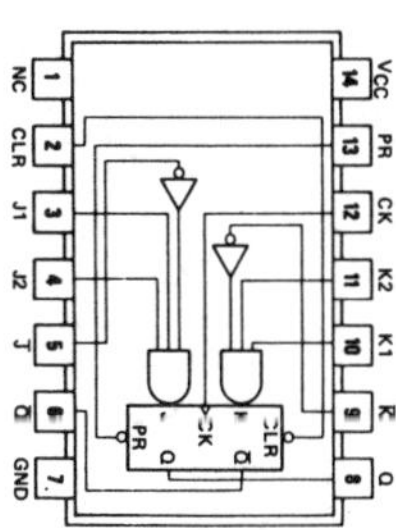

7472
AND-Gated J-K Master-Slave Flip-Flop

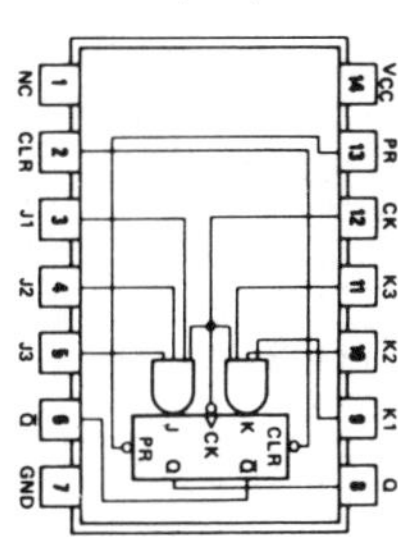

7474
Dual D-Type Flip-Flop

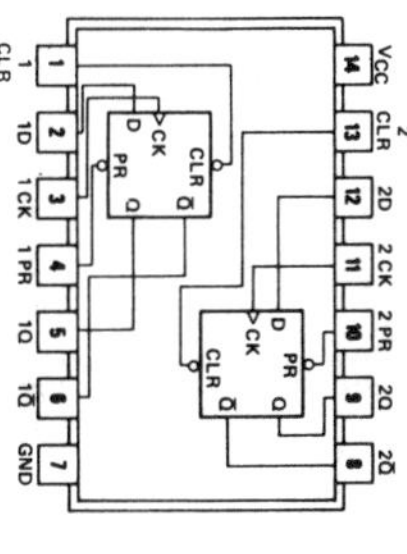

7475
4-Bit Bistable Latch

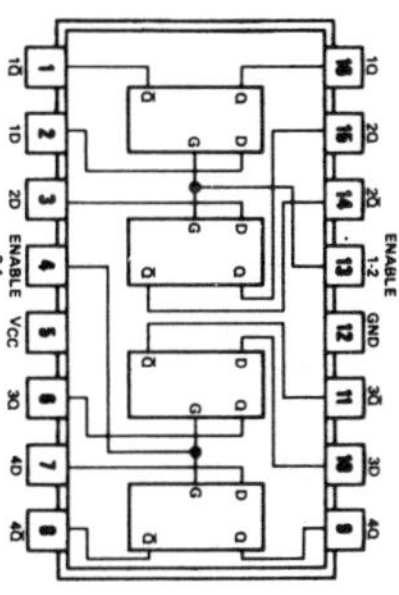

7476
Dual J-K Flip-Flop

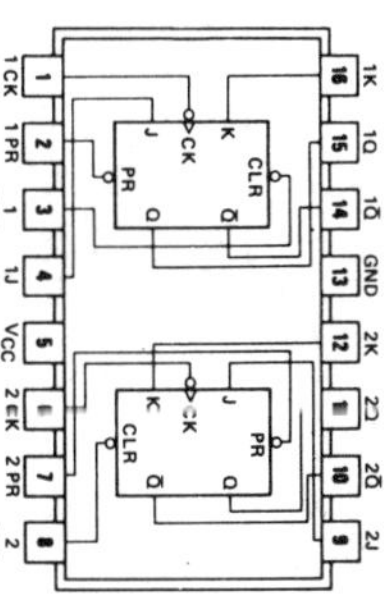

7483
4-Bit Binary Full Adder

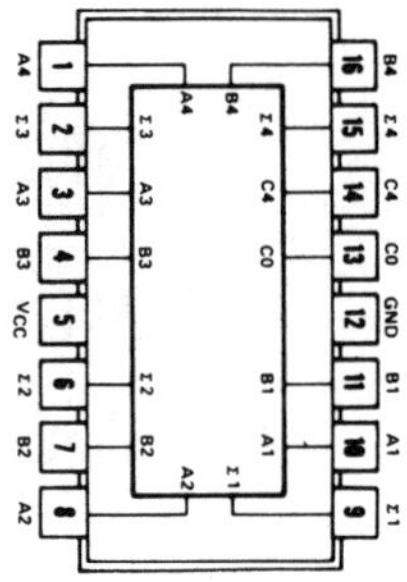

7492
Divide-by-12 Counter

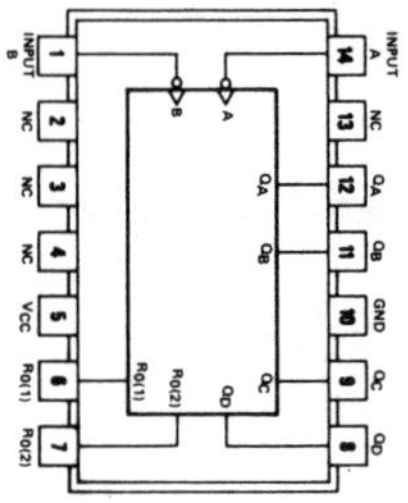

74122
Retriggerable Monostable Multivibrator

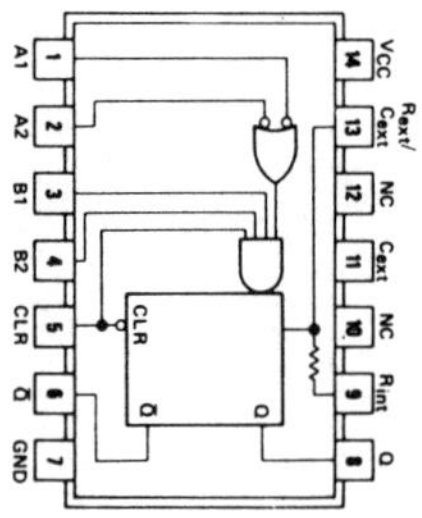

7486
Quad Two-Input Exclusive-OR Gate

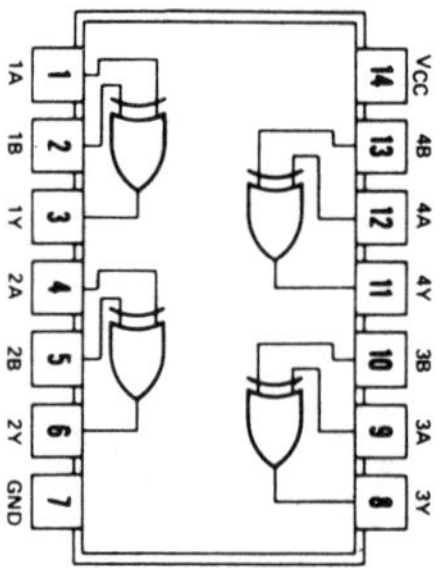

7493
Divide-by-8 and Divide-by-2 Binary Counter

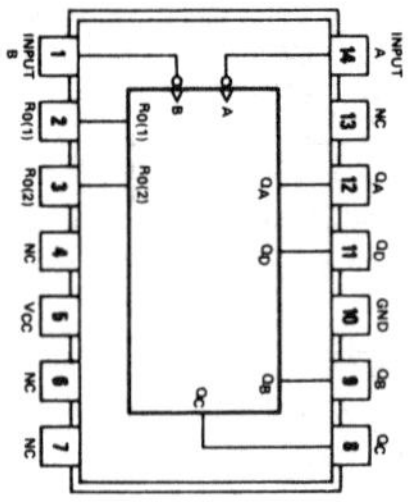

74123
Dual Retriggerable Monostable Multivibrator

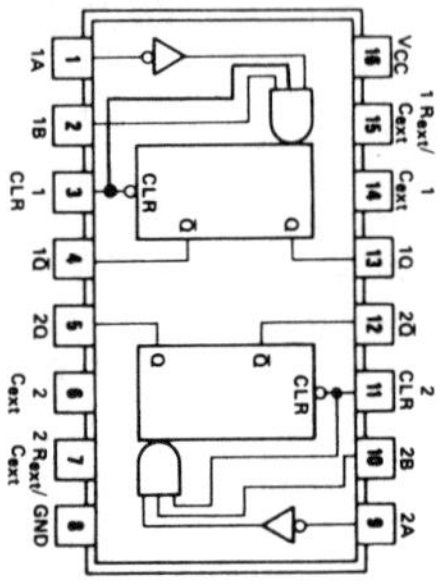

7490
Divide-by-2 and Divide-by-5 Decade Counter

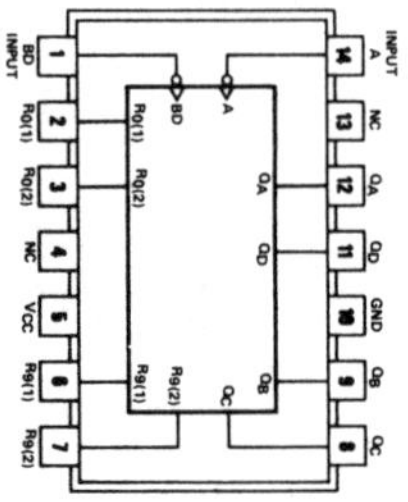

74107
Dual J-K Flip-Flop

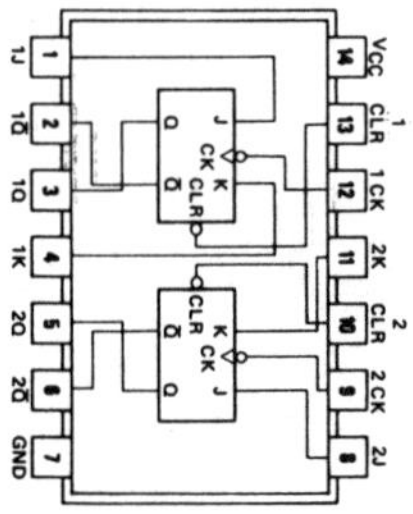

74142
Counter/Latch/Decoder/Driver

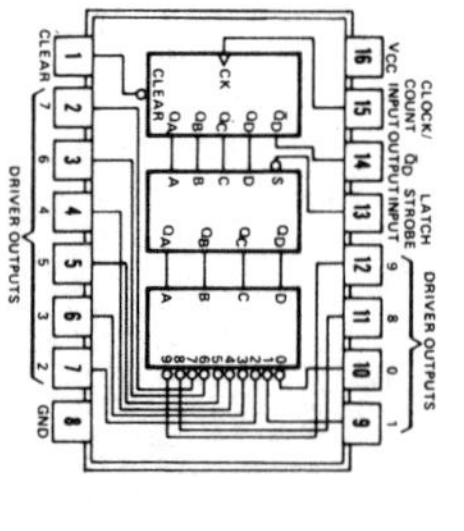

7491A
8-Bit Shift Register

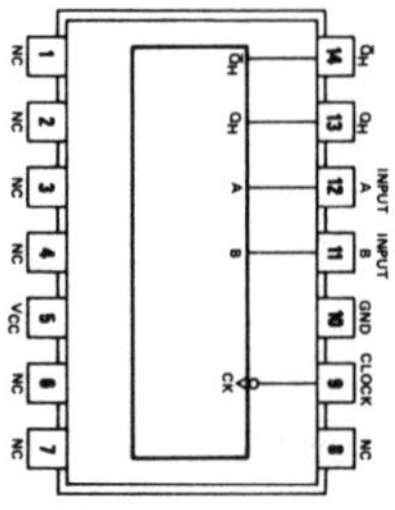

74121
Monostable Multivibrator

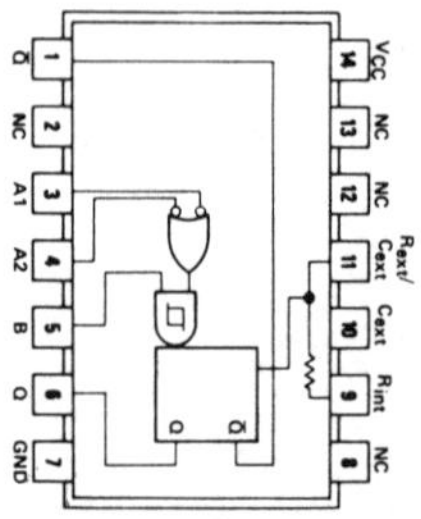

74150
1-of-16 Data Selector/Multiplexer

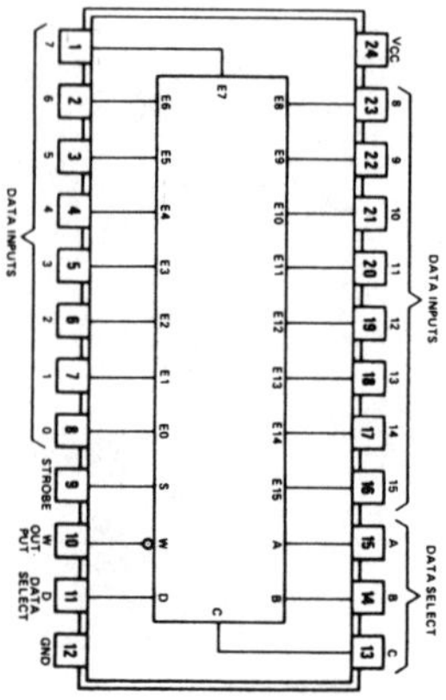

74151
1-of-8
Data Selector/Multiplexer

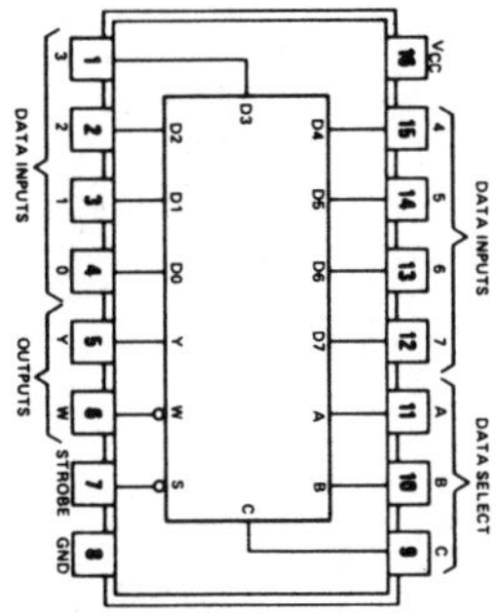

74154
4-Line to 16-Line
Decoder/Demultiplexer

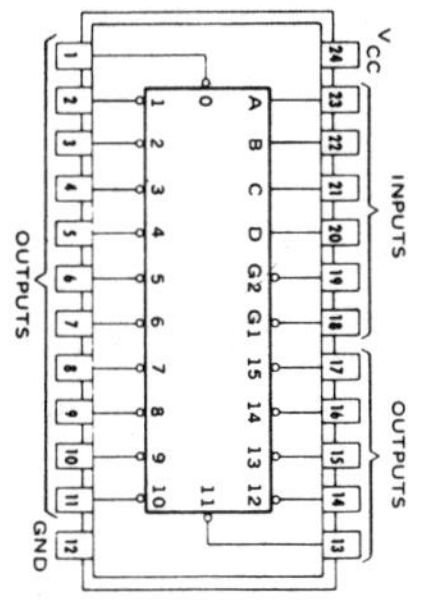

74155
Dual 1-of-4 Data Distributor

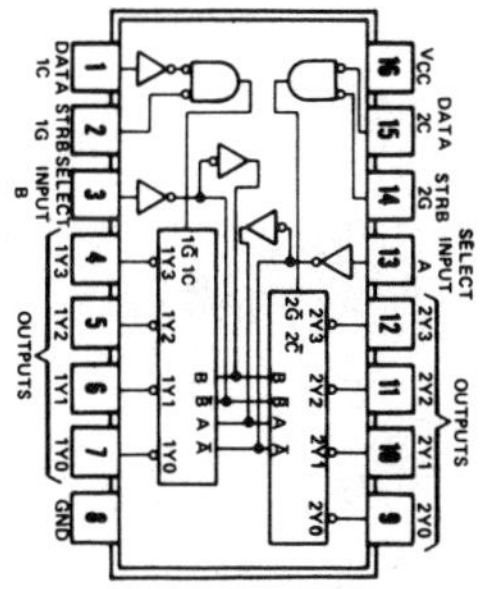

74157
Quad 1-of-2 Data Selector

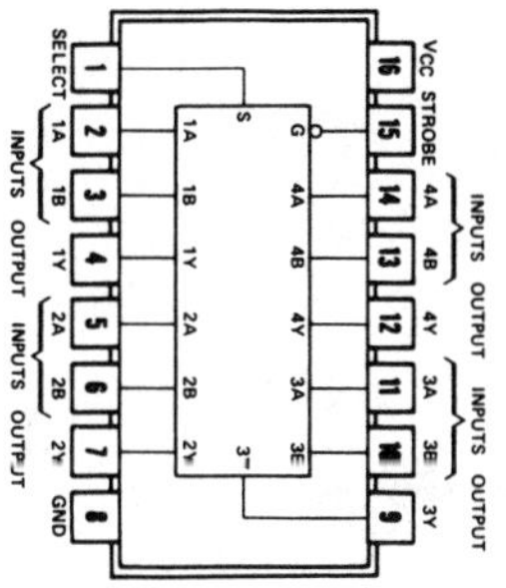

74164
8-Bit POSI Shift Register

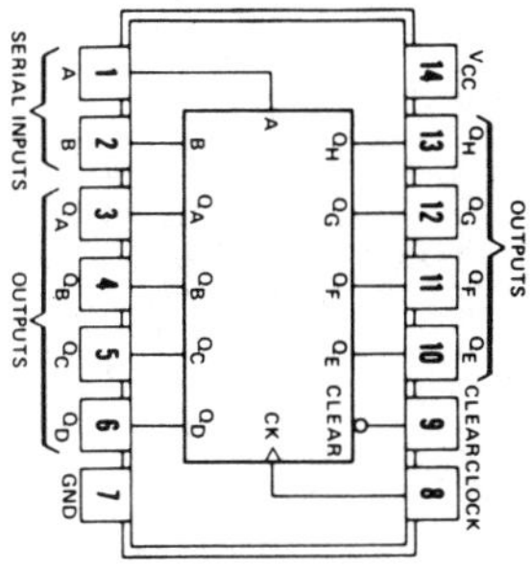

74165
8-Bit PISO Shift Register

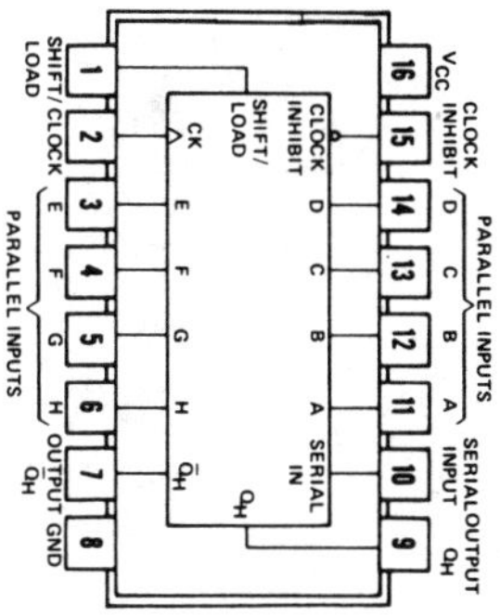

74173
4-Bit D-Type Register
with Tri-State Outputs

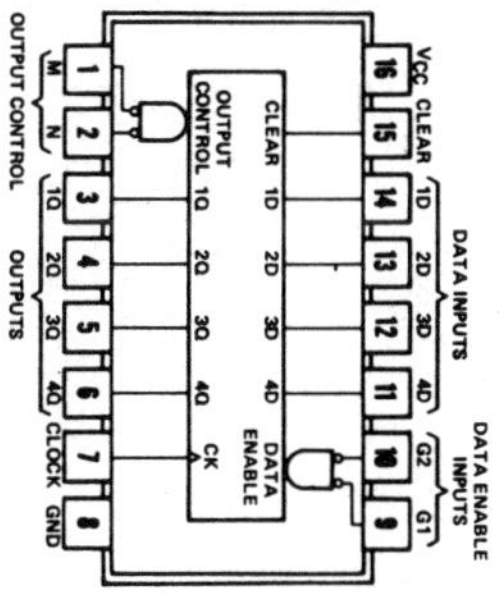

74180
9-Bit Odd/Even Parity
Generator/Checker

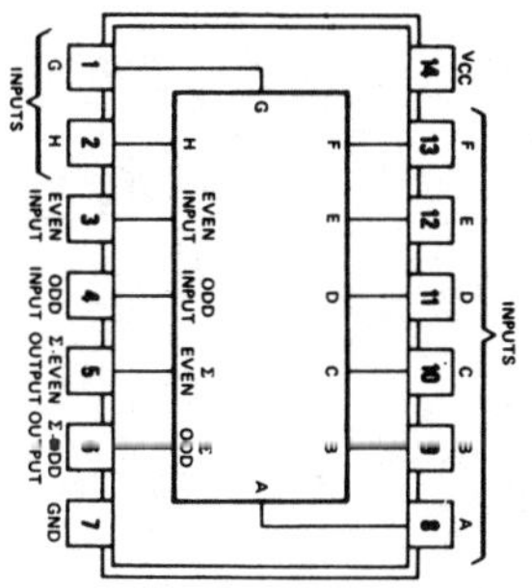

74181
Arithmetic Logic
Unit/Function Generator

74182
Look-Ahead Carry Generator

74191
Synchronous Up/Down Counter
with Down/Up Mode Control

74194
4-Bit Bidirectional
Universal Shift Register

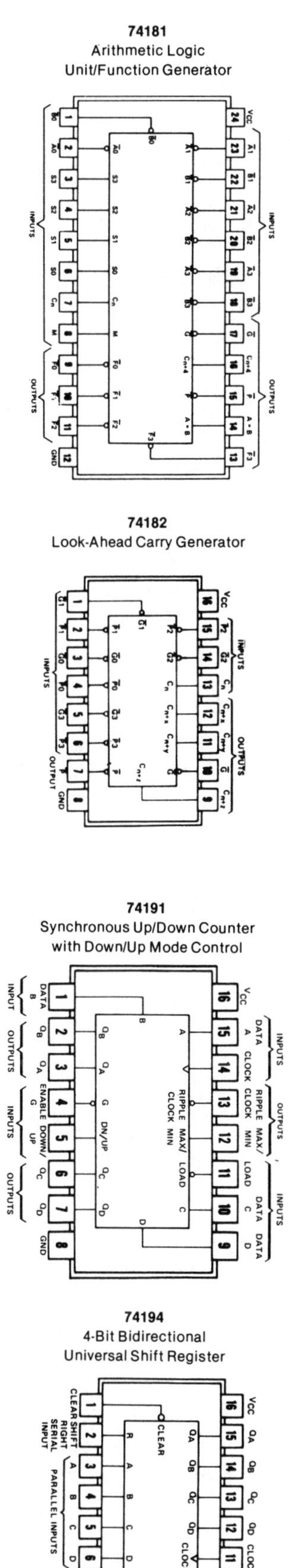

74198
8-Bit Bidirectional
Universal Shift Register

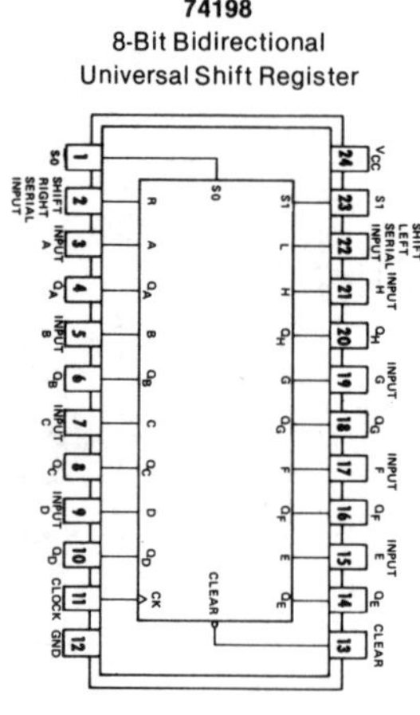

74LS245
Octal Bus Transceiver
with Tri-State Outputs

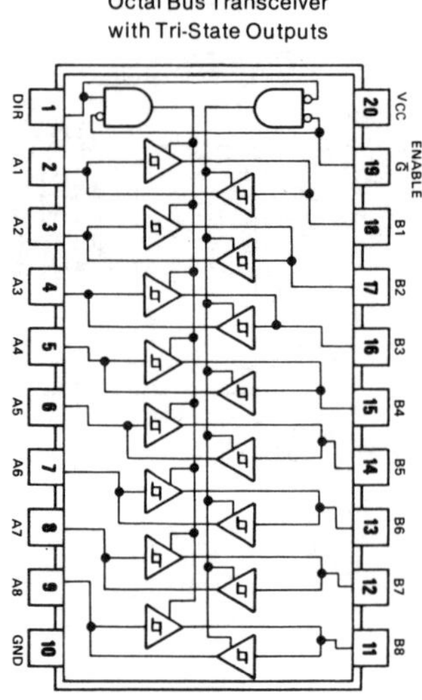

74S428
Controller and Bus Driver
for 8080A Systems

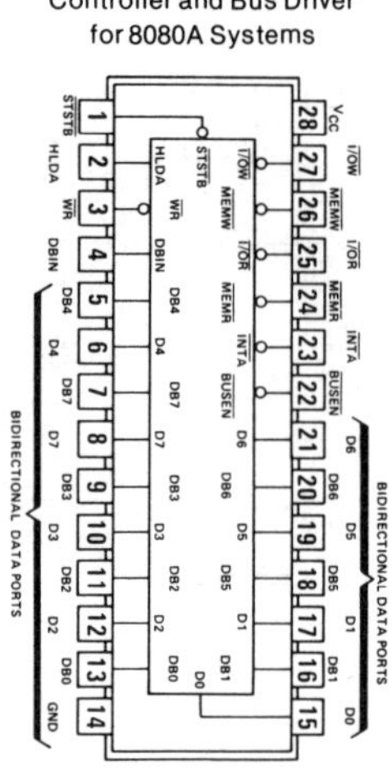

74298
Quad Two-Input Multiplexer
with Storage

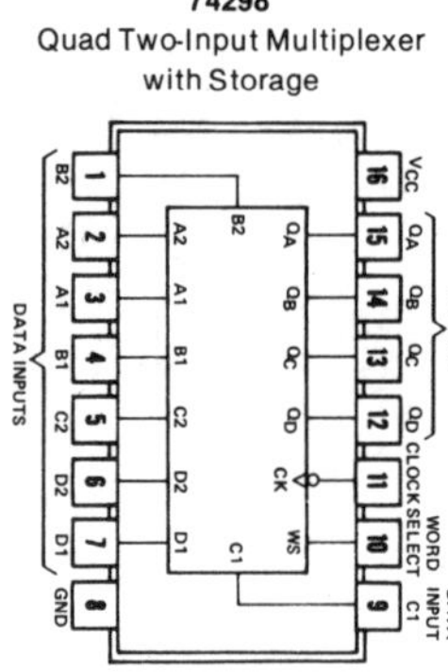

74LS670
4-by-4 Register File
with Tri-State Outputs

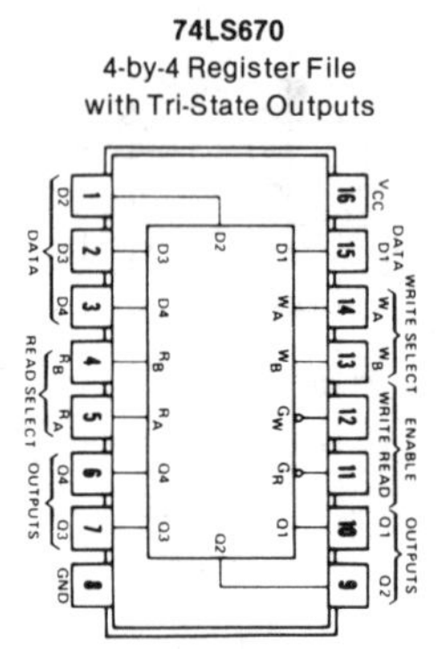

CMOS Pin-Outs

4001
Quad Two-Input NOR Gate

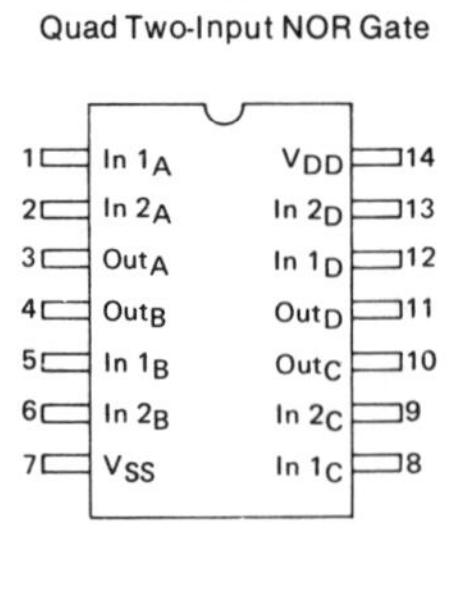

4006
Shift Register (Variable)

Pin			Pin
1	DP1	VDD	14
2	NC	Q4	13
3	C	Q9	12
4	DP5	Q8	11
5	DP10	Q13	10
6	DP14	Q18	9
7	VSS	Q17	8

4002
Dual Four-Input NOR Gate

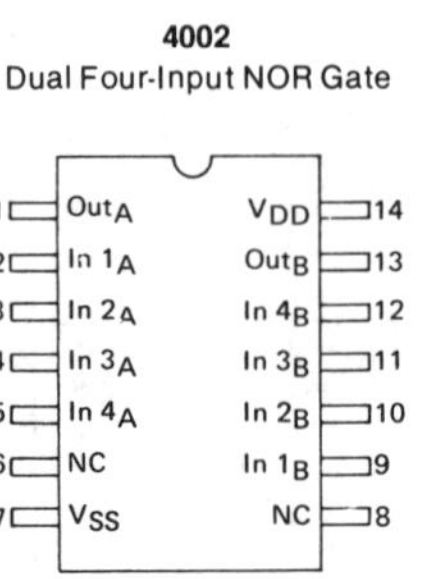

4009
Hex Inverter

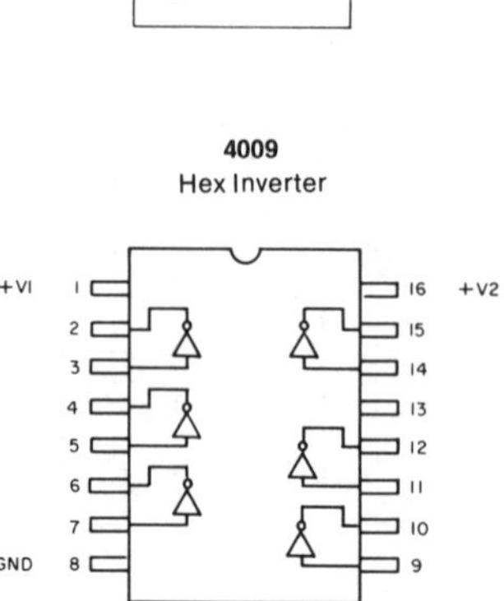

4010
Hex Buffer/Driver

4011
Quad Two-Input NAND Gate

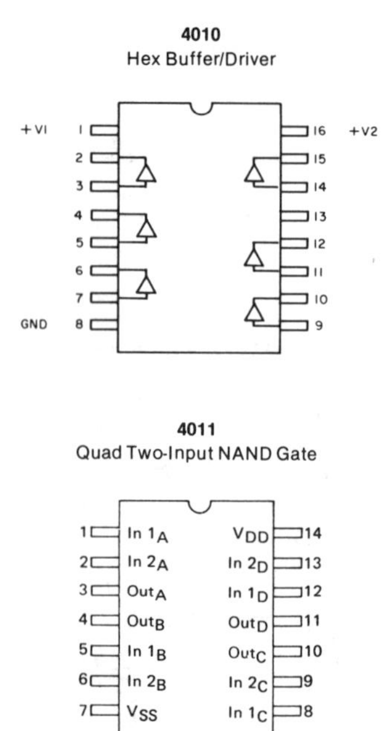

4012
Dual Four-Input NAND Gate

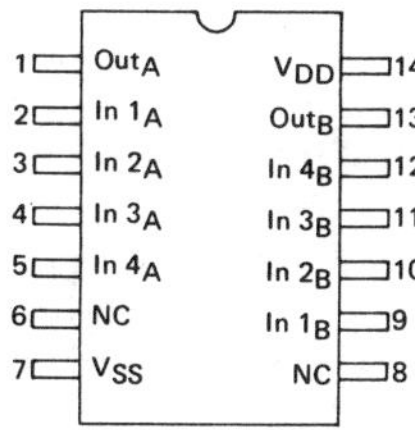

4018
Synchronous
Divide-by-2-through-10 Counter

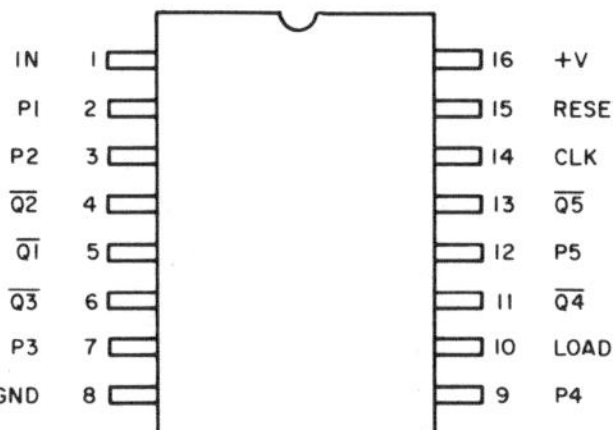

4028
BCD-to-Decimal Decoder

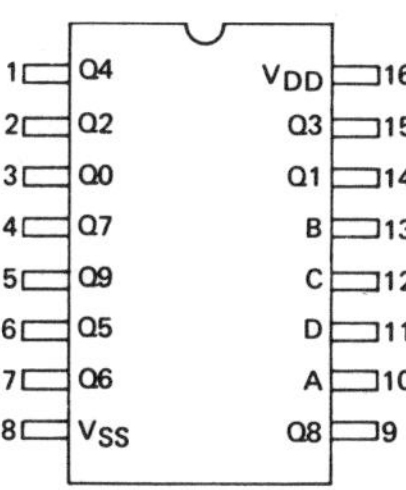

4013
Dual D-Type Flip-Flop

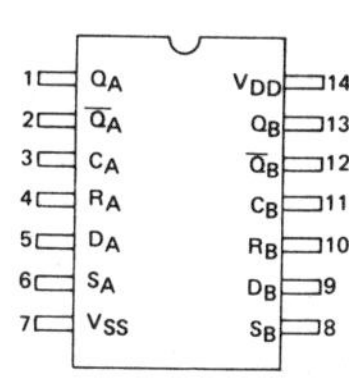

4020
14-Bit Ripple-Carry
Binary Counter

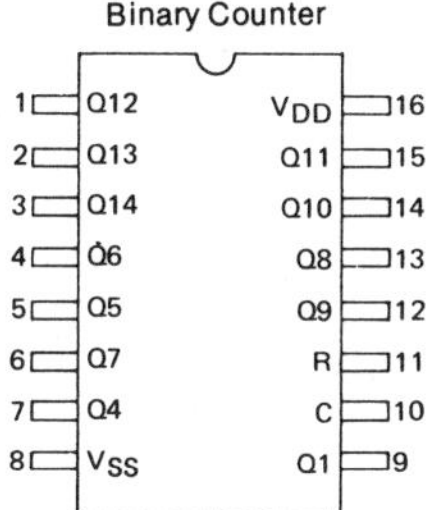

4030
Quad Exclusive-OR Gate

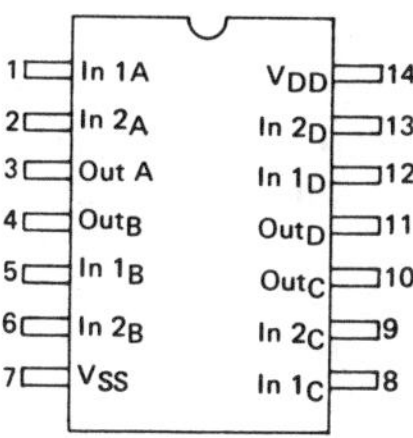

4015
Dual Four-State Shift Register

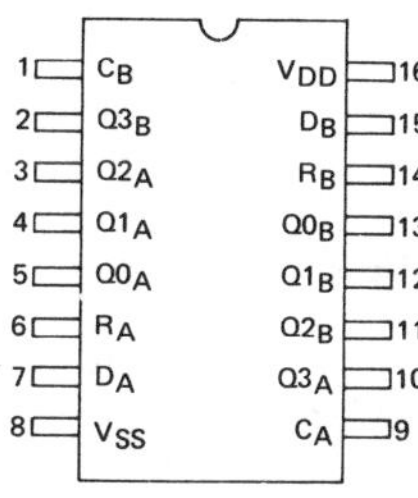

4023
Triple Three-Input NAND Gate

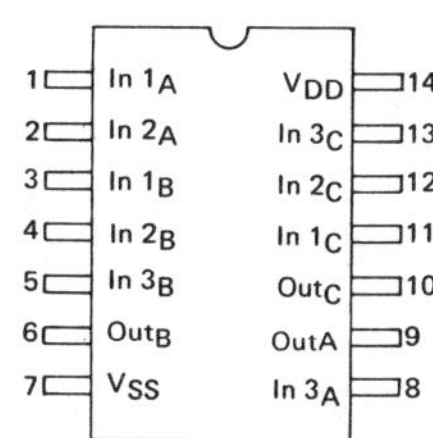

4031
64-Stage SISO Shift Register

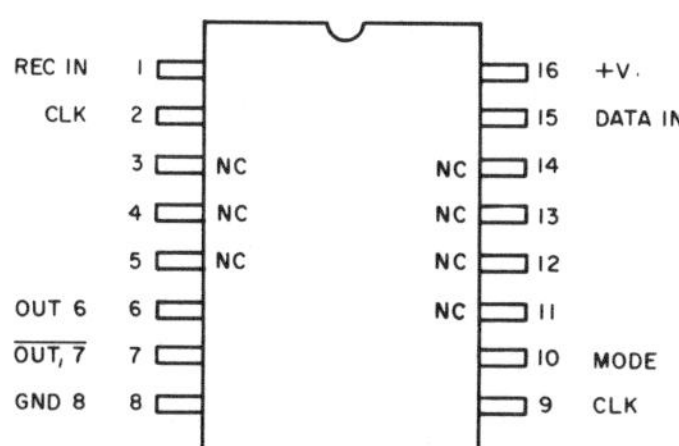

4017
Synchronous
Divide-by-10 Counter

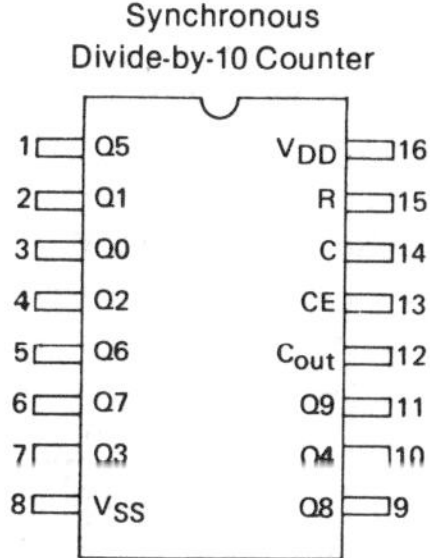

4024
7-Bit Binary Counter

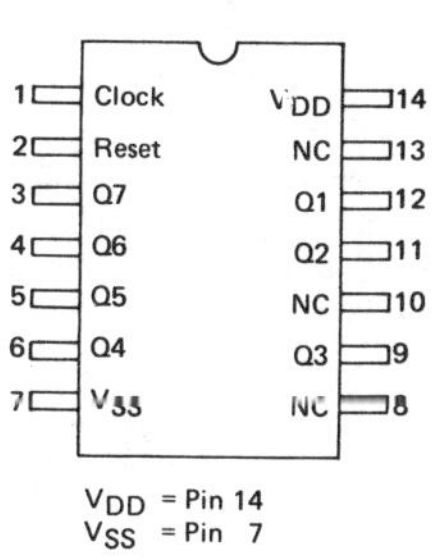

V_{DD} = Pin 14
V_{SS} = Pin 7
NC = No Connection

4035
Four-Stage PIPO Shift Register

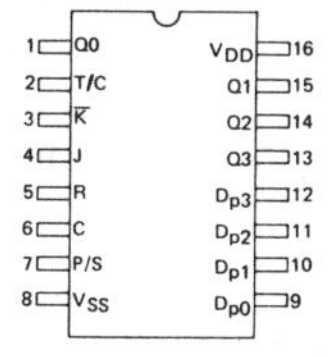

4040
12-Bit Binary/Ripple Counter

Pin	Signal	Pin	Signal
1	Q12	16	V_{DD}
2	Q6	15	Q11
3	Q5	14	Q10
4	Q7	13	Q8
5	Q4	12	Q9
6	Q3	11	R
7	Q2	10	C
8	V_{SS}	9	Q1

4052
Differential Four-Channel Analog Multiplexer/Demultiplexer

Pin	Signal	Pin	Signal
1	Y0	16	V_{DD}
2	Y2	15	X2
3	Y	14	X1
4	Y3	13	X
5	Y1	12	X0
6	Inh	11	X3
7	V_{EE}	10	A
8	V_{SS}	9	B

4071
Quad Two-Input OR Gate

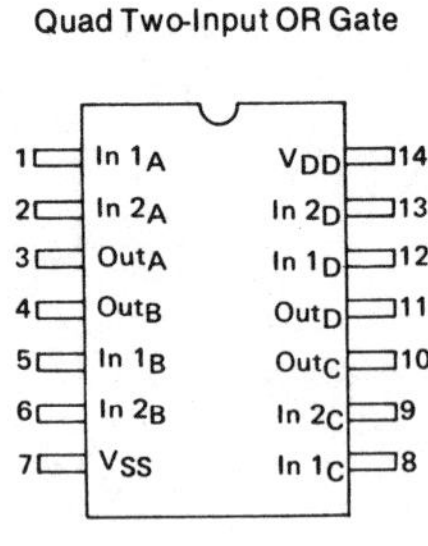

4043
Quad Three-State NOR R-S Latch

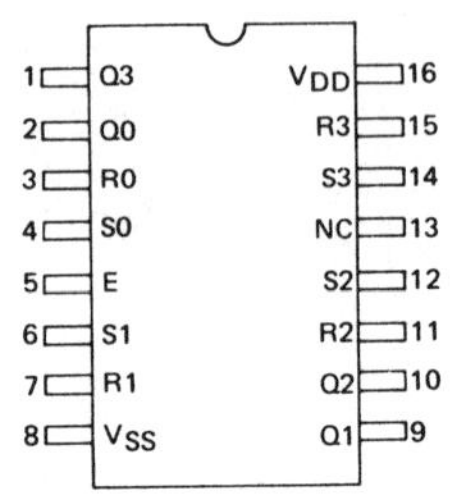

4053
Triple Two-Channel Analog Multiplexer/Demultiplexer

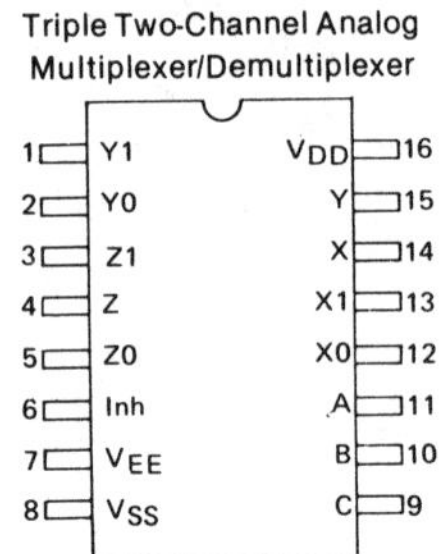

4072
Dual Four-Input OR Gate

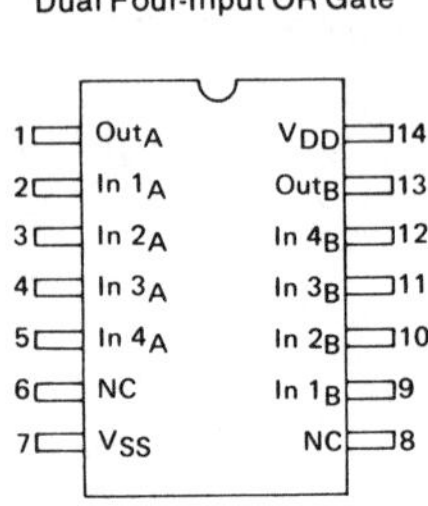

4044
Quad Three-State NAND R-S Latch

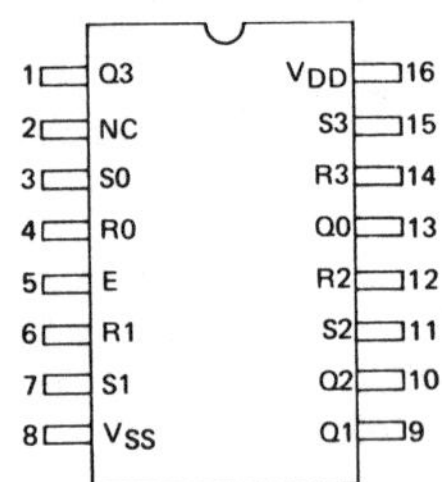

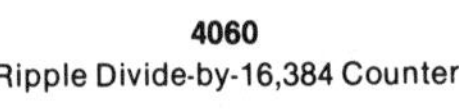
4060
Ripple Divide-by-16,384 Counter

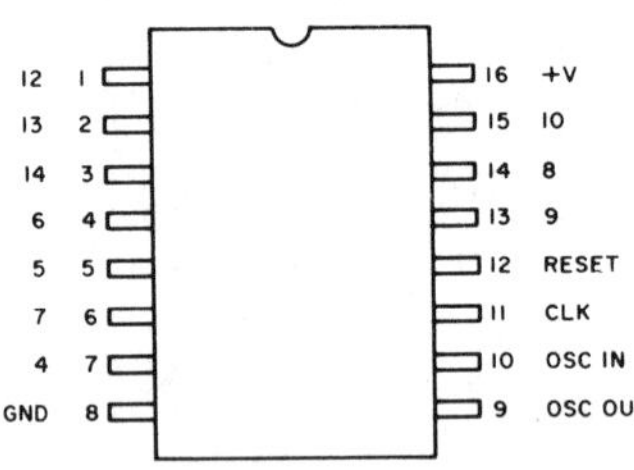

4073
Triple Three-Input AND Gate

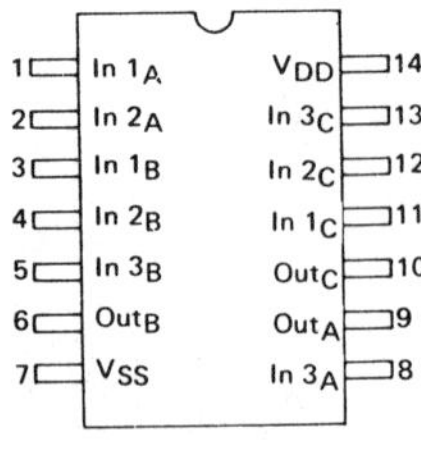

4051
Single Eight-Channel Analog Multiplexer/Demultiplexer

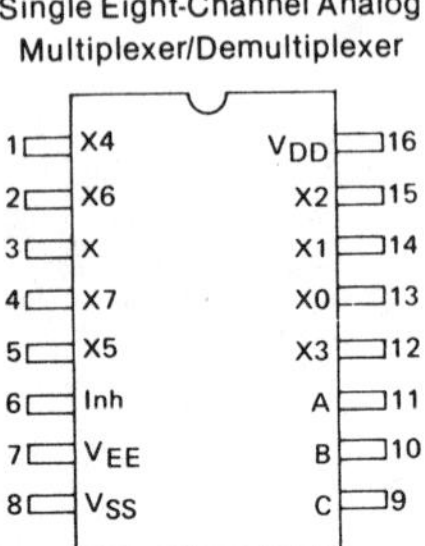

4070
Quad Two-Input Exclusive-OR Gate

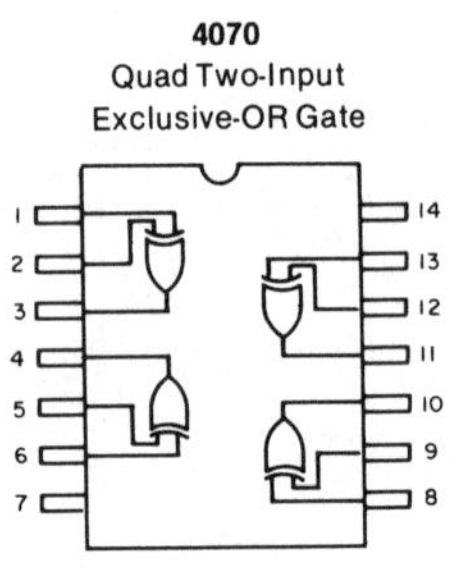

4075
Triple Three-Input OR Gate

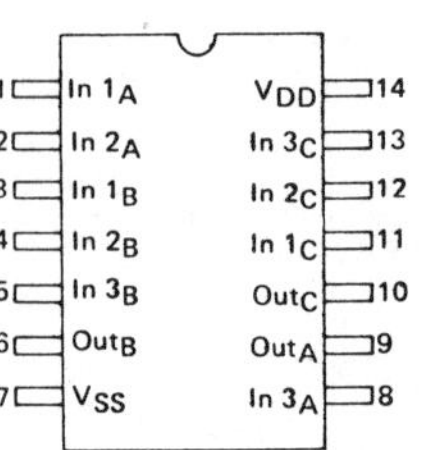

4077
Quad Two-Input
Exclusive-NOR Gate

1 2 3 4 5 6 GND 7
14 +V 13 12 11 10 9 8

4510
BCD Up/Down Counter

1 PE
2 Q4
3 P4
4 P1
5 Carry In
6 Q1
7 Carry Out
8 VSS
VDD 16
C 15
Q3 14
P3 13
P2 12
Q2 11
U/D 10
R 9

4522
BCD Programmable
Divide-by-4 Bit Counter

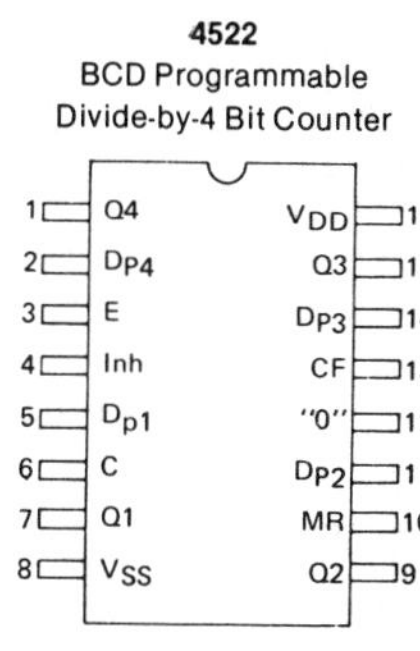

4081
Quad Two-Input AND Gate

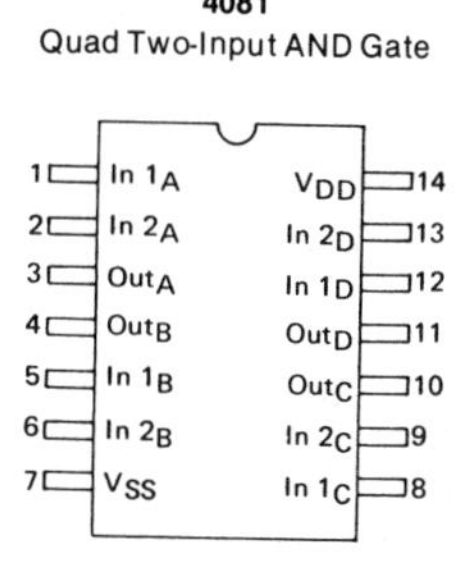

4514
4-Bit Latch
4-to-16 Line Decoder

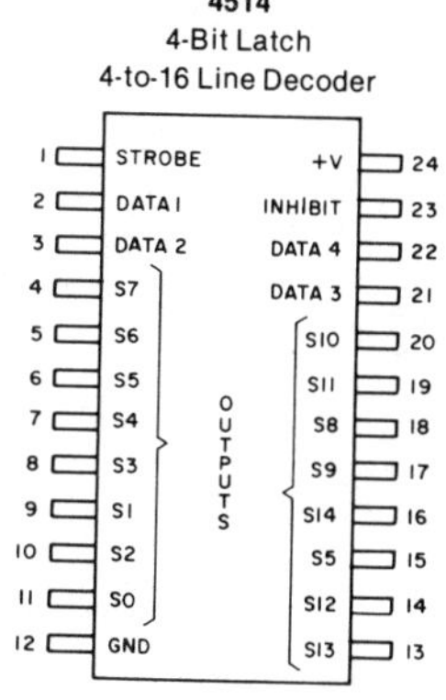

4526
Binary Programmable
Divide-by-4 Bit Counter

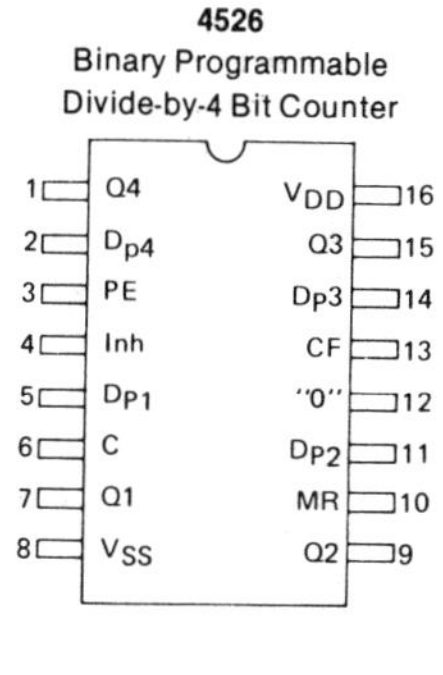

4082
Dual Four-Input AND Gate

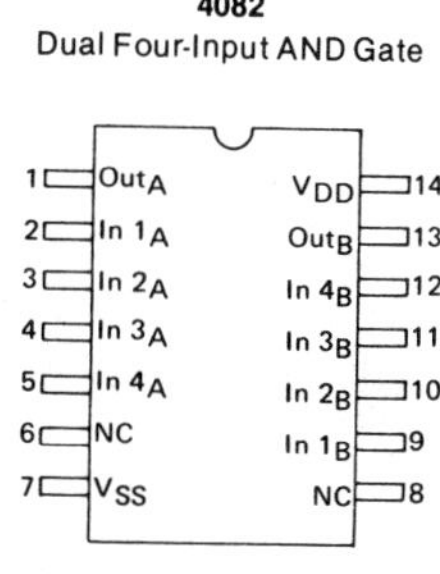

4516
Binary Up/Down Counter

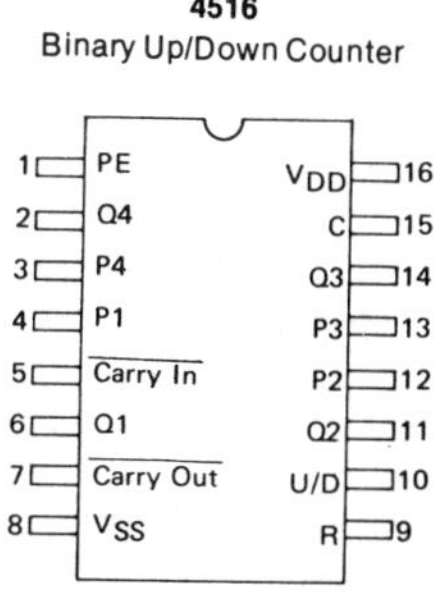

4539
Dual Four-Channel
Data Selector/Multiplexer

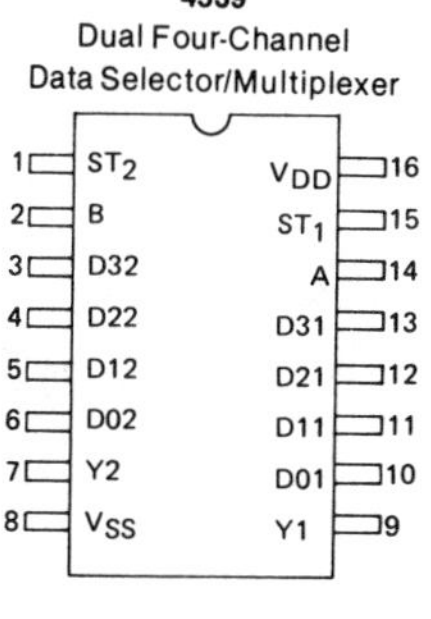

4503
Hex Buffer Driver

1 DisA
2 In 1
3 Out 1
4 In 2
5 Out 2
6 In 3
7 Out 3
8 VSS
VDD 16
Dis B 15
In 6 14
Out 6 13
In 5 12
Out 5 11
In 4 10
Out 4 9

4520
Dual Binary Up Counter

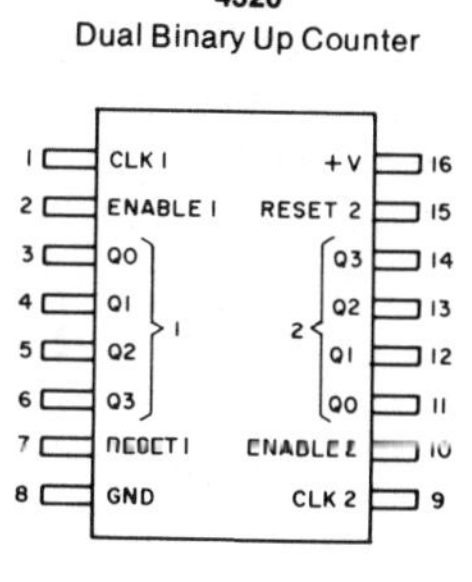

MC14043B

QUAD "NOR" R-S LATCH

MC14044B

QUAD "NAND" R-S LATCH

CMOS MSI

QUAD R-S LATCHES

The MC14043B and MC14044B quad R-S latches are constructed with MOS P-channel and N-channel enhancement mode devices in a single monolithic structure. Each latch has an independent Q output and set and reset inputs. The Q outputs are gated through three-state buffers having a common enable input. The outputs are enabled with a logical "1" or high on the enable input; a logical "0" or low disconnects the latch from the Q outputs, resulting in an open circuit at the Q outputs.

- Quiescent Current = 4.0 nA/pkg typical @ 10 Vdc
- Double Diode Input Protection
- Three-State Outputs with Common Enable
- Outputs Capable of Driving Two Low-Power TTL Loads, One Low-Power Schottky TTL Load, or Two HTL Loads Over the Rated Temperature Range
- Supply Voltage Range = 3.0 Vdc to 18 Vdc

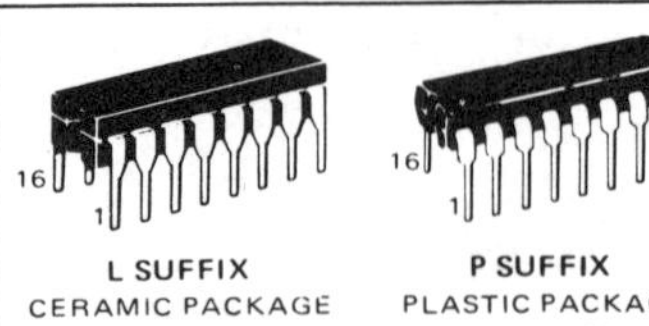

L SUFFIX
CERAMIC PACKAGE
CASE 620

P SUFFIX
PLASTIC PACKAGE
CASE 648

ORDERING INFORMATION

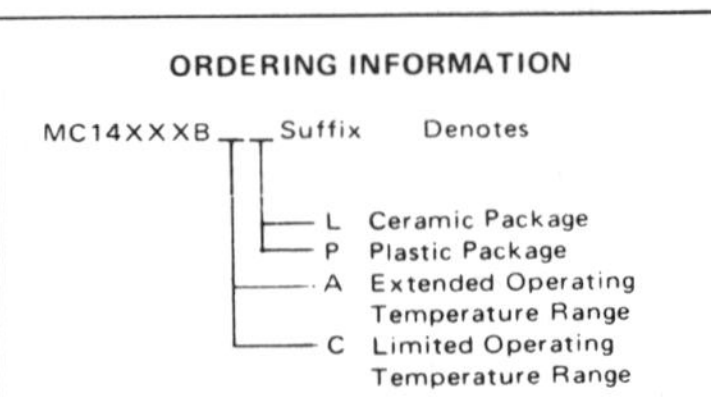

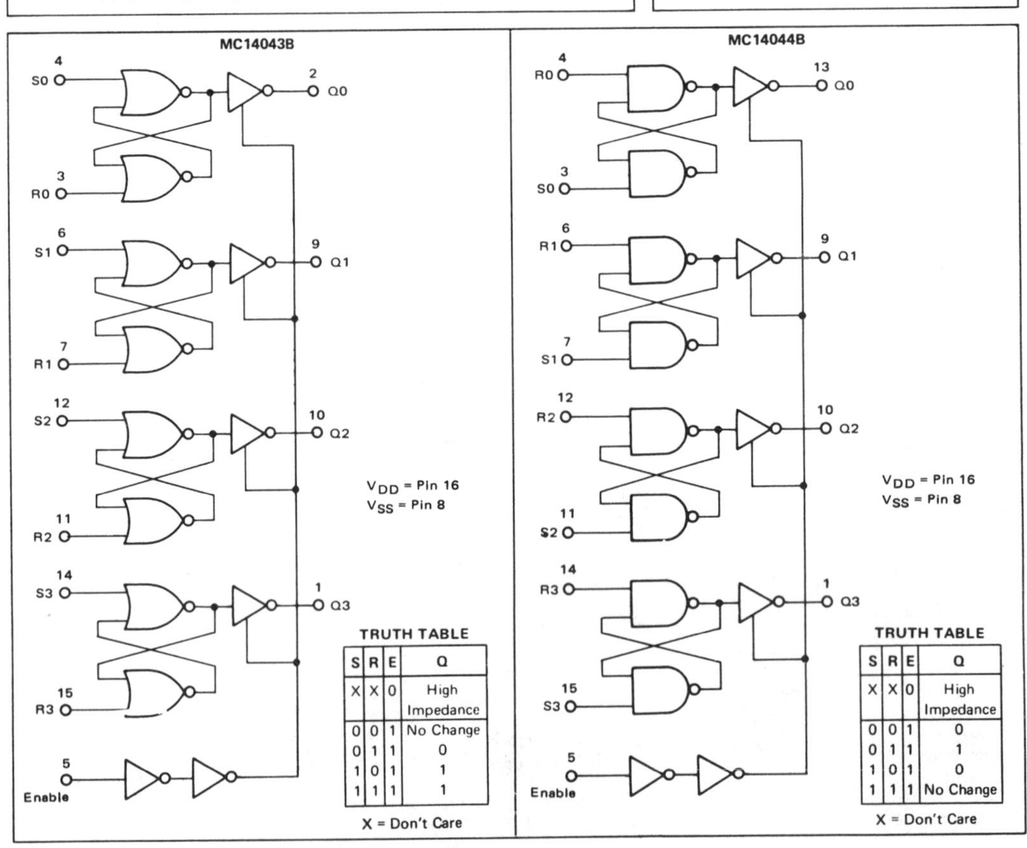

MC14043B TRUTH TABLE

S	R	E	Q
X	X	0	High Impedance
0	0	1	No Change
0	1	1	0
1	0	1	1
1	1	1	1

X = Don't Care

MC14044B TRUTH TABLE

S	R	E	Q
X	X	0	High Impedance
0	0	1	0
0	1	1	1
1	0	1	0
1	1	1	No Change

X = Don't Care

Other IC Pin-Outs

1702A
MOS EPROM

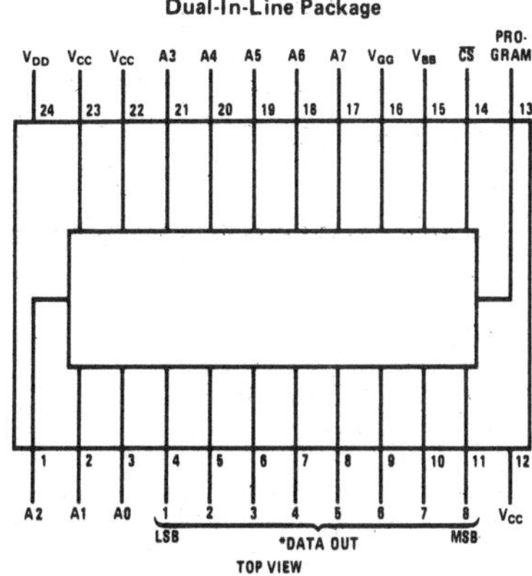

2102
1,024-Bit Decoded Static RAM

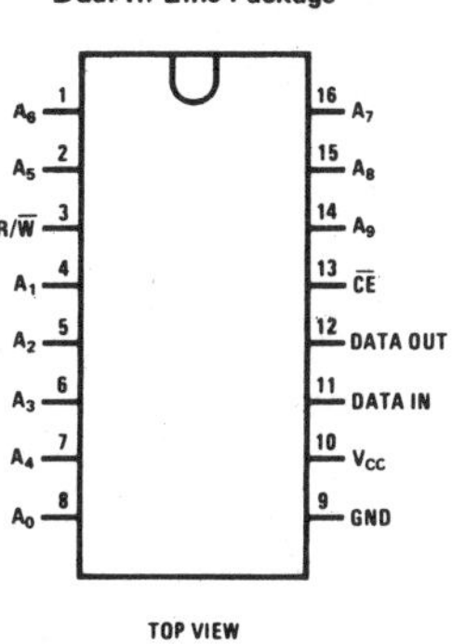

2111
1,024-Bit Static MOS RAM
with COMMON I/O
and Output Disable

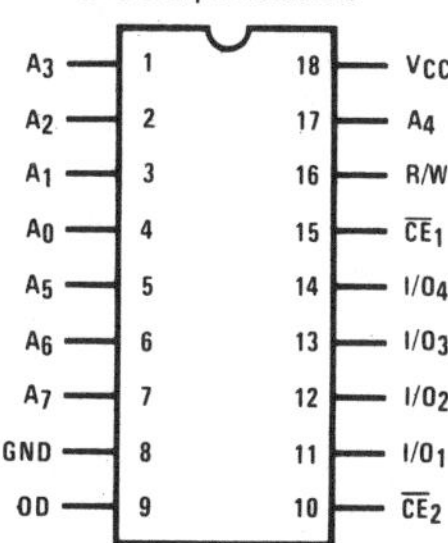

PIN NAMES

A_0-A_7	ADDRESS INPUTS
OD	OUTPUT DISABLE
R/W	READ/WRITE INPUT
$\overline{CE}1$	CHIP ENABLE 1
CE2	CHIP ENABLE 2
I/O_1-I/O_4	DATA INPUT/OUTPUT

6800
Monolithic
8-Bit Microprocessor

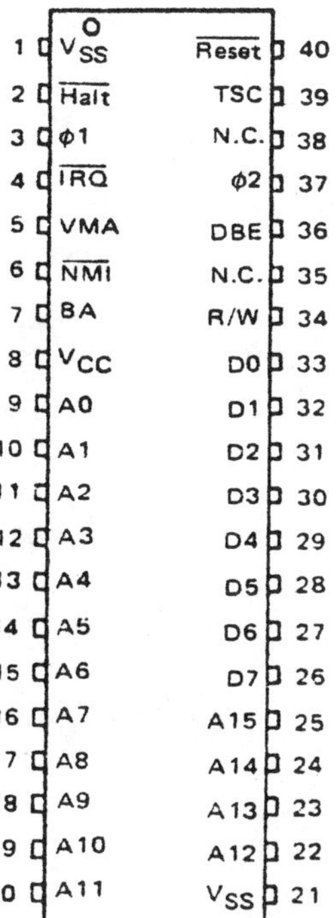

6802
Microprocessor with Clock
and RAM

Pin	Name	Name	Pin
1	V_{SS}	$\overline{Reset}$	40
2	$\overline{Halt}$	Xtal	39
3	MR	EXtal	38
4	$\overline{IRQ}$	E	37
5	VMA	RE	36
6	$\overline{NMI}$	V_{CC} Standby	35
7	BA	R/$\overline{W}$	34
8	V_{CC}	D0	33
9	A0	D1	32
10	A1	D2	31
11	A2	D3	30
12	A3	D4	29
13	A4	D5	28
14	A5	D6	27
15	A6	D7	26
16	A7	A15	25
17	A8	A14	24
18	A9	A13	23
19	A10	A12	22
20	A11	V_{SS}	21

6820
Peripheral Interface
Adapter (PIA)

Pin	Name	Name	Pin
1	V_{SS}	CA1	40
2	PA0	CA2	39
3	PA1	$\overline{IRQA}$	38
4	PA2	$\overline{IRQB}$	37
5	PA3	RS0	36
6	PA4	RS1	35
7	PA5	$\overline{Reset}$	34
8	PA6	D0	33
9	PA7	D1	32
10	PB0	D2	31
11	PB1	D3	30
12	PB2	D4	29
13	PB3	D5	28
14	PB4	D6	27
15	PB5	D7	26
16	PB6	E	25
17	PB7	CS1	24
18	CB1	$\overline{CS2}$	23
19	CB2	CS0	22
20	V_{CC}	R/W	21

8224
Timer

Name	Pin	Pin	Name
RESET OUTPUT	1	16	+5 V
$\overline{RESET}$ INPUT	2	15	XTAL 1
READY INPUT	3	14	XTAL 2
READY	4	13	TANK
SYNC	5	12	OSC
φ2 (TTL)	6	11	φ1
STATUS STROBE	7	10	φ2
GND	8	9	+12 V

Index